NOVEL THERAPEUTIC APPROACHES TARGETING OXIDATIVE STRESS

NOVEL THERAPEUTIC APPROACHES TARGETING OXIDATIVE STRESS

Edited by

Pawan Kumar Maurya

Professor and Head, Department of Biochemistry, Central University of Haryana, Mahendragarh, Haryana, India

Imteyaz Qamar

Assistant Professor, School of Biotechnology, Gautam Buddha University, Greater Noida, Uttar Pradesh, India

ELSEVIER

ACADEMIC PRESS

An imprint of Elsevier

Academic Press is an imprint of Elsevier
125 London Wall, London EC2Y 5AS, United Kingdom
525 B Street, Suite 1650, San Diego, CA 92101, United States
50 Hampshire Street, 5th Floor, Cambridge, MA 02139, United States
The Boulevard, Langford Lane, Kidlington, Oxford OX5 1GB, United Kingdom

Library of Congress Cataloging-in-Publication Data
A catalog record for this book is available from the Library of Congress

British Library Cataloguing-in-Publication Data
A catalogue record for this book is available from the British Library

ISBN 978-0-323-90905-1

For information on all Academic Press publications
visit our website at https://www.elsevier.com/books-and-journals

Publisher: Andre G. Wolff
Acquisitions Editor: Michelle Fisher
Editorial Project Manager: Tracy I. Tufaga
Production Project Manager: Kiruthika Govindaraju
Cover Designer: Miles Hitchen

Typeset by STRAIVE, India

Dedication

The publication of this book was finalized during the Coronavirus (COVID-19) pandemic. We would like to dedicate this book to all those who were affected by the pandemic, and in particular, to our teachers and health workforce around the world for their dedication and care during this difficult time.

Contents

4. Novel therapeutic approaches targeting oxidative stress in aging

Akshay Kumar and Pawan Kumar Maurya

5. Novel therapeutic approaches targeting oxidative stress in neurodegenerative diseases

Somu Yadav, Akshay Kumar, Deepika, and Pawan Kumar Maurya

6. The role of oxidative stress in kidney diseases

Vikas Saini

7. The role of oxidative stress and antioxidants across the spectrum of acute coronary syndrome

Wahid Ali, Neha Srivastava, and Priyanka Sharma

8. Biosensors in the detection of oxidative stress using discovered biomarkers

Swaraj Mohanty and Yasmin Ahmad

9. Biochemical approaches/methods in the detection of free radicals

Imteyaz Qamar and Uditi

Contributors

Rizwan Ahmad College of Medicine, Imam Abdulrahman bin Faisal University, Dammam, Saudi Arabia

Yasmin Ahmad Peptide and Proteomics Division, Defence Institute of Physiology & Allied Sciences (DIPAS), Defence Research & Development Organization (DRDO), Delhi, India

Haseeb Ahsan Department of Biochemistry, Faculty of Dentistry, Jamia Millia Islamia, New Delhi, India

Wahid Ali Department of Pathology, King George's Medical University, Chowk, Lucknow, Uttar Pradesh, India

Alexander Bambokian Centre for Neuroscience Studies (CNS), Queen's University, Kingston, ON, Canada

Elisa Brietzke Centre for Neuroscience Studies (CNS), Queen's University; Department of Psychiatry, Queen's University School of Medicine, Kingston, ON, Canada

Deepika Department of Biochemistry, Central University of Haryana, Mahendergarh, India

Archna Dhasmana Department of Biotechnology, School of Applied and Life Sciences, Uttaranchal University Dehradun, Uttarakhand, India

Fabiano A. Gomes Centre for Neuroscience Studies (CNS), Queen's University; Department of Psychiatry, Queen's University School of Medicine, Kingston, ON, Canada

Mohammad Yusuf Hasan School of Biomedical Sciences, Newcastle University, Newcastle upon Tyne, United Kingdom

Medha Kaushik Department of Toxicology, School of Chemical and Life Sciences, Jamia Hamdard, New Delhi, India

Pooja Kaushik Department of Toxicology, School of Chemical and Life Sciences, Jamia Hamdard, New Delhi, India

Akshay Kumar Department of Biochemistry, Central University of Haryana, Mahendergarh, India

Sumira Malik Amity Institute of Biotechnology, Amity University Jharkhand, Ranchi, Jharkhand, India

Pawan Kumar Maurya Department of Biochemistry, Central University of Haryana, Mahendergarh, India

Roumen Milev Centre for Neuroscience Studies (CNS), Queen's University; Department of Psychiatry, Queen's University School of Medicine, Kingston, ON, Canada

Swaraj Mohanty Peptide and Proteomics Division, Defence Institute of Physiology & Allied Sciences (DIPAS), Defence Research & Development Organization (DRDO), Delhi, India

Suhel Parvez Department of Toxicology, School of Chemical and Life Sciences, Jamia Hamdard, New Delhi, India

Shilpa Prasad Amity Institute of Biotechnology, Amity University Jharkhand, Ranchi, Jharkhand, India

Imteyaz Qamar School of Biotechnology, Gautam Buddha University, Greater Noida, UP, India

Vikas Saini Biomedical Sciences Program, Department of Vocational Studies & Skill Development, Central University of Haryana, Mahendergarh, India

Priyanka Sharma Department of Pathology, King George's Medical University, Chowk, Lucknow, Uttar Pradesh, India

Calvin Sjaarda Centre for Neuroscience Studies (CNS), Queen's University; Department of Psychiatry, Queen's University School of Medicine, Kingston, ON, Canada

Claudio N. Soares Centre for Neuroscience Studies (CNS), Queen's University; Department of Psychiatry, Queen's University School of Medicine, Kingston, ON, Canada

Neha Srivastava Department of Pathology, King George's Medical University, Chowk, Lucknow, Uttar Pradesh, India

Uditi School of Biotechnology, Gautam Buddha University, Greater Noida, UP, India

Mari Uyeda AC Camargo Cancer Centre, São Paulo, SP, Brazil

Somu Yadav Department of Biochemistry, Central University of Haryana, Mahendergarh, India

Preface

Interdisciplinary approaches in science have led into the development of therapeutic approaches targeting oxidative stress in various models for real clinical and biomedical applications. Oxidative stress is associated with diminished capacity of a biological system to overcome an overproduction of reactive oxygen species and other free radicals. Since oxidative stress is a leading cause of nucleic acid damage (DNA and RNA), genetic disorders, many environmental pollution-related diseases, aging, cardiovascular, cancer, diabetes, hypertension, and neurological disorders, there is an urgent need for novel therapeutic approaches which target oxidative stress for its prevention. There is growing evidence that oxidative stress may cause autism in children. Several biomarkers of oxidative stress have also been identified. The development of such kinds of methods requires versatile knowledge of biology, chemistry, molecular biology, immunology, and microbiology.

Oxidative stress has been suggested as the causative factor in aging and many diseases. Therefore, considerable interest in the development of rapid assay for biomarkers of these diseases has recently surfaced. These biomarkers are monitored using many biological samples such as blood, cell culture, saliva, and other biological fluids. Recent advances in technology lead to development of new therapeutic methods for targeting the oxidative stress status of organism. In this book, we will therefore also deal with novel therapeutic approaches targeting oxidative stress in aging, cardiovascular diseases, cancer, kidney disease, ferroptosis, and neurodegenerative and neuropsychiatric disorders. We will restrict ourselves to the more recent developments in this field.

In view of such clinical and healthcare importance of oxidative stress, this book has been proposed. In it, we start with a brief introduction about the role of oxidative stress in major disease conditions; we go on to focus mainly on the therapeutic approaches targeting oxidative stress in various diseases, which are very significant toward onsite clinical applications and can be used in the prevention of diseases.

After numerous deliberations, we came up with the idea to explore the possibility of developing a book on novel therapeutic approaches targeting oxidative stress to partially (if not completely) fill this void. Finally, we decided to develop a book by inviting chapters from experts in the field who have relevant research experience and an understanding of the intricacies of the subject. We had in mind a book that would help to alleviate most of the worries of both students and instructors. We discussed, argued, and disagreed until we concluded that a resource book would be a reasonable format, as it could provide sufficient information and literature for instructors to teach the subject, while providing students with ample information to gain better insight about the subject. Once we formulated these thoughts to develop a resource book, the ball started rolling, and we identified various experts and convinced them to contribute chapters.

This is our maiden effort to produce a book to help students and instructors focusing on oxidative stress and its impact in medical science. We hope that we will receive support from the readers of this book. We are always open to criticism, suggestions, and recommendations that can help to improve the content and presentation of the book. Your suggestions and criticisms will give us an opportunity to explore other aspects of oxidative stress in our future ventures and endeavors.

Acknowledgments

As Editors, we would like to express our gratitude and thanks to all the contributing authors, whose expertise and experience are now available to us in the form of book chapters in their respective fields. Our contributors have worked hard to update, upgrade, and improve their chapters. This book would never have been possible without their valuable work and timely support.

We would like to thank Prof. Tankeshwar Kumar, Hon'ble Vice-Chancellor, Central University of Haryana and Prof. Bhagwati Prakash Sharma, Hon'ble Vice-Chancellor, Gautam Buddha University for providing necessary facilities.

This book could never have been completed without Elsevier personnel working on this project. We are grateful to Ms. Tracy Tufaga (Editorial Project Manager), Ms. Swapna Praveen (Sr. Copyrights Coordinator, Copyrights Team), and Mr. Jaganathan Subramaniam who kept supporting, motivating, and pushing us throughout the project duration. We convey our heartfelt thanks to everyone who has contributed directly or indirectly to this book.

Finally, all this would never have been possible without "The Almighty" GOD to whom we owe our existence. The one who gifted us (human beings) a brain to hypothesize and analyze, courage to dream, and motivation to achieve, and hence we thank you for making our book instrumental in spreading knowledge and fueling creativity.

Pawan Kumar Maurya
Imteyaz Qumar

CHAPTER

1

Novel therapeutic approaches targeting oxidative stress in mood disorders

Alexander Bambokian[a], *Fabiano A. Gomes*[a,b], *Calvin Sjaarda*[a,b], *Claudio N. Soares*[a,b], *Roumen Milev*[a,b], *and Elisa Brietzke*[a,b]

[a]Centre for Neuroscience Studies (CNS), Queen's University, Kingston, ON, Canada, [b]Department of Psychiatry, Queen's University School of Medicine, Kingston, ON, Canada

1 Introduction

Mood disorders, including major depressive disorder (MDD) and bipolar disorder (BD), have been treated with a wide range of pharmacological therapies, including but not restricted to antidepressants, mood stabilizers, and second-generation antipsychotics [1]. However, these interventions are not always effective and may be associated with side effects, reducing the therapeutic options for a large group of individuals. For example, current gold standard antidepressant monotherapies only improve the severity of depressive symptoms in approximately 50% of patients, with only 30% achieving remission [1]. A possible explanation for this limitation is that the main target for all these medications is the monoaminergic brain systems.

Over the past decade, the main efforts of the scientific community for developing new treatments for mood disorders have focused on going beyond the monoaminergic system, concentrating mainly on three main pathways: modulation of neuroplasticity, mitochondrial dysfunction, and oxidative stress. Modulation of neuroplasticity includes changes in

https://doi.org/10.1016/B978-0-323-90905-1.00001-8

synaptic plasticity, dendritic arborization, and cell resilience [2]. Indeed, neuroplasticity impairments became the most prevalent theoretical paradigm to reconcile findings from different lines of evidence. For example, it is well known that neuroplasticity is compromised by hypothalamus-pituitary-adrenal-axis (HPA) activation due to stress, changes in gonadal hormones (e.g., testosterone, estrogen, progesterone), thyroid hormones, persistent long-term inflammatory activation, exposure to alcohol and drugs, and sleep deprivation [3]. In addition, neuroimaging studies have demonstrated reduction of hippocampus volume in MDD and some reduction of cortical thickness and hippocampus volume in BD [4–6].

More recently, treatments have focused on abnormalities in the process by which neurons and glia use energy [7]. For example, there is replicated evidence of mitochondrial dysfunction in both MDD [8] and BD [9,10]. From a more clinical point of view, several studies report a positive association between MDD and BD and systemic energy regulation dysfunction, clinically expressed by a disproportionally high prevalence of obesity and metabolic syndrome in individuals with mood disorders, which happens even in drug-naïve individuals [11,12]. Mansur et al. [13] proposed a model that centered on energy expenditure regulation as a disease model for BD. According to the authors, disruption of energy expenditure is a primary and sufficient cause of BD, with changes in neuroplasticity and neurotransmission being secondary to energy imbalances and general medical comorbidities as compensatory mechanisms [13].

Finally, reducing oxidative stress markers is promising to be an effective therapy in individuals with MDD and BD. Oxidative stress results from an imbalance between reactive oxygen species (ROS) production and their degradation by antioxidant defenses [14]. Physiological brain activity generates ROS, which antioxidant defenses should counterbalance. Over the past 15 years, mounting evidence on the role of oxidative stress has affirmed that both manic and depressive episodes are associated with oxidative stress [15,16]. Most of these studies quantified changes in oxidative stress markers throughout the disease, and during posttherapy periods, in individuals with MDD and BD when compared to healthy controls [17].

Increasing our understanding of the role that oxidative stress plays in the onset of mood disorders like MDD and BD, and development of novel therapeutic approaches targeting these mechanisms, may facilitate (and has already begun to do so) alternative therapies for treatment-resistant patients. However, to understand the therapeutic implications of targeting oxidative stress, it is critical first to recognize the essential role and sources of ROS across the central nervous system (CNS), as well as the physiological properties and cellular structures of the brain that make neurons highly susceptible to oxidative damage.

2 The production of ROS in the brain

The processes by which ROS are produced within neurons are similar to those recorded in other cell types. Oxidative phosphorylation, the mitochondrial metabolic process that provides aerobic organisms with most of their cellular adenosine trisphosphate (ATP), is a significant source of cellular ROS production within neurons. Occasionally during this process, oxygen (O_2) is subjected to a single-electron reduction, forming the superoxide anion radical (O_2^-), which most commonly occurs at Complex III of the electron transport chain (ETC)[2]. Due to its extremely high reactivity, the superoxide anion can immediately react with, and subsequently damage, nearby subcellular structures. Typically, this superoxide anion is rapidly converted into hydrogen peroxide (H_2O_2) via a superoxide dismutase-2 (SOD2) catalyzed dismutation. Despite its greater stability, hydrogen peroxide is also a ROS capable of initiating widespread neuronal damage due to its propensity to diffuse freely across both cellular and subcellular membranes [18].

Though oxidative phosphorylation generates the greatest concentration of intracellular ROS, various other enzymes also contribute to ROS production within the brain, though to a lesser degree. Neuronal nitric oxide synthase (nNOS) is an isoform of the NOS enzyme family that catalyzes L-arginine oxidation, which produces nitric oxide (NO) in both the central and peripheral nervous systems. Though NO production functions to mediate smooth muscle relaxation, regulate blood pressure, and provide synaptic plasticity to the CNS, NO and nNOS are responsible for contributing minor concentrations of ROS[3]. During periods where L-arginine concentrations are reduced, hydrogen peroxide and superoxide are produced in place of NO [18]. Additionally, any preexisting intracellular superoxide anions can interact with NO, creating the highly reactive peroxynitrite (NO_3^-), a reactive nitrogen species that can modify and damage intracellular proteins and disrupt the structural and functional integrity of the neuron. nNOS activity is indirectly regulated via intracellular calcium (Ca_2^+) concentrations, specifically through interactions with the Ca_2^+-binding protein calmodulin. In the brain, nNOS is found primarily anchored to the postsynaptic membrane of neurons, located in close proximity to Ca_2^+-transporting *N*-methyl-D-aspartate (NMDA) receptors [19]. This distribution enables efficient regulation of NO and ROS production based on postsynaptic neuronal depolarization.

The monoamine oxidase (MAO) enzyme family, specifically the MAO-A and MAO-B isoforms, is an additional ROS production source within neurons. MAO utilizes O_2 to catalyze the oxidative deamination of biogenic monoamines, producing H_2O_2 as a by-product. These isoforms function to inactivate monoamine neurotransmitters across the CNS and are found naturally distributed across different neuronal cell types. Though each catalyzes the same reaction, MAO-A is preferentially localized within catecholaminergic neurons, while MOA-B is predominantly situated in serotonergic and histaminergic neurons and glial cells [19,20]. Each form of MAO is bound to the outer mitochondrial membrane and contributes to oxidative stress in congruence with ROS produced through oxidative phosphorylation [21,22].

A third source of neuronal ROS may evolve from nicotinamide adenine dinucleotide phosphate (NADPH) oxidase (NOX), a family of plasma membrane-bound enzyme complexes whose expression across the CNS has only recently been described. Whereas the previous sources that have been discussed produce ROS as a by-product, NOX functions to specifically produce O_2^- via a single-electron transfer from NADPH to molecular oxygen [19]. Superoxide plays critical physiological roles within the brain, such as aiding immune protection against pathogens. However, if not regulated effectively, the overproduction of NOX-generated superoxide can damage nearby subcellular structures. Although the association of NOX with the pathogenesis of several neurodegenerative diseases, including Alzheimer's, Huntington's, and Parkinson's disease is not clearly understood, each disease displays increased NOX activity and heightened levels of oxidative stress markers [23]. The identification of other sources of neuronal ROS, and their role in neurodegenerative diseases and mood disorders, is a highly emphasized field in neuroscience research.

3 The elimination of ROS in the brain

Whether the ROS will have a detrimental or beneficial effect on the neuron is heavily reliant upon the activity of several key enzymes associated with ROS homeostasis. Numerous antioxidant systems ensure that any excess ROS, produced from the sources mentioned previously, is rapidly removed prior to cellular damage. This includes superoxide dismutase (SOD), a group of metalloenzymes that function to provide the first line of defense against ROS-inflicted cellular damage. SOD catalyzes the dismutation of superoxide radical anions into hydrogen peroxide and molecular oxygen, thus comprising an antioxidant mechanism that provides effective regulation over superoxide production [24]. Three SOD isotypes have been identified in humans, each distinguished based on their subcellular or extracellular location. These include SOD-1, SOD-2, and SOD-3, localized specifically in the cytoplasm, mitochondria, and extracellular matrix, respectively [24]. The distribution of the SOD isoforms throughout the cellular interior, and exterior, ensures rapid removal of superoxide, regardless of where it is produced within the cell. Given the significant production of superoxide originating from mitochondrial oxidative phosphorylation, SOD-2 operates as a critical component associated with regulating and thus preventing ROS overproduction. The rise in oxidative stress markers recorded in neurons often implicates SOD-2 dysfunction as a critical constituent associated with mood disorder pathology, though further investigation is required. Though total neuronal superoxide concentrations are heavily reliant upon the activity of mitochondrial SOD-2, SOD-1 and SOD-3 are also critical to maintaining neuronal integrity and preventing additional ROS production outside the mitochondria. SOD-1 maintains appropriate levels of superoxide within the cytoplasm of neurons and SOD-3 regulates extracellular superoxide in the extracellular space surrounding neurons across the CNS. Though not directly associated with neuronal protection, SOD-3 aids the metabolic regulation of neurons throughout the CNS, specifically by altering vascular tone and blood flow to the brain [24]. In unity, all three SOD isoforms provide neurons with an effective antioxidant mechanism that ensures tight regulation over the highly reactive superoxide anion radical.

Glutathione, a cysteine, glycine, and glutamic acid tripeptide, also protects neurons against oxidative stress through its antioxidant properties. Glutathione exists within cells in one of two redox states: a single molecule of reduced (GSH), or as an oxidized glutathione dimer (GSSG), consisting of the two molecules of glutathione bound together via a disulfide bond between cysteine residues. GSH provides antioxidant protection by reducing/neutralizing ROS, such as free radicals and peroxides, into oxygen species that pose a lower threat to cellular damage. This has established glutathione as a widely recognized "ROS scavenger." During this process, the ROS act as an oxidizing agent, converting two GSH molecules into GSSG. Though glutathione comprises a wide variety of physiological functions, including xenobiotic metabolism through conjugation, its antioxidant properties play a significant role in neuroprotection against oxidative stress. Cellular ratios of GSH:GSSG have been effectively used to predict the redox state of an individual cell where a low GSH:GSSG ratio indicates low oxidative stress and a high GSH:GSSG ratio indicates high oxidative stress, thus providing a metric for determining intracellular levels of oxidative stress [25]. Evidence suggests that healthy cells in their resting state have a GSH:GSSG ratio greater than 100:1, compared to 1:10 during periods of oxidative stress [25]. Once depleted, intraneuronal sources of GSH can become restored through one of three methods: the regeneration of oxidized glutathione (GSSG) back to its reduced form (GSH) by glutathione reductase, de novo synthesis from its constituent amino acids via a two-step catalyzed reaction, and the recycling of cysteine residues from conjugated glutathione [25]. The cystine residue in GSH regulates de novo synthesis of glutathione based on its intracellular availability, and thus encompasses a rate-limiting step that can be manipulated when targeting oxidative stress with respect to treating mood disorders.

4 Oxidative stress in the brain

Ultimately, oxidative stress will occur when the damaging effects of the ROS exceed the biological system's ability to neutralize the ROS and repair cellular damage. As seen across many cell types, oxidative conditions that overwhelm repair mechanisms across the CNS can cause potentially irreversible damage to neuronal proteins, lipids, and DNA. Amino acid side chains and protein backbones can become oxidized by ROS, eliciting detrimental effects on structural cytoskeletal and functional neuronal proteins. The loss of cellular integrity induced by protein modification can result in neuronal dysfunction, an outcome that has a significant implication for many neurodegenerative diseases and mood disorders. Base modification/damage and strand breaks by ROS, in both DNA and mitochondrial DNA (mtDNA), are responsible for a majority of ROS-inflicted neuronal dysfunction. More than 20 variations of oxidized bases have been discovered, though guanine oxidation to 8-oxoguanine has been recorded most frequently [26]. Polyunsaturated fatty acids are also common oxidative substrates, often subject to free radical or singlet oxygen oxidation.

Oxidative damage to mtDNA, proteins, and lipids can impair mitochondrial ATP production, as metabolic processes such as oxidative phosphorylation, fatty acid oxidation, and the tricarboxylic acid (TCA) cycle, among others, become disrupted. If sufficient damage has occurred such that the structural integrity of the mitochondria has been compromised, apoptotic factors, including cytochrome *c* and apoptosis-inducing factor (AIF), may leak from the mitochondrial intermembrane space into the cytosol, prompting apoptosis [27]. Alternatively, if significant concentrations of ROS are present, necrosis may result. Regardless of the macromolecular damage sustained by ROS, the loss of structural and functional neuronal integrity can result in cell death.

Considering the anatomical features of cells across the CNS, neurons are very vulnerable to oxidative damage compared to other cell types that compose different systems. The brain consumes 20% of the body's oxygen supply through aerobic processes such as oxidative phosphorylation. The significant utilization of oxygen throughout the CNS promotes ROS formation through metabolic processes at a rate that exceeds other cell types. Thus, if antioxidant mechanisms become overwhelmed by the large concentration of ROS produced throughout the CNS, the neurons can become subject to widespread macromolecular damage. Furthermore, the brain is particularly enriched with polyunsaturated fatty acids (PUFAs), such as arachidonic and docosahexaenoic acids, which can act as a suitable oxidative substrate [28]. PUFAs have several essential functions across the nervous system, including participation in axonal growth, neuronal development, memory, inflammatory response, neuronal membranes, and myelin formation [28]. These critical functions can become impaired following oxidative modification/damage, promoting neurodegeneration, cell death, and possibly the onset of affective disorders.

However, current research suggests that neurodegeneration may not only occur as a result of ROS-inflicted damage, but also have an effect on the activation of inflammatory pathways [29]. Oxidative damage of neuronal proteins and lipids can result in significant modification of cellular structures, generating novel, highly immunogenic epitopes [29]. Lipid membrane components that are modified by oxidative stress can be recognized by numerous immune cells, which can elicit a robust inflammatory response against the neuron. The interactions between oxidative stress and inflammatory pathway-mediated neurodegeneration are currently under review, particularly to determine their role in the pathogenesis of numerous mood disorders.

5 Markers of oxidative stress and mood disorders

Several converging bodies of evidence support the role of oxidative stress in mood disorders [30]. To date, more than 400 studies support the association of oxidative stress and MDD or BD [30]. This research includes data from both animal and human studies, especially those evaluating the levels of different oxidative mediators in peripheral blood. It is thought that oxidative mediators are secondary messengers for the transcriptional regulation of genes involved in the regulation of neuroplasticity [31]. Here, we summarize the results of the human studies supporting the association between products of oxidative stress and mood disorders.

There is no consensus about the best biomarkers to evaluate the severity of oxidative stress. Most oxidative species have a short half-life and thus are difficult to measure directly. Consequently, different biomarkers including lipid peroxidation, protein oxidation, and DNA oxidative damage have been used.

Several comparative studies and a meta-analysis of different oxidative stress markers suggest that oxidative stress is increased and antioxidant defenses are reduced in MDD [32]. For example, different studies support the theory that increased levels of malondialdehyde (MDA), evaluated by the levels of thiobarbituric acidic reactive substance (TBARS), are a main indicator of cell membrane lipid peroxidation, in both MDD [33] and BD [34]. Peripheral levels of TBARS are not specific to any of these conditions and are similar between MDD and BD [33]. TBARS has some well-known limitations, including the facts that MDA is not only a product of lipid peroxidation, but also a by-product of prostaglandin synthesis, and that its levels can be overestimated by the assays that are commonly used. Therefore, MDA and TBARS may not be the best indicators of oxidative stress in vivo [35,36].

F2-isoprostanes, oxidized derivatives of arachidonic acid, have also been evaluated in mood disorder. It has been postulated that these molecules are precise biomarkers of lipid peroxidation within the brain [37]. Different studies have found elevated peripheral levels of F2-isoprostanes in MDD [38,39].

Protein oxidation may be another potential marker for quantifying oxidative stress in the brain. Protein oxidation is defined by a modification in the protein's structure induced by the direct or indirect action of ROS—for example, by a relative lack of antioxidant defenses. The primary process by which ROS prompts protein oxidation is the oxidation of amino acids (especially cysteine and methionine) and also in the bridges between them. These modifications will induce changes in the physical structure of proteins, or their chemical properties, including their susceptibility to the action of proteolytic enzymes [40]. Oxidation of proteins has been evaluated by the level of protein carbonyl content (PCC). Although protein oxidation has been less studied compared to lipid peroxidation, there is also some evidence of protein oxidation in MDD [15] and BD [41].

In the same way that ROS may cause damage to lipids and proteins, DNA can also be susceptible to these mediators [42]. The molecule that is the index of damage to DNA is the 8-oxoG, a guanine analogue. The presence of 8-oxoG in the DNA impairs the latter's structure and function, as it tends to pair with adenine rather than cytosine during transcription; this can result in a protein with a different amino acid. Increased levels of 8-oxoG were described in urine, plasma, and leucocytes in individuals with MDD [42]. DNA damage was also described in BD, although this damage was evaluated using other methods, such as comet assay [43].

The accumulation of these biomarkers of oxidative stress can have detrimental effects on neuronal integrity, creating an association between oxidative stress and the development of mood disorders. Oxidative modifications of cellular structures can impair critical neuronal processes, such as mitochondrial signaling and neuronal excitability [44]. The adverse effect on neuronal phenotype that is prompted by oxidative damage contributes to the pathology of numerous mood disorders. Interneuron deficits, characterized by the presence of these previously described markers of oxidation, have been observed in the brains of patients with both schizophrenia and various other mood disorders [44]. To determine the exact mechanism of oxidative stress and neurological impairment in the development of mood disorders, animal models of oxidative stress are continuously being studied.

6 Antioxidant treatments in mood disorders

As oxidative stress is involved in the pathophysiology of MDD and BD, a natural next step is to target this system to obtain an improvement of those conditions. A first approach to the implications of oxidative stress in mood disorders is related to the potential antioxidant action of already commonly used pharmacological and nonpharmacological agents. A second method is the use of antioxidant molecules as adjunctive treatments in mood disorders. Finally, a third approach is the manipulation of oxidative stress through lifestyle changes such as diet and exercise. In this section, these three strategies will be described and explained.

6.1 Antioxidant action of conventional treatments for mood disorders

Antidepressants, such as selective serotonin reuptake inhibitors (SSRIs), are considered the first-line treatment in MDD [45]. In vitro, SSRIs have the potential to inhibit the enzymatic system involved in the production of ROS [46]. Treatment with SSRIs such as fluoxetine and citalopram is associated with increased levels of SOD activity, reduction in lipid peroxidation, and increased levels of ascorbic acid, all indicators of an enhanced antioxidant defense system. These benefits were demonstrated in a clinical trial in humans [47]. Interestingly, the potential antioxidant effect of the SSRIs, together with an antiinflammatory effect, was the basis of the proposal of a potential utility of these agents in COVID-19 treatment [48,49].

Another antidepressant with antioxidant properties is lithium. This has been found to prevent or antagonize DNA damage, free radical formation, and lipid peroxidation in diverse models of BD [50]. Lithium has also been shown to reduce the levels of SOD in the prefrontal cortex in amphetamine models of mania [51] and to improve peripheral markers of oxidative stress in a proof of concept twin-study [52]. These results were replicated in human studies, both in healthy subjects and in individuals with BD. Khairova et al. [53] investigated the effect of lithium administration in the oxidative stress system of healthy volunteers and found that the administration of lithium for 2–4 weeks was associated with a significant reduction in SOD levels with a decrease in the ratio of SOD:CAT, which provoked a decrease in cell hydrogen peroxide levels. Lithium has also been associated with a decrease in lipid peroxidation, evident by a reduction of TBARS peripheral levels, and an association between a reduction of antioxidant mechanisms and clinical response to lithium in BD [54]. These studies support an antioxidant role for lithium and suggest that the beneficial neuroprotective action of lithium in BD may be due in part to antioxidant action.

Finally, a more contemporary approach proposes that antioxidant properties of antidepressants and mood stabilizers may be partially responsible for the benefit of those agents in the clinical trajectory of BD and multiepisodic recurrent MDD [55]. The concept of neuroprogression systematizes a robust body of evidence supporting the idea that—at least for a large proportion of patients with BD and MDD—the clinical course is related to progressive clinical deterioration, reduction in the probability of response to treatment, and progressive worsening of cognitive performance and functional status [56]. It has been proposed that, together with neurotrophic factors and a low grade and persistent proinflammatory activation, oxidative stress imbalances can be critical factors in neuroprogression. If this theory is proven to be true in longitudinal studies, targeting oxidative stress will become a promising way to produce a beneficial impact in the clinical course of BD and MDD [57].

6.2 Antioxidants as an add-on therapy in mood disorders

Different antioxidant molecules have been explored as potential antidepressants and mood stabilizers. The main rationale for their introduction is that although traditionally used psychotropic agents might have antioxidant properties, they are probably not robust enough to guarantee an effective antidepressant and antimanic action. In this section, the most investigated approaches are explained.

6.2.1 N-acetylcysteine (NAC)

One of the most investigated antioxidant substances in mood disorders is NAC. This is a synthetic derivate of the endogenous amino acid L-cysteine and a precursor of one of the most important antioxidant molecules, glutathione. NAC has been available for years for the treatment of cough and as an antidote to paracetamol overdose. Michael Berk and collaborators conducted several studies with NAC in the treatment of psychiatric disorders, including MDD and BD [58,59]. Its main mechanism of action is the improvement of the antioxidant defense system, but it also has been suggested that NAC may affect the modulation of glutamatergic transmission. NAC effect in MDD has been investigated in both preclinical and clinical studies. Animal studies support the antiglutamatergic and antioxidant actions of NAC in different brain areas, including the hippocampus [60,61]. Clinical trials with this substance suggest a potential of NAC as an add-on treatment in depression. Although most trials did not show a difference between the substance and placebo in the severity of depressive symptoms assessed by standardized instruments in the short term, there is some evidence in favor of NAC at the 16-week postdiscontinuation outcomes [58]. A beneficial effect of NAC in secondary outcomes such as quality of life has also been described. NAC was also investigated in BD, mainly as a treatment for bipolar depression, by the same group. The authors did not find a strong effect of NAC supplementation in bipolar depression, although it was possible to detect some positive secondary signals related to a late response [59]. Similarly, the most recent meta-analysis involving NAC in MDD and BD did not find a robust antimanic effect, although a larger effect was found in the treatment of schizophrenia. Indeed, the neuroprotective properties of NAC did not seem to be specific to mood disorders, and were also described in other mental illnesses such as addictions, anxiety disorders, autism spectrum disorders, impulse control disorders, attention deficit and hyperactivity disorder, obsessive-compulsive disorder, and schizophrenia [62].

Antioxidant vitamins (vitamin E/alpha-tocopherol; vitamin C/ascorbic acid and beta-carotene) and trace elements (zinc): It is well known that any situation of stress and homeostatic imbalance will be related to an increase in adrenal production and mobilization, and utilization of vitamins and minerals, such as vitamin C, vitamin E, magnesium, potassium, and micronutrients, which accelerates the metabolism of proteins, fats, and carbohydrates to produce readily available energy sources to overcome a real or perceived threat [63]. Specific vitamins and zinc are being considered as a secondary antioxidant defense as nonenzymatic antioxidant molecules. There is no consistent evidence that individuals with MDD or BD have a higher chance than the general population of presenting deficiencies of antioxidant vitamins. However, there are some studies in humans suggesting that supplementation of those elements can be associated with a potential benefit in the treatment of mood disorders. For example, there is preliminary evidence in favor of the supplementation of vitamin C in depression [64]. Regarding zinc, this trace mineral has been implicated in the pathophysiology of depression for a long time. Zinc is found sequestered by a family of proteins called metallothioneins (MTs), and is released during inflammation and neuronal damage. Metallothioneins are also a well-known antioxidant group of molecules and have cytoprotective action [65].

6.2.2 Medications with some antioxidant action but not as the main mechanism of action

Some drugs that are already in use for the treatment of nonpsychiatric conditions have been recently repurposed as antidepressants and mood stabilizers. This is the case of metformin, for example. Metformin is the most used medication in treating type 2 diabetes mellitus. It has been recognized as having pleiotropic actions, some of them potentially beneficial in depression, a common comorbidity in diabetes. Metformin has recently been shown to possess antioxidant properties and has the capacity to enhance synaptic plasticity and serotonin activity [66].

6.3 Manipulation of oxidative stress through lifestyle interventions

It is generally accepted that traditional dietary patterns, such as the Mediterranean diet and the consumption of specific foods such as fish, fresh vegetables, and fruits, have been associated with a lower risk of depression or lower severity of depressive symptoms. In contrast, high-fat Western diets, rich in ultra-processed foods and sugar-sweetened beverages, have been associated with a higher risk of depressive symptoms in all age groups [67,68]. Different studies have shown that diet and nutrition are involved in the pathophysiology of depression, the risk of the emergence of depressive symptoms in the general population, and the severity of depressive symptoms [69,70]. Supplementation with antioxidant foods such as green tea polyphenols or isoflavonoid intake has also been shown to have a positive impact on the reduction of the severity of depressive symptoms [71]. In the same way, dietetic interventions are among the most promising approaches in the treatment of mood disorders [72].

Exercise is another important lifestyle intervention that has been increasingly studied in depression. Higher levels of physical activities are associated with reductions in both the incidence and prevalence of common mental disorders [73,74]. Recent meta-analyses have documented the positive effect of exercise in reducing depressive symptoms as both a monotherapy and adjunctive treatment [75]. Several potential neurobiological mechanisms have been implicated in this association, including changes in cardiovascular and endocrine mediators, neuroplasticity, inflammation, and oxidative stress [76,77].

7 Conclusions

Mood disorders, including MDD and BD, remain a significant challenge in public health, not only because they are highly prevalent, but also because currently available treatment options are limited in providing relief for a significant percentage of patients. Improved understanding of the actions of oxidative stress in the brain opens a new avenue of investigation, allowing science to expand the knowledge of the regulation of neuroplasticity processes, but also generating insights for very novel interventions. In parallel, this new knowledge also increases the interest in using simple interventions such as a healthy diet for the prevention and treatment of depression.

Although the state of the art of mood disorders includes the relevance of oxidative stress imbalances among the abnormalities involved in the pathophysiology of mood disorders, this new knowledge was not fully translated to clinical studies with real-life patients, and there has been no approved antioxidant intervention in the treatment of mood disorders. Patients' and doctors' hopes are for a quick expansion of research on these topics and the incorporation of new alternatives to the therapeutic arsenal that can be used for the treatment of mood disorders.

References

[1] R.S. McIntyre, M. Berk, E. Brietzke, B.I. Goldstein, C. López-Jaramillo, L.V. Kessing, G.S. Malhi, A.A. Nierenberg, J.D. Rosenblat, A. Majeed, E. Vieta, M. Vinberg, A.H. Young, R.B. Mansur, Bipolar disorders, Lancet 396 (10265) (2020) 1841–1856.

[2] P.N. Moretti, V.K. Ota, E.S. Gouvea, M. Pedrini, M.L. Santoro, F. Talarico, L.M. Spindola, C.M. Carvalho, C. Noto, G. Xavier, E. Brietzke, A. Gadelha, R. Bressan, J. Mari, S. Belangero, Accessing gene expression in treatment-resistant schizophrenia, Mol. Neurobiol. 55 (8) (2018) 7000–7008.

[3] J.D. Rosenblat, E. Brietzke, R.B. Mansur, N.A. Maruschak, Y. Lee, R.S. McIntyre, Inflammation as a neurobiological substrate of cognitive impairment in bipolar disorder: evidence, pathophysiology and treatment implications, J. Affect. Disord. 188 (2015) 149–159.

[4] D.P. Hibar, L.T. Westlye, N.T. Doan, N. Jahanshad, J.W. Cheung, C.R.K. Ching, A. Versace, A.C. Bilderbeck, A. Uhlmann, B. Mwangi, B. Krämer, B. Overs, C.B. Hartberg, C. Abé, D. Dima, D. Grotegerd, E. Sprooten, E. Bøen, E. Jimenez, F.M. Howells, G. Delvecchio, H. Temmingh, J. Starke, J.R.C. Almeida, J.M. Goikolea, J. Houenou, L.M. Beard, L. Rauer, L. Abramovic, M. Bonnin, M.F. Ponteduro, M. Keil, M.M. Rive, N. Yao, N. Yalin, P. Najt, P.G. Rosa, R. Redlich, S. Trost, S. Hagenaars, S.C. Fears, S. Alonso-Lana, T.G.M. van Erp, T. Nickson, T.M. Chaim-Avancini, T.B. Meier, T. Elvsåshagen, U.K. Haukvik, W.H. Lee, A.H. Schene, A.J. Lloyd, A.H. Young, A. Nugent, A.M. Dale, A. Pfennig, A.M. McIntosh, B. Lafer, B.T. Baune, C.J. Ekman, C.A. Zarate, C.E. Bearden, C. Henry, C. Simhandl, C. McDonald, C. Bourne, D.J. Stein, D.H. Wolf, D.M. Cannon, D.C. Glahn, D.J. Veltman, E. Pomarol-Clotet, E. Vieta, E.J. Canales-Rodriguez, F.G. Nery, F.L.S. Duran, G.F. Busatto, G. Roberts, G.D. Pearlson, G.M. Goodwin, H. Kugel, H.C. Whalley, H.G. Ruhe, J.C. Soares, J.M. Fullerton, J.K. Rybakowski, J. Savitz, K.T. Chaim, M. Fatjó-Vilas, M.G. Soeiro-de-Souza, M.P. Boks, M.V. Zanetti, M.C.G. Otaduy, M.S. Schaufelberger, M. Alda, M. Ingvar, M.L. Phillips, M.J. Kempton, M. Bauer, M. Landén, N.S. Lawrence, N.E.M. van Haren, N.R. Horn, N.B. Freimer, O. Gruber, P.R. Schofield, P.B. Mitchell, R.S. Kahn, R. Lenroot, R. Machado-Vieira, R.A. Ophoff, S. Sarró, S. Frangou, T.D. Satterthwaite, T. Hajek, U. Dannlowski, U.F. Malt, V. Arolt, W.F. Gattaz, W.C. Drevets, X. Caseras, I. Agartz, P.M. Thompson, O.A. Andreassen, Cortical abnormalities in bipolar disorder: an MRI analysis of 6503 individuals from the ENIGMA bipolar disorder working group, Mol. Psychiatry 23 (4) (2018) 932–942.

[5] J. Sato, Y. Hirano, N. Hirakawa, J. Takahashi, N. Oribe, H. Kuga, I. Nakamura, S. Hirano, T. Ueno, O. Togao, A. Hiwatashi, T. Nakao, T. Onitsuka, Lower hippocampal volume in patients with schizophrenia and bipolar disorder: a quantitative MRI study, J Pers Med. 11 (2) (2021) 121.

[6] L. Schmaal, D.P. Hibar, P.G. Sämann, G.B. Hall, B.T. Baune, N. Jahanshad, J.W. Cheung, T.G.M. van Erp, D. Bos, M.A. Ikram, M.W. Vernooij, W.J. Niessen, H. Tiemeier, A. Hofman, K. Wittfeld, H.J. Grabe, D. Janowitz, R. Bülow, M. Selonke, H. Völzke, D. Grotegerd, U. Dannlowski, V. Arolt, N. Opel, W. Heindel, H. Kugel, D. Hoehn, M. Czisch, B. Couvy-Duchesne, M.E. Rentería, L.T. Strike, M.J. Wright, N.T. Mills, G.I. de Zubicaray, K.L. McMahon, S.E. Medland, N.G. Martin, N.A. Gillespie, R. Goya-Maldonado, O. Gruber, B. Krämer, S.N. Hatton, J. Lagopoulos, I.B. Hickie, T. Frodl, A. Carballedo, E.M. Frey, L.S. van Velzen, B.W.J.H. Penninx, M.J. van Tol, N.J. van der Wee, C.G. Davey, B.J. Harrison, B. Mwangi, B. Cao, J.C. Soares, I.M. Veer, H. Walter, D. Schoepf, B. Zurowski, C. Konrad, E. Schramm, C. Normann, K. Schnell, M.D. Sacchet, I.H. Gotlib, G.M. Mac Queen, B.R. Godlewska, T. Nickson, A.M. McIntosh, M. Papmeyer, H.C. Whalley, J. Hall, J.E. Sussmann, M. Li, M. Walter, L. Aftanas, I. Brack, N.A. Bokhan, P.M. Thompson, D.J. Veltman, Cortical abnormalities in adults and adolescents with major depression based on brain scans from 20 cohorts worldwide in the ENIGMA major depressive disorder working group, Mol. Psychiatry 22 (6) (2017) 900–909.

[7] J. Sacramento, A. Wichert, M.C. van Rossum, Energy efficient sparse connectivity from imbalanced synaptic plasticity rules, PLoS Comput. Biol. 11 (6) (2015), e1004265.
[8] E. Tobe, Mitochondrial dysfunction, oxidative stress, and major depressive disorder, Neuropsychiatr. Dis. Treat. 567 (2013).
[9] H. Jeong, M.K. Dimick, A. Sultan, A. Duong, S.S. Park, E.S. El Sabbagh, D, Goldstein BI, Andreazza AC., Peripheral biomarkers of mitochondrial dysfunction in adolescents with bipolar disorder, J. Psychiatr. Res. 123 (2020) 187–193.
[10] M. Lundberg, V. Millischer, L. Backlund, L. Martinsson, P. Stenvinkel, C.M. Sellgren, C. Lavebratt, M. Schalling, Lithium and the interplay between telomeres and mitochondria in bipolar disorder, Front. Psych. 11 (2020), 586083.
[11] R.B. Grigolon, E. Brietzke, R.B. Mansur, M.A. Idzikowski, F. Gerchman, F.G. De Felice, R.S. McIntyre, Association between diabetes and mood disorders and the potential use of anti-hyperglycemic agents as antidepressants, Prog. Neuropsychopharmacol. Biol. Psychiatry 95 (2019), 109720.
[12] R.B. Mansur, E. Brietzke, R.S. McIntyre, Is there a "metabolic-mood syndrome"? A review of the relationship between obesity and mood disorders, Neurosci. Biobehav. Rev. 52 (2015) 89–104.
[13] R.B. Mansur, Y. Lee, R.S. McIntyre, E. Brietzke, What is bipolar disorder? A disease model of dysregulated energy expenditure, Neurosci. Biobehav. Rev. 113 (2020) 529–545.
[14] G. Pizzino, N. Irrera, M. Cucinotta, G. Pallio, F. Mannino, V. Arcoraci, F. Squadrito, D. Altavilla, A. Britto, Oxidative stress: harms and benefits for human health, Oxid. Med. Cell. Longev. 2017 (2017) 8416763.
[15] B.S. Diniz, A.P. Mendes-Silva, L.B. Silva, L. Bertola, M.C. Vieira, J.D. Ferreira, M. Nicolau, G. Bristot, E.D. da Rosa, A.L. Teixeira, F. Kapczinski, Oxidative stress markers imbalance in late-life depression, J. Psychiatr. Res. 102 (2018) 29–33.
[16] B. Pfaffenseller, G.R. Fries, B. Wollenhaupt-Aguiar, G.D. Colpo, L. Stertz, B. Panizzutti, P.V. Magalhães, F. Kapczinski, Neurotrophins, inflammation and oxidative stress as illness activity biomarkers in bipolar disorder, Expert Rev. Neurother. 13 (7) (2013) 827–842.
[17] T. Liu, S. Zhong, X. Liao, J. Chen, T. He, S. Lai, Y. Jia, A meta-analysis of oxidative stress markers in depression, PLoS One 10 (10) (2015), e0138904.
[18] Y. Xu, C. Wang, J.J. Klabnik, J.M. O'Donnell, Novel therapeutic targets in depression and anxiety: antioxidants as a candidate treatment, Curr. Neuropharmacol. 12 (2) (2014) 108–119, https://doi.org/10.2174/1570159X11666131120231448.
[19] T.F. Beckhauser, J. Francis-Oliveira, R. De Pasquale, Reactive oxygen species: physiological and physiopathological effects on synaptic plasticity, J. Exp. Neurosci. 10 (Suppl. 1) (2016) 23–48, https://doi.org/10.4137/JEN.S39887.
[20] Q. Chen, E.J. Vazquez, S. Moghaddas, C.L. Hoppel, E.J. Lesnefsky, Production of reactive oxygen species by mitochondria: central role of complex III, J. Biol. Chem. 278 (38) (2003) 36027–36031.
[21] B. Heinzel, M. John, P. Klatt, E. Böhme, B. Mayer, Ca2+/calmodulin-dependent formation of hydrogen peroxide by brain nitric oxide synthase, Biochem. J. 281 (Pt 3) (1992) 627–630.
[22] J.C. Shih, K. Chen, M.J. Ridd, Monoamine oxidase: from genes to behavior, Annu. Rev. Neurosci. 22 (1999) 197–217.
[23] M.W. Ma, J. Wang, Q. Zhang, R. Wang, K.M. Dhandapani, R.K. Vadlamudi, D.W. Brann, NADPH oxidase in brain injury and neurodegenerative disorders, Mol. Neurodegener. 12 (1) (2017) 7.
[24] J.M. Flynn, S. Melov, SOD2 in mitochondrial dysfunction and neurodegeneration, Free Radic. Biol. Med. 62 (2013) 4–12.
[25] J. Pizzorno, Glutathione! Integr. Med. (Encinitas). 13 (1) (2014) 8–12.
[26] M.S. Cooke, M.D. Evans, M. Dizdaroglu, J. Lunec, Oxidative DNA damage: mechanisms, mutation, and disease, FASEB J. 17 (10) (2003) 1195–1214.

[27] M.P. Murphy, How mitochondria produce reactive oxygen species, Biochem. J. 417 (1) (2009) 1–13.
[28] L.J. Falomir-Lockhart, G.F. Cavazzutti, E. Giménez, A.M. Toscani, Fatty acid signaling mechanisms in neural cells: fatty acid receptors, Front. Cell. Neurosci. 13 (2019) 162.
[29] M. Siwek, M. Sowa-Kućma, D. Dudek, K. Styczeń, B. Szewczyk, K. Kotarska, P. Misztakk, A. Pilc, M. Wolak, G. Nowak, Oxidative stress markers in affective disorders, Pharmacol. Rep. 65 (6) (2013) 1558–1571.
[30] T.M. Michel, D. Pulschen, J. Thome, The role of oxidative stress in depressive disorders, Curr. Pharm. Des. 18 (36) (2012) 5890–5899.
[31] D. Lindqvist, F.S. Dhabhar, S.J. James, C.M. Hough, F.A. Jain, F. Saverio Bersani, V.I. Reus, J.E. Verhoeven, E.S. Espel, L. Mahan, R. Rosser, O.M. Wolkowitz, S.H. Mellon, Oxidative stress, inflammation and treatment response in major depression, Psychoendocrinology 76 (2017) 197–205.
[32] C.N. Black, M. Bot, P.G. Scheffer, P. Cuijpers, B.W.J.H. Penninx, Is depression associated with increased oxidative stress? A systematic review and meta-analysis, Psychoneuroendocrinology 51 (2015) 164–175.
[33] M. Sowa-Kucma, K. Styczen, M. Siwek, P. Misztak, R.J. Nowak, D. Dudek, J.K. Rybakowski, G. Nowak, M. Maes, Lipid peroxidation and immune biomarkers are associated with major depression and its phenotypes, including treatment-resistant depression and melancholia, Neurotox. Res. 33 (2) (2018) 448–460.
[34] M. Siwek, M. Sowa-Kucma, K. Styczen, P. Misztak, B. Szewczyk, R. Topor-Madry, G. Nowak, D. Dudek, J.K. Rybakowski, Thiobarbituric acid-reactive substances: markers of an acute episode and a late stage of bipolar disorder, Neuropsychobiology 73 (2) (2016) 116–122.
[35] I. Dalle-Donne, R. Rossi, R. Colombo, D. Giustarini, A. Milzani, Biomarkers of oxidative damage in human disease, Clin. Chem. 52 (2006) 601–623.
[36] E.A. Meagher, G.A. Fitzgerald, Indices of lipid peroxidation in vivo: strengths and limitations, Free Radic. Biol. Med. 28 (2000) 1745–1750.
[37] E. Niki, Biomarkers of lipid peroxidation in clinical material, Biochim. Biophys. Acta 1840 (2014) 809–817.
[38] C.P. Chung, D. Schmidt, C.M. Stein, J.D. Morrow, R.M. Salomon, Increased oxidative stress in patients with depression and its relationship to treatment, Psychiatry Res. 206 (2013) 213–216.
[39] N. Dimopoulus, C. Piperi, V. Psarra, R.W. Lea, A. Kalofoutis, Increased plasma levels of 8-iso-PGF2alpha and IL-6 in an elderly population with depression, Psychiatry Res. 161 (2008) 59–66.
[40] W. Zhang, S. Xiao, D.U. Ahn, Protein oxidation: basic principles and implications for meat quality, Crit. Rev. Food Sci. Nutr. 53 (11) (2013) 1191–1201.
[41] T.C. Mondin, T.A. Cardoso, F.P. Moreira, C. Wiener, J.P. Oses, L.D.M. Souza, K. Jansen, P.V.S. Magalhaes, F. Kapczinski, R.A. Silva, Circadian preferences, oxidative stress and inflammatory cytokines in bipolar disorder: a community study, J. Neuroimmunol. 301 (2016) 23–29.
[42] P. Czarny, P. Wigner, P. Galecki, T. Sliwinski, The interplay between inflammation, oxidative stress, DNA damage, DNA repair and mitochondrial dysfunction in depression, Prog. Neuropsychopharmacol. Biol. Psychiatry 80 (Pt C) (2018) 309–321.
[43] A.C. Andreazza, B.N. Frey, B. Erdtmann, M. Salvador, F. Rombaldi, A. Santin, C.A. Goncalves, F. Kapczinski, DNA damage in bipolar disorder, Psychiatry Res. 153 (1) (2007) 27–32.
[44] S. Salim, Oxidative stress and psychological disorders, Curr. Neuropharmacol. 12 (2) (2014) 140–147.
[45] C. Otte, S.M. Gold, B.W. Penninx, C.M. Pariante, A. Etkin, M. Fava, D.C. Mohr, A.F. Schatzberg, Major depressive disorder, Nat. Rev. Dis. Primers. 2 (2016) 16065.
[46] O. Dalmizrak, K. Teralı, E.B. Asuquo, I.H. Ogus, N. Ozer, The relevance of glutathione reductase inhibition by fluoxetine to human health and disease: insights derived from a combined kinetic and docking study, Protein J. 38 (5) (2019) 515–524.

[47] S.D. Khanzode, G.N. Dakhale, S.S. Khanzode, A. Saoji, R. Palasodkar, Oxidative damage and major depression: the potential antioxidant action of selective serotonin reuptake inhibitors, Redox Rep. 8 (6) (2003) 365–370.
[48] M.G.M. Hamed, R.S. Hagag, The possible immunoregulatory and antiinflammatory effects of selective serotonin reuptake inhibitors in coronavirus disease patients, Med. Hypotheses 144 (2020), 110140.
[49] E.J. Lenze, C. Mattar, C.F. Zorumski, A. Stevens, J. Schweiger, G.E. Nicol, J.P. Miller, L. Yang, M. Yingling, M.S. Avidan, A.M. Reiersen, Fluvoxamine vs placebo and clinical deterioration in outpatients with symptomatic COVID-19: a randomized clinical trial, JAMA 324 (22) (2020) 2292–2300.
[50] L. Shao, L.T. Young, J.F. Wang, Chronic treatment with mood stabilizers lithium and valproate prevents excitotoxicity by inhibiting oxidative stress in rat cerebral cortical cells, Biol. Psychiatry 58 (11) (2005) 879–884.
[51] B.N. Frey, S.S. Valvassori, G.Z. Réus, M.R. Martins, F.C. Petronilho, K. Bardini, F. Dal-Pizzol, F. Kapczinski, J. Quevedo, Effects of lithium and valproate on amphetamine-induced oxidative stress generation in an animal model of mania, J. Psychiatry Neurosci. 31 (5) (2006) 326–332.
[52] B.N. Frey, A.C. Andreazza, M. Kunz, F.A. Gomes, J. Quevedo, M. Salvador, C.A. Gonçalves, F. Kapczinski, Increased oxidative stress and DNA damage in bipolar disorder: a twin-case report, Prog. Neuropsychopharmacol. Biol. Psychiatry 31 (1) (2007) 283–285.
[53] R. Khairova, R. Pawar, G. Salvadore, M.F. Juruena, R.T. de Sousa, M.G. Soeiro-de-Souza, M. Salvador, C.A. Zarate, W.F. Gattaz, R. Machado-Vieira, Effects of lithium on oxidative stress parameters in healthy subjects, Mol. Med. Rep. 5 (3) (2012) 680–682.
[54] R.T. de Sousa, C.A. Zarate Jr., M.V. Zanetti, A.C. Costa, L.L. Talib, W.F. Gattaz, R. Machado-Vieira, Oxidative stress in early stage bipolar disorder and the association with response to lithium, J. Psychiatr. Res. 50 (2014) 36–41.
[55] B. Wollenhaupt-Aguiar, F. Kapczinski, B. Pfaffenseller, Biological pathways associated withNeuroprogressionin bipolar disorder, Brain Sci. 11 (2) (2021) 228.
[56] G. Morris, B.K. Puri, A.J. Walker, M. Maes, A.F. Carvalho, C.C. Bortolasci, K. Walder, M. Berk, Shared pathways for neuroprogression and somatoprogression in neuropsychiatric disorders, Neurosci. Biobehav. Rev. 107 (2019) 862–882.
[57] R. Kupka, A. Duffy, J. Scott, J. Almeida, V. Balanzá-Martinez, B. Birmaher, D.J. Bond, E. Brietzke, I. Chendo, B.N. Frey, I. Grande, D. Hafeman, T. Hajek, M. Hillegers, M. Kauer-Sant'Anna, R.B. Mansur, A. van der Markt, R. Post, M. Tohen, H. Tremain, G.H. Vazquez, E. Vieta, L. Yatham, M. Berk, M. Alda, F. Kapczinski, Consensus on nomenclature for clinical staging models in bipolar disorder: a narrative review from the International Society for Bipolar Disorders (ISBD) staging task force, Bipolar Disord. 23 (7) (2021) 659–678, https://doi.org/10.1111/bdi.13105.
[58] M. Berk, O.M. Dean, S.M. Cotton, S. Jeavons, M. Tanious, K. Kohlmann, K. Hewitt, K. Moss, C. Allwang, I. Schapkaitz, J. Robbins, H. Cobb, F. Ng, S. Dodd, A.I. Bush, G.S. Malhi, The efficacy of adjunctive N-acetylcysteine in major depressive disorder: a double-blind, randomized, placebo-controlled trial, J. Clin. Psychiatry 75 (6) (2014) 628–636.
[59] M. Berk, A. Turner, G.S. Malhi, C.H. Ng, S.M. Cotton, S. Dodd, Y. Samuni, M. Tanious, C. McAulay, N. Dowling, J. Sarris, L. Owen, A. Waterdrinker, D. Smith, O.M. Dean, A randomised controlled trial of a mitochondrial therapeutic target for bipolar depression: mitochondrial agents, N-acetylcysteine, and placebo, BMC Med. 17 (1) (2019) 18.
[60] J. Fernandes, G.L. Gupta, N-acetylcysteine attenuates neuroinflammation associated depressive behavior induced by chronic unpredictable mild stress in rat, Behav. Brain Res. 364 (2019) 356–365.
[61] I. Smaga, B. Pomierny, W. Krzyżanowska, L. Pomierny-Chamioło, J. Miszkiel, E. Niedzielska, A. Ogórka, M. Filip, N-acetylcysteine possesses antidepressant-like activity through reduction of oxidative stress: behavioral and biochemical analyses in rats, Prog. Neuropsychopharmacol. Biol. Psychiatry 39 (2) (2012) 280–287.

[62] W. Zheng, Q.E. Zhang, D.B. Cai, X.H. Yang, Y. Qiu, G.S. Ungvari, C.H. Ng, M. Berk, Y.P. Ning, Y.T. Xiang, N-acetylcysteine for major mental disorders: a systematic review and meta-analysis of randomized controlled trials, Acta Psychiatr. Scand. 137 (5) (2018) 391–400.

[63] M. Gautam, M. Agrawal, M. Gautam, P. Sharma, A.S. Gautam, S. Gautam, Role of antioxidants in generalised anxiety disorder and depression, Indian J. Psychiatry. 54 (3) (2012) 244–247.

[64] Q.Q. Han, T.T. Shen, F. Wang, P.F. Wu, J.G. Chen, Preventive and therapeutic potential of vitamin C in mental disorders, Curr. Med. Sci. 38 (1) (2018) 1–10.

[65] N.N. Omar, R.F. Tash, Fluoxetine coupled with zinc in a chronic mild stress model of depression: providing a reservoir for optimum zinc signaling and neuronal remodeling, Pharmacol. Biochem. Behav. 160 (2017) 30–38.

[66] S. Papachristou, N. Papanas, Reduction of depression in diabetes: a new pleiotropicactionof metformin? Diabetes Ther. (2021) 1–4.

[67] A. O'Neil, S.E. Quirk, S. Housden, S.L. Brennan, L.J. Williams, J.A. Pasco, M. Berk, F.N. Jacka, Relationship between diet and mental health in children and adolescents: a systematic review, Am. J. Public Health 104 (10) (2014) e31–e42.

[68] S.E. Quirk, L.J. Williams, A. O'Neil, J.A. Pasco, F.N. Jacka, S. Housden, M. Berk, S.L. Brennan, The association between diet quality, dietary patterns and depression in adults: a systematic review, BMC Psychiatry 13 (2013) 175.

[69] L.B. Martins, J.R. Braga Tibães, M. Sanches, F. Jacka, M. Berk, A.L. Teixeira, Nutrition-based interventions for mood disorders, Expert Rev. Neurother. (2021) 1–13.

[70] W. Marx, M. Lane, M. Hockey, H. Aslam, M. Berk, K. Walder, A. Borsini, J. Firth, C.M. Pariante, K. Berding, J.F. Cryan, G. Clarke, J.M. Craig, K.P. Su, D. Mischoulon, F. Gomez-Pinilla, J.A. Foster, P.D. Cani, S. Thuret, H.M. Staudacher, A. Sánchez-Villegas, H. Arshad, T. Akbaraly, A. O'Neil, T. Segasby, F.N. Jacka, Diet and depression: exploring the biological mechanisms of action, Mol. Psychiatry 26 (1) (2021) 134–150.

[71] Q. Huang, H. Liu, K. Suzuki, S. Ma, C. Liu, Linking what we eat to our mood: a review of diet, dietary antioxidants, and depression, Antioxidants (Basel). 8 (9) (2019) 376.

[72] F.N. Jacka, A. O'Neil, R. Opie, C. Itsiopoulos, S. Cotton, M. Mohebbi, D. Castle, S. Dash, C. Mihalopoulos, M.L. Chatterton, L. Brazionis, O.M. Dean, A.M. Hodge, M. Berk, A randomised controlled trial of dietary improvement for adults with major depression (the 'SMILES' trial), BMC Med. 15 (1) (2017) 23.

[73] R.K. Dishman, C.P. McDowell, M.P. Herring, Customary physical activity and odds of depression: a systematic review and meta-analysis of 111 prospective cohort studies, Br. J. Sports Med. (2021) (Jan 5:bjsports-2020-103140).

[74] F.B. Schuch, D. Vancampfort, J. Firth, S. Rosenbaum, P.B. Ward, E.S. Silva, M. Hallgren, A. Ponce De Leon, A.L. Dunn, A.C. Deslandes, M.P. Fleck, A.F. Carvalho, B. Stubbs, Physical activity and incident depression: a Meta-analysis of prospective cohort studies, Am. J. Psychiatry 175 (7) (2018) 631–648.

[75] F.B. Schuch, B. Stubbs, The role of exercise in preventing and treating depression, Curr. Sports Med. Rep. 18 (8) (2019) 299–304.

[76] A. Kandola, G. Ashdown-Franks, J. Hendrikse, C.M. Sabiston, B. Stubbs, Physical activity and depression: towards understanding the antidepressant mechanisms of physical activity, Neurosci. Biobehav. Rev. 107 (2019) 525–539.

[77] F.B. Schuch, A.C. Deslandes, B. Stubbs, N.P. Gosmann, C.T. Silva, M.P. Fleck, Neurobiological effects of exercise on major depressive disorder: a systematic review, Neurosci. Biobehav. Rev. 61 (2016) 1–11.

CHAPTER

2

Ferroptosis: Oxidative stress and pathophysiology

Haseeb Ahsan[a], *Mohammad Yusuf Hasan*[b], *and Rizwan Ahmad*[c]

[a]Department of Biochemistry, Faculty of Dentistry, Jamia Millia Islamia, New Delhi, India, [b]School of Biomedical Sciences, Newcastle University, Newcastle upon Tyne, United Kingdom, [c]College of Medicine, Imam Abdulrahman bin Faisal University, Dammam, Saudi Arabia

1 Introduction

Cell death is critical in mammalian development, homeostasis, and disease, and it is highly regulated with other biological processes. The regulated form of cell death (RCD) is under the control of specific intrinsic cellular mechanisms. Programmed cell death (PCD) is a subset of RCD that is predestined to occur in normal physiological processes [1]. Apoptosis through a caspase-dependent mechanism was the first RCD to be characterized at the molecular level. In recent years, there is a growing interest in the molecular mechanisms responsible for apoptosis and RCD that control cell death [2]. Ferroptosis is another form of RCD that involves iron-catalyzed lipid damage and is modulated by lipid repair systems involving the antioxidant glutathione (GSH) and enzyme glutathione peroxidase 4 (GPx) (Fig. 1).

An increasing body of evidence suggests that nonapoptotic cell death is actually a class of genetically regulated cell death termed "regulated necrosis," which includes necroptosis, pyroptosis, netosis, parthanatos, and ferroptosis. Ferroptosis is a new form of cell death that involves iron accumulation and lipid peroxidation in cells. It is triggered by iron-catalyzed lipid peroxidation initiated by non-enzymatic (Fenton reactions) and enzymatic mechanisms (lipoxygenases) [2–5]. The Fenton chemistry is the reaction between Fe(II) and hydrogen peroxide (H_2O_2) to form hydroxyl

https://doi.org/10.1016/B978-0-323-90905-1.00002-X

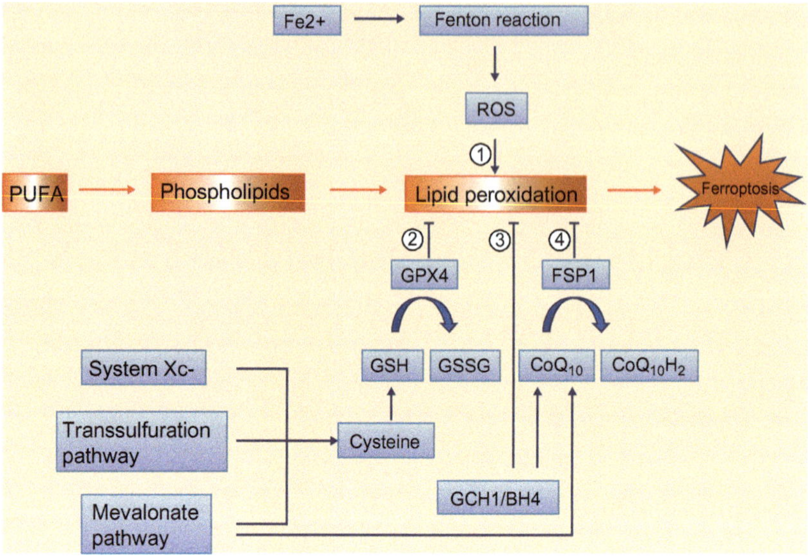

FIG. 1 Molecular mechanisms of involved in the process of ferroptosis. 1. Iron metabolism; 2. GSH/GPX4 pathway; 3. GCH1/BH4 pathway; 4. FSP1/CoQ10/NAD(P)H pathway. *Reproduced from Y. Wu, C. Yu, M. Luo, C. Cen, J. Qiu, S. Zhang, K. Hu, Ferroptosis in cancer treatment: another way to Rome, Front. Oncol. 10 (2020) 571127, https://doi.org/10.3389/fonc.2020.571127.*

radicals (OH˙). The term "ferroptosis" is derived from the Greek word *ptosis* (falling) and the Latin word *ferrum* (iron). It was first discovered as a unique form of cell death induced by the small molecule erastin in 2012. Cells undergoing ferroptosis have a typical necrotic morphology with the mitochondria consisting of decreased cristae and condensed membrane, and the damaged outer plasma membrane [6,7]. Ferroptosis is mediated by mitochondrial voltage-dependent anion-selective channels (VDACs), erastin-induced VDAC2/3 opening leading to mitochondrial iron uptake, generation of reactive oxygen species (ROS), increased mitochondrial potential, and oxidative stress [8,9]. Mitochondria play a crucial role in ferroptosis through mitochondrial metabolism and other regulatory pathways [10], although other organelles are also involved in ferroptosis. Endoplasmic reticulum (ER)-related oxidative stress, Golgi complex associated lipid peroxidation and lysosomes contribute to the induction of ferroptosis [11]. In the presence of the antioxidant GSH, GPx inhibits lipid peroxidation and protects cells from ferroptosis. Since this discovery, a complex interaction between iron, cysteine, and lipid metabolism has emerged as an important regulator of ferroptosis [12]. Moreover, cancer cells exhibit an increased iron demand, which makes them more vulnerable to ferroptosis [13].

Ferroptosis is the form of cell death that is induced by the small molecule erastin, which inhibits the import of cystine, leading to GSH depletion and inactivation of the phospholipid GPX4. Inactivation of GPX4 through depletion of GSH with erastin, or through the GPX4 inhibitor Ras-selective lethal (RSL)3, ultimately results in lipid peroxidation, which causes cell death. Thus, suppression of lipid peroxidation is required in mammalian cells to prevent the onset of ferroptosis, which may occur by default under conditions where GPX4 is inactivated. This process has been implicated in a variety of pathological conditions and therapeutic interventions [14] (Fig. 2).

The cell death occurring exclusively by ferroptosis may be suppressed by iron chelators, lipophilic antioxidants, inhibitors of lipid peroxidation, and depletion of polyunsaturated fatty acids (PUFAs). For example, some degenerative diseases appear to be caused by defective repair of peroxidized lipids, resulting in cell death. The emerging evidence also suggests that ferroptosis may have a tumor suppressor function in removing cells that lack essential nutrients in their environment, or that have been compromised by infection or environmental stress. It is conceivable that ferroptosis could be triggered during development or normal homeostatic tissue turnover by the accumulation of glutamate, iron, and polyunsaturated phospholipids or by depletion of endogenous inhibitors of ferroptosis, such as GSH, reduced nicotinamide adenine dinucleotide phosphate (NADPH), GPX4, or vitamin E [14]. It is dependent on enzymatic reactions, including biosynthesis of polyunsaturated phospholipids, which are the substrates of lipid peroxidation products, acyl-CoA-synthetase long chain (ACSL) 4 and lysophosphatidylcholine

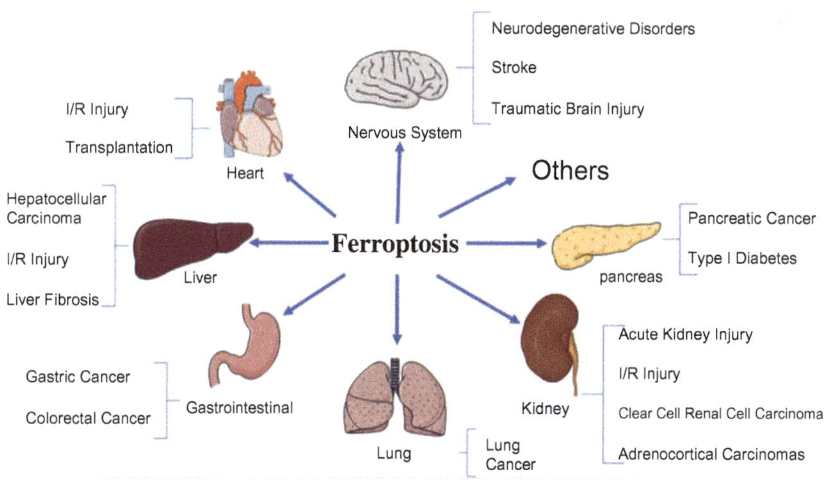

FIG. 2 Role of ferroptosis in systemic diseases. *Reproduced from J. Li, F. Cao, HI Yin, et al., Ferroptosis: past, present and future, Cell Death Dis. 11 (2020) 88.*

acyltransferase (LPCAT) 3, and selective oxygenation of polyunsaturated-phosphatidylethanolamines (PE) by lipoxygenases (LOXS) [14].

2 Immunology

Ferroptosis plays an important role in diseases along with other types of cell death [15–17]. The activation of autophagy may lead to changes in ferritin leading to the process of ferritin-associated autophagy that is mediated by the nuclear receptor coactivator (NCOA) 4 and cause an increase in the content of unstable iron in cells thereby promoting ferroptosis [18]. Therefore, there are some common features in the regulation of ferroptosis, apoptosis, autophagy, and other modes of cell death pathways. Changes in the iron content in cells plays an important role in the occurrence and development of ferroptosis. Free intracellular Fe^{+} can produce the $OH^{\cdot}$ or peroxide radicals ($ROO^{\cdot}$) through the Fenton reaction, further oxidizing lipids. However, copper is also involved in the redox metabolism of biological systems and has similar effects on glutamate-induced oxidation and erastin-mediated ferroptosis in immortalized mouse hippocampal neuronal cell line (HT22). Therefore, in addition to iron, other metal ions can also regulate the occurrence of ferroptosis [19,20]. Increasing evidence shows that ferroptosis may be accompanied by inflammation and the cells activate the innate immune system through ferroptosis by regulating inflammation, signal transduction, and cell growth [21,22]. Some studies have found that GPX4 activation may be used as an antiinflammatory or cytoprotective therapy. However, there are still no specific biomarkers of ferroptosis, such as those that for apoptosis (caspase activation) and autophagy (lysosome formation) [23].

Cell death and immunity maintain homeostasis through complex molecular and cellular mechanisms. On one hand, the effective elimination of cellular debris through phagocytosis is essential to prevent inflammatory and autoimmune diseases. On the other hand, the release or exposure of intracellular molecules from dead or dying cells can elicit adaptive immunity, which leads to an immune response against intracellular pathogens and tumor-associated antigens. Cell death can be divided into two different types: tolerogenic and immunogenic cell death (ICD) [24,25]. While the concept of ICD was originally described for chemotherapy-induced apoptotic cell death, it is now believed that ICD can occur in various types of non-apoptotic cell death caused by chemotherapy, radiotherapy, or other anticancer treatments. The type and activity of damage-associated molecular patterns (DAMPs) during the course of cell death plays key roles in determining ICD [26–28].

Efimova et al. reported that RSL3 stimulates ferroptosis that results in ICD, which occurs in an ATP- and P2X7-dependent manner [29]. Cancer

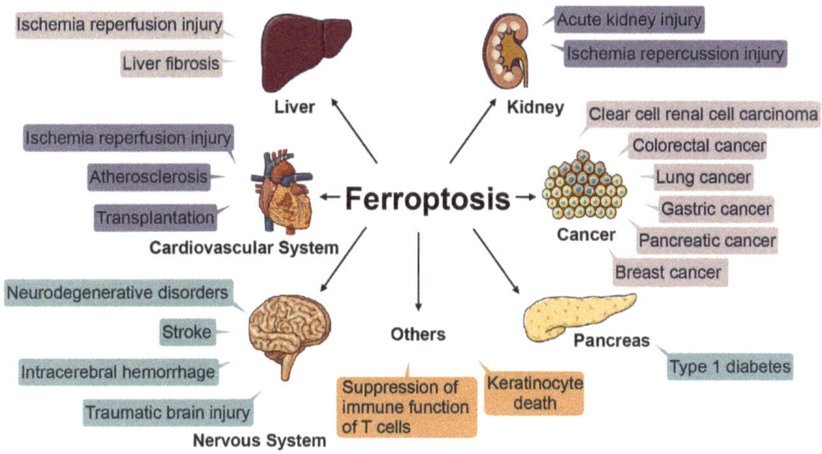

FIG. 3 Ferroptosis in various pathological processes including cancer. *Reproduced from Z. Li, L. Chen, C. Chen, et al., Targeting ferroptosis in breast cancer, Biomark. Res. 8 (2020) 58, https://doi.org/10.1186/s40364-020-00230-3.*

cell immunogenicity depends on two factors: adjuvanticity (which is dictated by DAMPs) and antigenicity (which depends on tumor-associated antigens), many of which may be due to oncogenic stress [26,30]. Hence, further studies are required to ascertain whether different cell death modalities including apoptosis, ferroptosis, necroptosis, and pyroptosis may be responsible to induce ICD through distinct mechanisms. With respect to antigenicity, the specific biochemical context of ferroptosis with oxidative damage to the plasma membrane may generate an "altered self" that favors the immune recognition of tumor antigens. Therefore, it will be interesting to determine the different ICD modalities on the induction of an immune response against distinct tumor-associated antigens. Indeed, it might be possible to conceive "mixed" vaccines composed of cancer cells that succumb to distinct ICD types and may achieve superior anticancer immune response [28] (Fig. 3).

3 Conclusion

Ferroptosis is a new form of cell death that results from iron accumulation and lipid peroxidation in cells. It involves depletion in the antioxidant enzymes resulting in lipid peroxidation and oxidative stress. It is an iron-dependent and oxidative damage induced cell death involving cytological changes such as decreased mitochondrial cristae and damaged mitochondrial membranes. These cell abnormalities resulted from the loss of selective permeability of plasma membrane due to intense membrane lipid peroxidation and the occurrence of oxidative stress. Ferroptosis is an exceptional metabolic process involving amino acids, lipids, NADPH, and microelements. However, cysteine depletion, GPX4 inactivation, and iron overload causes cells to undergo metabolic stress or ferroptotic cell death. Ferroptosis is characterized by metabolic imbalance and perturbations in redox homeostasis. Furthermore, the inactivation of GPX4 contributes to lipid peroxidation and results in the induction of ferroptosis. In cellular models, the morphology and characteristics of cells and mitochondria may be explored to provide evidence to distinguish ferroptosis from other types of cell death. Thus, further understanding is necessary to determine the link between different metabolic processes involved in the association between ferroptosis and autophagy and other forms of RCD and ferroptosis.

Acknowledgment

R.A. is grateful to Prof. Bassam Hassan Awary, Prof. Ali AlSultan, Prof. Mehdi Abumadni, Dr. Kholoud AlGhamdi, and Dr. Mohammad Madadin for their help and support during the preparation of the manuscript. H.A. wishes to thank his family for their endless, everlasting and eternal support in life and wellbeing. H.A. also acknowledges the inspiration, motivation, and encouragement of Prof. Fahim H. Khan and Prof. Waseem A. Siddiqui, Aligarh Muslim University, India.

Conflict of interest

The authors declare that there is no conflict of interest.

References

[1] L. Galluzzi, J.M. Bravo-San Pedro, I. Vitale, S.A. Aaronson, J.M. Abrams, D. Adam, E.S. Alnemri, L. Altucci, D. Andrews, M. Annicchiarico-Petruzzelli, et al., Essential versus accessory aspects of cell death: recommendations of the NCCD 2015, Cell Death Differ. 22 (2015) 58–73.

[2] M. Conrad, J.P. Angeli, P. Vandenabeele, B.R. Stockwell, Regulated necrosis: disease relevance and therapeutic opportunities, Nat. Rev. Drug Discov. 15 (2016) 348–366.

[3] J.G. Nirmala, M. Lopus, Cell death mechanisms in eukaryotes, Cell Biol. Toxicol. (2019).

[4] J.J. Peng, W.T. Song, F. Yao, X. Zhang, J. Peng, X.J. Luo, X.B. Xia, Involvement of regulated necrosis in blinding diseases: focus on necroptosis and ferroptosis, Exp. Eye Res. 191 (2020), 107922.

[5] D. Tang, R. Kang, T.V. Berghe, P. Vandenabeele, G. Kroemer, The molecular machinery of regulated cell death, Cell Res. 29 (5) (2019) 347–364.

[6] S.J. Dixon, K.M. Lemberg, M.R. Lamprecht, R. Skouta, E.M. Zaitsev, C.E. Gleason, D.N. Patel, A.J. Bauer, A.M. Cantley, W.S. Yang, B. Morrison, B.R. Stockwel, Ferroptosis: an iron-dependent form of nonapoptotic cell death, Cell 149 (5) (2012) 1060–1072.

[7] J.P. Friedmann Angeli, M. Schneider, B. Proneth, Y.Y. Tyurina, V.A. Tyurin, V.J. Hammond, N. Herbach, M. Aichler, A. Walch, E. Eggenhofer, D. Basavarajappa, O. Radmark, S. Kobayashi, T. Seibt, H. Beck, F. Neff, I. Esposito, R. Wanke, H. Forster, O. Yefremova, M. Heinrichmeyer, G.W. Bornkamm, E.K. Geissler, S.B. Thomas, B.R.

Stockwell, V.B. O'Donnell, V.E. Kagan, J.A. Schick, M. Conrad, Inactivation of the ferroptosis regulator Gpx4 triggers acute renal failure in mice, Nat. Cell Biol. 16 (12) (2014) 1180–1191.

[8] E.N. Maldonado, K.L. Sheldon, D.N. DeHart, J. Patnaik, Y. Manevich, D.M. Townsend, S.M. Bezrukov, T.K. Rostovtseva, J.J. Lemasters, Voltage-dependent anion channels modulate mitochondrial metabolism in cancer cells: regulation by free tubulin and erastin, J. Biol. Chem. 288 (17) (2013) 11920–11929.

[9] N. Yagoda, M. von Rechenberg, E. Zaganjor, A.J. Bauer, W.S. Yang, D.J. Fridman, A.J. Wolpaw, I. Smukste, J.M. Peltier, J.J. Boniface, R. Smith, S.L. Lessnick, S. Sahasrabudhe, B.R. Stockwell, RAS-RAF-MEK-dependent oxidative cell death involving voltage-dependent anion channels, Nature 447 (7146) (2007) 864–868.

[10] H. Wang, C. Liu, Y. Zhao, G. Gao, Mitochondria regulation in ferroptosis, Eur. J. Cell Biol. 99 (1) (2020), 151058.

[11] Y. Wu, S. Zhang, X. Gong, S. Tam, D. Xiao, S. Liu, Y. Tao, The epigenetic regulators and metabolic changes in ferroptosis-associated cancer progression, Mol. Cancer 19 (1) (2020).

[12] J.P. Friedmann Angeli, D.V. Krysko, M. Conrad, Ferroptosis at the crossroads of cancer-acquired drug resistance and immune evasion, Nat. Rev. Cancer 19 (7) (2019) 405–414.

[13] Y. Sun, P. Chen, B. Zhai, M. Zhang, Y. Xiang, J. Fang, S. Xu, Y. Gao, X. Chen, X. Sui, G. Li, The emerging role of ferroptosis in inflammation, Biomed. Pharmacother. 127 (2020), 110108.

[14] B.R. Stockwell, J.P. Friedmann Angeli, H. Bayir, A.I. Bush, M. Conrad, S.J. Dixon, S. Fulda, S. Gascón, S.K. Hatzios, V.E. Kagan, K. Noel, X. Jiang, A. Linkermann, M.E. Murphy, M. Overholtzer, A. Oyagi, G.C. Pagnussat, J. Park, Q. Ran, C.S. Rosenfeld, K. Salnikow, D. Tang, F.M. Torti, S.V. Torti, S. Toyokuni, K.A. Woerpel, D.D. Zhang, Ferroptosis: a regulated cell death Nexus linking metabolism, redox biology, and disease, Cell 171 (2) (2017) 273–285.

[15] L. Jiang, et al., Ferroptosis as a p53-mediated activity during tumour suppression, Nature 520 (2015) 57–62.

[16] L. Jiang, J.H. Hickman, S.J. Wang, W. Gu, Dynamic roles of p53-mediated metabolic activities in ROS-induced stress responses, Cell Cycle 14 (2015) 2881–2885.

[17] A. Tarangelo, et al., p53 suppresses metabolic stress-induced ferroptosis in cancer cells, Cell Rep. 22 (2018) 569–575.

[18] W. Hou, et al., Autophagy promotes ferroptosis by degradation of ferritin, Autophagy 12 (2016) 1425–1428.

[19] J. Lewerenz, G. Ates, A. Methner, M. Conrad, P. Maher, Oxytosis/ferroptosis-(re-) emerging roles for oxidative stress-dependent nonapoptotic cell death in diseases of the central nervous system, Front. Neurosci. 12 (2018) 214.

[20] P. Maher, Potentiation of glutathione loss and nerve cell death by the transition metals iron and copper: implications for age-related neurodegenerative diseases, Free Radic. Biol. Med. 115 (2017) 92–104.

[21] C. Li, et al., Activation of glutathione peroxidase 4 as a novel anti-inflammatory strategy, Front. Pharmacol. 9 (2018) 1120.

[22] A. von Mässenhausen, W. Tonnus, A. Linkermann, Cell death pathways drive necroinflammation during acute kidney injury, Nephron 140 (2018) 144–147.

[23] J. Li, F. Cao, Y. Hl, et al., Ferroptosis: past, present and future, Cell Death Dis. 11 (2020) 88.

[24] D.R. Green, T. Ferguson, L. Zitvogel, G. Kroemer, Immunogenic and tolerogenic cell death, Nat. Rev. Immunol. 9 (2009) 353–363, https://doi.org/10.1038/nri2545.

[25] N. Yatim, S. Cullen, M.L. Albert, Dying cells actively regulate adaptive immune responses, Nat. Rev. Immunol. 17 (2017) 262–275, https://doi.org/10.1038/nri.2017.9.

[26] L. Galluzzi, I. Vitale, S. Warren, S. Adjemian, P. Agostinis, A.B. Martinez, T.A. Chan, G. Coukos, S. Demaria, E. Deutsch, et al., Consensus guidelines for the definition, detection and interpretation of immunogenic cell death, J. Immunother. Cancer 8 (2020), e000337.

[27] O. Kepp, A. Marabelle, L. Zitvogel, G. Kroemer, Oncolysis without viruses—inducing systemic anticancer immune responses with local therapies, Nat. Rev. Clin. Oncol. 17 (49–64) (2020) 49–64.
[28] D. Tang, O. Kepp, G. Kroemer, Ferroptosis becomes immunogenic: implications for anticancer treatments, Oncoimmunology 10 (1) (2020) 1862949.
[29] I. Efimova, E. Catanzaro, L. Van der Meeren, V.D. Turubanova, H. Hammad, T.A. Mishchenko, M.V. Vedunova, C. Fimognari, C. Bachert, F. Coppieters, et al., Vaccination with early ferroptotic cancer cells induces efficient antitumor immunity, J. Immunother. Cancer 8 (2020), e001369.
[30] Q. Wen, J. Liu, R. Kang, B. Zhou, D. Tang, The release and activity of HMGB1 in ferroptosis, Biochem. Biophys. Res. Commun. 510 (2019) 278–283.

CHAPTER

3

Oxidative stress and its biological significance

Mari Uyeda

AC Camargo Cancer Centre, São Paulo, SP, Brazil

1 Introduction

Cell membranes contain a high concentration of polyunsaturated fatty acids, which are often subjected to peroxidation, leading to inhibition of cell growth and death. The oxidation of phospholipids within cells can trigger the path of apoptosis mediated by mitochondria, reactive oxygen species (ROS) or lipid peroxidation by-products that react mainly with cardiolipin molecules and that are linked to cytochrome C [1], inducing disturbances in interactions of cytochrome C-cardiolipin, and its dissociation and release in the cytoplasm induce biochemical reactions that result in the activation of caspase and subsequent cell death [2]. This can be regulated by the protein Bcl-2 for presenting pro- and anti-apoptotic activities, linked to the external cellular environment, and modulate cellular permeabilization [3].

Bax and Bak are examples of anti-apoptotic proteins that can be activated by breaking their connection with anti-apoptotic proteins or by interacting with activating proteins that induce their conformational changes. Inactivated Bax proteins can be located as monomers in the cytosol or associated with the extracellular medium during the activation process, Bax forms homo-oligomers and is inserted both in the extracellular medium and in Bak, leading to the formation of membrane pores and permeabilization, promoting the release of cytochrome C in the cytosol [4] (Fig. 1).

https://doi.org/10.1016/B978-0-323-90905-1.00003-1

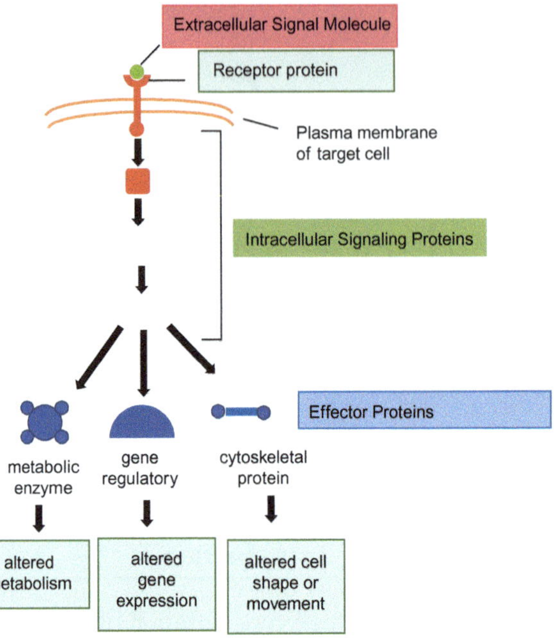

FIG. 1 Cellular response to external injuries.

2 Oxidative stress in 4-hydroxynonenal (HNE) leading to apoptosis

Overexpression of Bcl-2 inhibits the release of cytochrome C from mitochondria and the apoptotic response becomes blocked, such as the HNE-induced caspase activation that is suppressed in transfected bcl-2 colorectal carcinoma cells. Cytosol cytochrome C and binds to the apoptotic protease activating factor of adapter protein 1 (Apaf-1) and induces an apoptosome assembly in the presence of ATP/dATP, activating pro-caspase-9 directly within the apoptosome complex; pro-caspase-9 is then cut into active caspase-9, which activates caspases-3, -6, and -7, leading to DNA fragmentation and cell death [5]. If the cell level of ATP/dATP is depleted, cell death is redirected from apoptosis to necrosis. The release of cytochrome C and the formation of apoptosis can also be triggered through the extrinsic pathway of apoptosis. In addition, the accumulation of damage directly in the mitochondria can also cause increased production of oxidants and degenerative events, and the HNE can be generated directly through the oxidation of the mitochondrial phospholipid cardiolipin, as well as through other oxidation products, in this case reacting with the surrounding molecules close to the site of their formation, promoting apoptosis chain reactions derived again from mitochondria in the pheochromocytoma cell line (PC12), involved in colorectal carcinogenesis [6].

Stimulation of HNE in human osteosarcoma cells MG63 could activate caspase-3 and alter the Bax/Bcl-2 ratio, inducing cell death by increasing the growth of breast cancer cells and promoting their angiogenesis and invasion. In addition, elevated levels of malondealdehyde (MDA) can be found in the plasma and serum of patients with breast, lung, ovarian, thyroid, and oral cancer, and pre-cancer states. The levels of MDA in patients with lung cancer are related to the stage of the disease, and high levels of salivary MDA can be found in patients with squamous cell carcinoma and with precursor lesions [7].

3 Antioxidant supplementation increasing tumorigenesis

We have always believed that a high concentration of ROS leads to tumorigenesis, and its suppression using antioxidants could be prophylactic if it were not therapeutic. anTo reach these conclusions, many studies have been carried out since 1978, involving monolayers, immortalized cell lines showing the ectopic expression of oncogenesis, which can lead to increased ROS generation, causing damage to nucleic acids. However, on the other hand, we can see that the dietary supplementation of some antioxidants can modify this reality, such as the following:

- *Supplementation of N-acetylcysteine (NAC) or vitamin E:* significantly accelerates tumor development and mortality in studies involving animals with lung cancer induced by Kras and Braf, in addition to promoting NAC metastasis in human melanoma xenografts.
- *Folate supplementation:* promotes the development and progression of breast cancer in murine models.
- *Supplementation of beta-carotene, NAC, vitamin A, or vitamin E:* did not reduce the incidence of head and neck, lung, colorectal, or prostate cancer.

ROS are produced by enzymatic and non-enzymatic systems within eukaryotic cells and play important roles in cell physiology and pathophysiology. Although physiological concentrations are crucial to ensure cell survival, overproduction of ROS is detrimental to cells and is considered to be key to the development of various diseases, such as neurodegenerative diseases, cardiovascular diseases, and cancer. Cancer cells are generally subjected to higher levels of ROS that further stimulate the malignant phenotype by stimulating sustained proliferation, avoiding death, angiogenesis, invasiveness, and metastasis. Cancer stem cells (CSCs), also known as cancer cells tumor initiators (TICs), are a small subpopulation of cells within a heterogeneous tumor that have the unique ability to lead to tumor initiation, progression, and metastasis, and are resistant to current anticancer therapies, enriched in residual tumors after chemotherapy and/or radiation.

3.1 Generation of mitochondrial ROS and cancer formation

Mitochondria generate ROS during their normal metabolism and attack mitochondrial membranes and mtDNA directly near the site of their formation, leading to mitochondrial dysfunction and further production of ROS. They interact extensively with nuclear DNA and proteins and cell membrane lipids. Damage with large amounts of mutations or DNA arrangements, genomic instability, the formation and interruption of protein functions, altered metabolic and signal pathways induced by oxidative stress are situations found in cancer cells that have high levels of reactive oxygen species and are commonly exposed to stressful conditions such as hypoxia (low oxygen levels), loss of cell adhesion and extracellular matrix cells, imbalance in oxidative metabolism, and various environmental factors [8].

ROS is a regulator of several signal transduction pathways and can participate in various stages of tumor transformation and progression, such as:

- self-sufficiency in signs of growth;
- insensitivity to antiproliferative signs;
- evasion of apoptosis;
- unlimited replicative potential;
- sustained angiogenesis;
- invasion and metastasis; and.
- metabolism and inflammation.

The mitogen-activated protein kinase pathway (MAPK) and phosphorinositide 3 kinase (PI3K) can be considered the most important growth pathways regulated by ROS, with MAPK activating transcription factors such as Jun C Fos, which regulate the expression of genes involved in cell proliferation, differentiation, transformation, and inflammation, whereas Pi3K is related to the RAS pathway involved in cell survival [8].

4 Regulation of the cell cycle in the oxidative state

Before mitosis, the cell undergoes a DNA verification process to identify possible damage and, if necessary, perform its repair through mutated telangiectasia ataxia kinase (ATM), which is a key protein in this process and whose main function is to respond to the breakdown of the double strand of DNA. It can also be activated by oxidative stress, activating p53, which is related to numerous types of tumors, such as breast, lymphoma, sarcoma, and carcinoma (Fig. 2) [9].

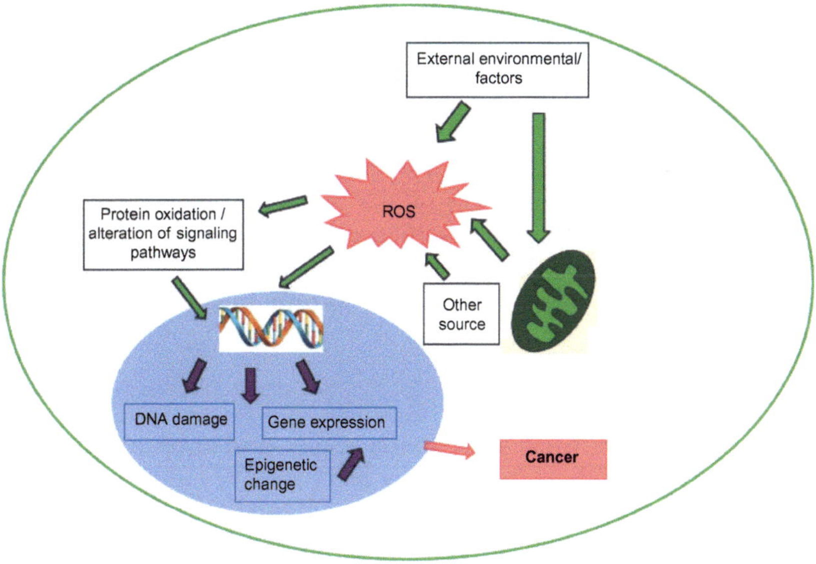

FIG. 2 Participation of ROSs in the development of cancer. External factors, mitochondria, among others, are a source of ROS. The intracellular increase in ROS can result in protein oxidation, changes in signaling pathways, DNA damage, epigenetic changes, and changes in gene expression, factors that can contribute to the development of cancer.

5 Viral infections and oxidative stress

The hepatitis C virus (HCV) triggers cellular responses mediated by ROS, representing a risk of hepatocellular carcinoma, through the previously described pathways (MAKP and PI3K). Oxidative stress could act in synergy with the human papillomavirus (HPV), creating ideal conditions for cervical tumor initiation and progression [8].

6 Epigenetic factors and carcinogenesis

Epigenetics is a chemical change in chromatin that does not involve changes in DNA nucleotide sequences, participating in the regulation of gene expression such as DNA methylation, post-translational modifications in histones, and non-coded RNAs. Epigenetic processes are potentially reversible, unlike genetic changes, and are therefore favorable to treatment [8].

DNA methylation is the most studied epigenetic mechanism and is characterized by the addition of a methyl group in carbon 5 of cytosines adjacent to guanines (cpg dinucleotides). Methylation in promoters is associated with gene silencing, and many genetic regions that are silenced by methylation in normal cells, such as transposons, often demethylate in cancer. Approximately 60% of human genes have a high concentration of cpg dinucleotides, islands of CpGs, in their promoters. In these regions, DNA methylation plays an important role, as it defines the status of genetic transcription [9].

In general, promoters that contain the unmethylated CpGs islands are capable of transcription, while methylated promoters are transcriptionally inactive. In tumor cells, many promoters of tumor suppressor genes become methylated, being silenced, and contributing to the loss of cell control [9].

7 Extracellular superoxide angles and apoptosis

Tumor cells produce extracellular superoxide anions with the participation of NADPH oxidase1 (NOX1) that serves as protection against apoptosis in these tumor cells, located on the cell membrane and necessary for the reduction of molecular oxygen. Extracellular superoxide anions are necessary to regulate the proliferation of cancer cells and to maintain their transformed state [10].

To maintain their transformed state, ROS signaling through the HOCl and NO/peroxytonitrite pathways spontaneously react with hydroxyl ions (H-) or are conducted by superoxide dismutase (SOD) to form unstable hydrogen peroxide [10]. It is used as a substrate to generate exogenous HOCl by the peroxidase domain (POD) encoded by the double oxidase (DUOX). Then, the HOCl can interact with superoxide anions, generating a hydroxyl radical, which penetrates the intracellular space, induces lipid peroxidation, and consequently promotes apoptosis derived from mitochondria, as shown in Fig. 3. However, tumor cells express the membrane-associated catalase, which can efficiently decompose H_2O_2, right after its generation. In addition, SOD can play the role of protective co-modulation through partial inhibition of HOCl/superoxide anion interactions. The generation of H_2O_2-derived SOD can also be compensated for by the interrelation of SOD with a catalase-mediated protective effect. In addition, in the presence of high concentrations of H_2O_2, compared to POD, HOCl can react with H_2O_2 to block HOCl signaling [10].

The NO/peroxytonitrite signaling pathway that modulates tumor cell death consists of the following:

(1) Inside cells, no synthase promotes NO synthesis derived from arginine.
(2) NO passes through the cell membrane.

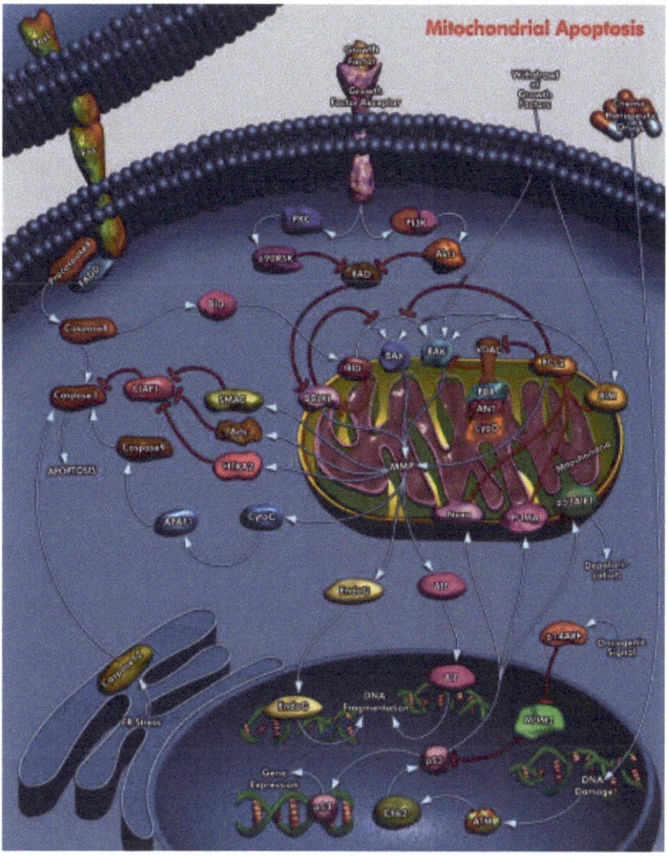

FIG. 3 Mitochondrial pathways of apoptosis. O: Death receptors that transmit apoptosis-inducing signals (ligand FAZ and FAZ/tumor necrosis factor (TNF)-α receptors and TNF-α receptors) and bait receptors of the same family. B: Intracellular proteins that monitor DNA damage, tumor suppressors. W: Anti-apoptotic members of the Bcl-2 family and other proteins that block cell death.

(3) The formation of extracellular interaction of peroxytonitrite NO with the superoxide anion and subsequent reactions lead to the generation of nitrogen dioxide (NO_2) and hydroxyl radical.
(4) These reactions initiate lipid peroxidation and apoptosis.

However, membrane-associated catalase can protect the tumor cell, preventing the generation of peroxytonitrite and NO oxidation. In addition, the inhibition of NO/superoxide anions by SOD, as in the previous case, modulates the negative effect on the apoptosis pathway. High concentrations of HNE can inactivate membrane-associated catalase, leading to the death of tumor cells through apoptosis derived from mitochondria, consequently conferring anti-cancer effects [10].

7.1 HNE and its adducts in the regulation of transcription factors

The signaling pathway of the transcription factors of the nuclear factor-kB (NF-kB) and the transcription factor of the activating protein 1 (AP-1) are involved in cancer, through the regulation that the HNE and its adducts exert on these factors and the expression of genes responsible for immunity, apoptosis, and cell cycle. The NF-κB family is composed of five proteins: p65 (RelA), RelB, c-Rel, p105/p50 (NF-κB1), and p100/52 (NF-κB2), which combine to form different κB NF-active dimers. In the cytosol, NF-κB dimers are found in the inactive form anchored by IκB. There are two ways to activate NF-kB-induced genetic transcription. These are triggered by α TNF and cytokines IL-1 (canonical signaling) or CD40 and BAFF antigen receptors (non-canonical/alternative signaling). There is the activation of the IKK complex (IKKα, IKKβ, and IKKγ) and the phosphorylation of IκB proteins, which are in interaction with the p50-RelA subunit, with cytokines during the canonical signaling pathway. This reaction leads to the degradation of IkB proteins and the release of the active dimer p50-RelA NF-κB [11].

Non-canonical NF-kB signaling consists of NIK (NF-κB kinase inducer) inducing the activation of IKKα, which phosphorylates the p100 NF-κB subunit of the p100-RelB NF-κB dimer directly in the cytosol, leading to the generation of an active p52-RelB NF-κB dimer, causing darkening active NF-κB can translocate to the nucleus and induce transcription of target genes. Target genes for NF-κB include apoptosis regulators (Bcl-XL and IAPs), cytokines (TNF-α, IL-1, IL-6, and IL-12), cyclins and growth factors (G-CSF and M-CSF), immunoregulatory proteins (VCAM, ICAM, and MHC I), and other NIK (kinase-inducing NF-κB) that induce the activation of IKKα, which phosphorylates the p100 NF-κB subunit of the p100-RelB NF-κB dimer directly on the cytosol. This leads to the generation of an active p52-RelB NF-κB dimer. In both cases, active NF-κB browning can translocate to the nucleus and induce transcription of target genes. Target genes for NF-κB include apoptosis regulators (Bcl-XL and IAPs), cytokines (TNF-α, IL-1, IL-6, and IL-12), cyclins and growth factors (G-CSF and M-CSF), immunoregulatory proteins (VCAM, ICAM, and MHC I), and other NIK (kinase-inducing NF-κB) that induce the activation of IKKα, which phosphorylates the p100 NF-κB subunit of the p100-RelB NF-κB dimer directly on the cytosol. This leads to the generation of an active p52-RelB NF-κB dimer. In both cases, active NF-κB browning can translocate to the nucleus and induce transcription of target genes. Target genes for NF-κB include apoptosis regulators (Bcl-XL and IAPs), cytokines (TNF-α, IL-1, IL-6, and IL-12), cyclins and growth factors (G-CSF and M-CSF), immunoregulatory proteins (VCAM, ICAM, and MHC I), among others [12].

The AP-1 complex is composed of members of the JUN, FOS, ATF, and MAF families, and determines genes that affect differentiation, proliferation, and apoptosis, such as Bcl2, EGFR, p53, and CD44. The protein proliferation family is associated with tumorigenesis, mainly the c-JUN and c-FOS oncoproteins, but in some cases can suppress the formation of tumors. Regulation of AP-1 activity occurs in several ways: differential expression of unique AP-1 components, interactions with auxiliary proteins, and regulation of transcription and post-translation, through activation of MAPK pathways (ERK1/ERK2, JNK/SAPK, and p38) by external stimuli, leading to the induction of transcription of the FOS and JUN genes and the formation of other AP-1 complexes.

The HNE can regulate the generation of the AP-1 complex by inducing the expression of the FOS and JUN genes or activate/inhibit the MAPK pathways. It can regulate the FAS/FADD-dependent apoptosis pathway (Fig. 3), where FAS belongs to the TNF receptor superfamily and FADD has a death domain, being able to interact with the TNF receptor and mediate signs of apoptosis [13].

The connection of the FasR with the Fas ligand (FasL) induces its trimerization, leading to the interaction with the adapter molecule FADD, forming the death-inducing signaling complex (COMPLEX DISC I). Activated caspase-8 can stimulate caspase-3 signaling in two ways:

(1) caspase-8 cuts off the Bcl-2 interaction protein (Bid), which alters the permeabilization of the mitochondrial membrane or induces Bak activation. It unleashes the release of cytochrome C, activating the mitochondrial-dependent apoptosis pathway.
(2) Caspase-8 directly activates caspase-3, -7, or -6, leading to apoptotic DNA fragmentation and cell death [14].

In another case, FasR can recruit death-associated protein (DAXX), which later binds to the kinase that regulates the apoptosis signal 1 (ASK1) and activates, followed by the induction of ask1 into the JNK/SAPK and p38 pathways. The interaction of FasR with protein kinase 1 that interacts with the receptor (RIP1) must be a component of the DISC II complex. This complex, depending on the combination of signaling proteins, can promote apoptosis, necrosis, and NF-kB, JNK/SAPK, and MAP38 p38 pathways.

HNE can induce FAS-mediated apoptosis in HepG2 cells by exposing these cells to sublethal concentrations of HNE that promote DAXX export from the nucleus to the cytoplasm and facilitate the binding of FAS-DAXX, but also activate ASK1, JNK, and caspase 3, leading to cell death, which also occurs in Jurkat and HLE B-3 cells [15].

The expression and functions of FAS can be modulated by the HNE in a manner dependent on time and concentration; this modulation is essential for the execution of apoptosis, also being involved in the apoptosis pathway p53 in HepG2 cells, resulting in the induction of p53 expression, phosphorylation, and activation of targets downstream Bax and p21. HNE treatment has been shown to increase protein expression in the p53 family and its targets in the SK-N-BE neuroblastoma cell line, leading to neuronal cell death [16].

7.2 How oxidative DNA damage is critical to cancer

The ROS and lipids from peroxidation can affect genomic and mitochondrial DNA, leading to several types of DNA damage, such as single- and double-strand breaks, intro and DNA interconnection, DNA adduct formation, DNA-based changes, and deoxyribose. Subsequently, the replication of the damaged DNA before repair results in mutations in the DNA and genomic instability. Double-stranded DNA breaks (DSBs) are the most dangerous deficiency. They cause severe genetic mutations that lead to various disorders and tumor progression [17].

Simple tape breaks (SSBs) are less harmful to cells if they are repaired quickly; if not, they result in serious injury and can contribute to cancer. Fig. 4 shows the damage caused to the DNA strands.

8-oxoguanine (8-OHG) is one of the multiple oxidation products generated in DNA through dG oxidation. 8-OHG may erroneously add adenine to make GT and CA substitutions in the genome. The nucleoside form of 8-OHG is 8-oxo-2′-deoxyguanosine (8-OHdG), which has been proposed as an indicator of oxidative damage to DNA in vivo and in vitro. 8-OHdG

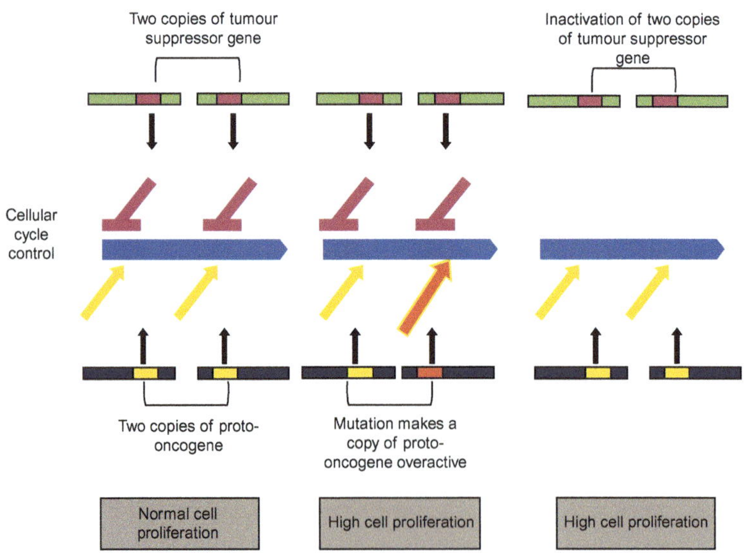

FIG. 4 Damage caused to the DNA chain.

can also be subjected to keto-enol tautomerism to favor the oxidized product 8-oxo-7,8-dihydro-2-deoxyguanosine (8-oxodG), which is also commonly used as a marker of DNA oxidative damage, being associated with cancer for contributing to spontaneous mutagenesis [18].

Mitochondria generate large amounts of ROS, directly exposing MTDNA to oxidative stress because it lacks "protective" histones and has a limited repertoire of available DNA repair pathways [19]. MTDNA injuries can have serious consequences, and the steps mentioned are extremely important: replication of MTDNA, error repair, and degradation:

Replication—The replication of mitochondrial DNA differs from that of nuclear DNA in terms of the cell cycle and can occur at any stage of the cycle. DNA γ polymerase is the enzyme responsible for the replication process.

DNA repair—This process protects the genome from damage that can lead to harmful mutations. DNA can undergo base oxidation and base incompatibility. Once the error is located, specific molecules are "sent" to the site to perform their repair through base excision (base excision repair, or BER), nucleotide excision (nucleotide excision repair, or NER) or incompatibility repair (repair of incompatibility of mismatch repair, MMR).

DNA degradation—When mitochondrial DNA has several lesions that are already considered irreversible, the only solution is to degrade it, and thus prevent mutagenesis. This degradation is the responsibility of endonuclease G.

7.3 Transcription factors that promote cancer

In sarcoma, overexpression of SOD suppresses radiation-induced transformation, increasing the production of fibroblasts. High amounts of ROS after RAS oncogene expression are not present in cells where RAS oncogene is expressed in the endogenous locus being artifacts, showing that the loss of tumor suppressor p53, in cancer cell lines through siRNA-mediated silencing or splenocytes and thymocytes for Trp53, increases ROS production.

Like p53, other transcription factors regulate the intracellular redox environment and can be considered as important aspects in the tumor suppression function, such as the FOXO family of proteins, whose activation leads to the positive regulation of antioxidant enzymes, such as MnSOD and catalase. Likewise, the BRCA1 genetic ablation of oxidative stress is mediated by NRF2, which increases ROS levels and the frequency of spontaneous mammary tumor formation [20].

Elevated ROS is a direct mutagen in DNA or a pathway to induce genomic instability after activation of topoisomerase II, being essential in tumor initiation and progression [21].

7.4 Tumor-promoting genetic changes induce endogenous antioxidants

NRF2 is essential for the transactivation of genes involved in the maintenance of redox homeostasis and its regulation is positive, with the proliferation of cancer cells, metabolic reprogramming, chemoresistance, serine biosynthesis, and translation of mRNA identified in pancreatic, skin, breast, prostate, lung cancer, and maintenance of redox homeostasis [22].

The XC system is essential for importing cysteine to support the synthesis of glutathione (GSH) and is related to the proliferation and resistance to multiple drugs of various types of cancer cells, including lung and ovarian neoplasms. The pharmacological inhibition of xCT inhibits tumor growth, including hepatocarcinoma, glioma, and prostate and breast cancer. Overexpression of SLC7A11 in breast and liver cancer cells decreases tumor growth mediated by p53 (3KR), maintaining its antioxidant capacity and ability to prevent ferroptosis [23]. Elevated levels of SGM should be considered necessary for the initiation and progression of some types of tumors, such as the pancreas, breast, and lung, as well as the loss of TIGAR (the gene that controls glucose metabolism in maintaining NADPH levels during regeneration of GSH) can impair the proliferation, and survival of cancer cells after oxidative stress, showing that endogenous antioxidants are essential in the tumor process.

7.5 Endogenous antioxidants promote cell migration and metastasis

As tumor cells detach themselves from the extracellular matrix and invade the membrane, ROS levels increase, and the reduction in GSH decreases. In contrast to ROS reported roles in promoting cell migration, folate-mediated NADPH-generating enzymes promote distant metastases in melanoma [24]. The administration of NAC raises the levels of these circulating cells, promoting distant metastases, showing that the epithelial cells undergo cell death after the detachment of the matrix in the culture due to reduced glucose absorption, ATP depletion, and high oxidative stress. Oncogenic signaling promotes the survival of cells separated from the matrix, increasing glucose absorption, and flow through the pentose phosphate pathway to generate NADPH equivalent reducing agents. Finally, the treatment of these cells with the antioxidants NAC or Trolox can substantially increase the number and the average size of the colonies, indicating a role of antioxidants in increasing the transforming activity of cells that suffer from oncogenic aggressions. IDH1-dependent reductive carboxylation, and promotes the growth of mitochondrial NADPH, and the increase in intracellular reduction equivalents through metabolic changes in favor of supporting the growth, survival, and migration of cancer cells [23].

7.6 *N*-acetylcysteine—More than an antioxidant

NAC increases the synthesis of GSH, enters cells, and within the cytosol, is converted into L-cysteine (precursor to GSH). It is a source of sulfhydryl groups in cells, and a rescue agent for a system under oxidative stress because it interacts with ROS, such as OH and H_2O_2 [25]. At low doses (0.1–1 mM), NAC functions as a redox state regulator, but at high doses (> 5 mM), it can directly alter the structure of TGF-beta [26], and can selectively induce p53-dependent apoptosis in genetically transformed cells, an activity that is not seen in chain-breaking antioxidants such as vitamin E and Trolox. It is important to note that an increase in GSH is not necessary for NAC triggered by apoptosis. The treatment of breast cancer cells with NAC has been shown to suppress the activation of NRF2 target genes in the presence of ROS-inducing agents, such as tert-butyl hydroquinone.

7.7 The NADPH relationship and the redox potential

NADPH provides reduction equivalents for GSH regeneration. Two of the main contributors to the NADPH cytosolic reduction equivalents are the pentose phosphate pathway and a serine carbon metabolism (folate cycle). Other NADPH generators include ME1, IDH1/2, and NNT.

NADPH is used as an anabolic pathway for lipid synthesis, cholesterol synthesis, and lengthening of the fatty acid chain (its most important function and not the redox function). The increased demand for NADPH-producing enzymes in cancer cells may reflect not necessarily an increase in redox support, but an increase in anabolic demands [27].

7.8 The ambiguous relationship between mitochondrial metabolism and ROS

Changes in the mitochondrial activity of cancer cells will inevitably lead to changes in cell metabolism and redox status. One example is the pharmacological inhibition of electron transport chain (ETC) complex III in cancer cells, showing that ROS is necessary for the stabilization of hypoxia-dependent HIF1, which may be a consequence of direct inhibition of prolyl hydroxylase (PHD) enzymes through the oxidation of essential iron not bound to heme. However, most pharmacological, and genetic interventions that alter ETC function. it induces changes in oxygen consumption and mitochondrial ROS production, and it is a challenge to distinguish the difference between the regulation of mitochondrial HIFL-dependent proline hydroxylation by oxygen depletion or increased reactive oxygen. Likewise, the inability to modulate ROS production without changing ATP synthesis represents a major obstacle in understanding the physiological roles of specific mitochondrial sites in the production of ROS [28].

7.9 ROS stimulates or inhibits tumorigenesis

ROS can have dual functions in tumorigenesis, where, on the one hand, they can suppress cell growth through genotoxic stress and interrupt protein translation, and, on the other hand, they can promote cell growth by activating mitogenic signaling cascades. The role of ROS in the cellular result is more diverse than expected. Cellular responses to ROS reflect a complex integration of ROS type, location, and levels. For example, mitochondrial ROS must be central in promoting damage and death [29], while membrane-generated ROS contributes to proliferation, tumorigenicity, migration, and metastasis [30]. However, these distinctions are not absolute, as mitochondrial ROS also plays a role in proliferation, tumorigenicity, migration, and metastasis, while the ROS generated by NOX in the membrane can also induce cell death via ferroptosis and necrosis. In this context, oxidized lipids on the membrane have also been shown to exhibit a variety of functions. For example, low-density oxidized lipoprotein can induce cellular apoptosis in vascular smooth muscle cells or activation of NF-KB in endothelial cells [31].

7.10 Conceptual and experimental limitations of ROS in the field of redox cancer biology

The involvement of ROS in cancer is not limited to indiscriminate macromolecular damage; its regulation is topological and temporal and functions as secondary intracellular messengers, with each reactive species performing unique signaling events, such as kinases, ubiquitilases, and acetyltransferases, among others. However, unlike these more familiar signaling events, target specificity in redox signaling occurs at the atomic level and not at the macromolecular level. Therefore, it is expected that the potential number of specific ROS effects will be massive and little explored. Just like kinase signaling, the output of redox signaling is the integration of mechanisms, specificity, and the functional consequence of each post-oxidative translation modification. Since total phosphate levels cannot be used as a metric to decipher the biological consequence of kinase signaling events, standard approaches to quantify relative changes in global ROS levels should not be expected to reveal adequately the biological consequence of specific events of redox signaling [31].

7.11 Cell senescence and tumor suppression

Cell senescence was first described more than 40 years ago, where normal diploid cells were shown to have limited replicative potential. Unlike cancer cells, normal cells at the end of their replicative life are in a process known as cell senescence, which is characterized by the irreversible arrest of the cell cycle, morphological changes, epigenetic changes, lack of response to growth factors, shortening and dysfunction of the telomer, sustained metabolic activity, and high DNA damage, activating the signaling pathways ASK1/JNK and ASK1/p38 in fibroblasts and cancer cells. This mediation can occur through the oxidation of TRX1 H_2O_2 and subsequent ASK1 dissociation, thus triggering the suppression of anti-apoptotic factors through the sustained activation of MAP kinase (MKK) 4/MKK7/JNK and MKK3/MKK6 downstream/p38 paths. In addition, cell senescence was considered a tumor suppressor mechanism. The activation of negative growth-regulating genes and proteins, which inhibit the initiation of DNA synthesis, during senescence, such as p53 and Rb, the tumor suppressors, have been shown to play a role in the regulation of cellular senescence [9]. The inactivation of the p53 and Rb genes contributes to extend the proliferative life span of normal fibroblasts [9]. However, for normal cells to transform into cancer cells, at least four or five mutations are required [9].

For colorectal tumors, this number of four or five genetic changes could also be considered in the development of cancer. The other growth-inhibitory genes, such as p21, p16, p33, p19, and p27, can also be considered the main effects of cell senescence [9]. In addition, ROS is related to the mediation and activation of the death receptor through the negative regulation of the cytoprotective FLIP protein. Oxidizing agents, such as buthionine sulfoximine, functionally inactivate the mitogenic signaling cascades conducted by AKT in pancreatic duct cells [22], as well as in skeletal muscle cells [32].

The oncogenes RAS and MYC are involved in the regulation of senescence induction pathways.

The oncogenic transformation of diploid fibroids by RAS results in a permanent arrest of the cell cycle, simultaneous induction of tumor suppressor proteins p53 and p16, and cellular senescence. The oncogenic transformation of cells by RAS requires a cooperating oncogene or the inactivation of tumor suppressors. However, escaping the RAS-induced arrest by interrupting the p53 or p16/Rb pathways can lead to cell transformation. In contrast to RAS, MYC overexpression results in apoptosis in the primary cells. This effect can be mediated by the activation of P19 ARF/p53 and p14 ARF/Vias E2F-1 [22]. The expression of other oncoproteins, such as RAF and MEK, also induces cell cycle arrest and activates senescence mediators (p53, p21 Cip1, and p16 Ink4a) in fibroblasts [22]. Overexpression of the Bmi-1 oncogene induces the primary immortalization of fibroblasts and the negatively regulated expression of tumor suppressors p16 and p19 ARF. Fibroblasts deficient in Bmi-1 show greater expression of p16 and p19 ARF and a phenotype of premature senescence [9].

Telomere shortening is one of the main mechanisms that induce cell senescence and inhibit tumorigenesis [9]. However, it can trigger not only the senescence response but also the telomere crisis. The progressive shortening of telomeres and the loss of tumor suppressor function results in massive chromosomal instability, secondary genetic alterations that facilitate carcinogenesis. The loss of telomere function and the consequent genomic instability can cooperate with p53 deficiency and promote tumorigenic initiation, in addition to senescence and apoptosis induced by telomere dysfunction and p53 activation in the suppression of tumors [33].

7.12 The mTOR pathway is involved in the senescent phenotype and cancer

The mTOR path leads to the process of converting proliferative arrest to irreversible senescence and is involved in longevity. On the other hand, it is activated by mutations in oncogenes such as RAF, RAS, and PI3K, and inhibition of several tumor suppressor genes, including p53. However, cells with mTOR-activated oncogenes are required to disable cell cycle checkpoints in order to proliferate [34].

Senescent cells are characterized by persistent signaling of DNA damage (DDR), which can be induced by mitochondrial dysfunction and oxidative stress. Detects DNA damage, signals the presence of damage to genomic DNA, and promotes its repair. In fact, DDR has been recognized as an anticancer mechanism that leads to cell cycle arrest followed by cell senescence or apoptosis. DDR markers, such as overexpression p21, activation of ATM and CHK2 kinase checkpoints, histone H2AX phosphorylation, and p53 accumulation or phosphorylation, commonly occur in the early stages of tumors and precursor lesions. These data indicate that, in an early stage of tumorigenesis, cells activate DDR to slow or prevent cancer. Thus, DDR can be considered as another tumor suppression mechanism that could be triggered by oxidative stress and in which senescence is involved [35].

Cancer progression is also slower in the elderly, and metastases are less frequent. The loss of genome "plasticity" during development and aging is a possible explanation for this decline. Deregulation of the potential for chromosomal recombination predisposes to malignant transformation, but a decline occurs with aging. Thus, during senescence, the loss of "plasticity" protects against deviations in cell proliferation, delaying tumor growth and metastases in the elderly. In addition, cell senescence is also mentioned as a mechanism for reverting tumor cells to normal. In the population of tumor cells, there are cells that do not divide with senescence [36].

Normal cells have a genetically programmed deadline for cell replication and are designed to prevent excessive proliferation, and senescence is one of the main defense mechanisms against malignant transformation.

However, in some conditions, normal cells can "escape senescence." Treatment of primary cells with viral oncogenes results in cell death or apoptosis. At a certain frequency, cells can become immortal. In vivo, during senescence, many changes, including oncogenic ones, can accumulate. In addition, senescent cells secrete factors that can promote the growth of other cells and tumorigenesis. Therefore, although senescence is recognized as a powerful tumor suppressor mechanism, it remains a risk factor for cancer.

8 Extreme oxidative stress with T cells

The treatment that elevates oxidative stress to the point that, instead of favoring the growth of cancer cells, cause their death. T cells are the patient's cells that are designed to better fight cancer. The therapy causes fatal double-strand DNA damage to cancer cells and the treatment is recognized as interfering with the production of glutathione, a natural antioxidant found in all cells, as it increases the production and accumulation of ROS within tumor cells, resulting in increased production of pro-cytokine T cells—inflammatory drugs that regulate many functions that cancer needs to control, such as cell proliferation, differentiation, and death. The tumor-alpha necrosis factor can act directly on tumor cells and induce ROS within them, with tumor regression.

For breast cancer and lymphoma, the tumor-alpha necrosis factor must act with chemotherapy to increase oxidative stress and cancer cell death, so drugs that are known to raise ROS levels can somehow replicate the benefit of killing adoptive T cell tumor cells.

9 Oxidative stress in lung cancer

Lung cancer is the leading cause of cancer-related deaths worldwide, with a discouraging 5-year survival rate of 17% [37]. There are two main types of lung cancer: small cell lung cancer (SCLC), which accounts for ~ 15% of lung cancer patients, and non-small cell lung cancer (NSCLC), comprising the remaining ~ 85% [38], as shown in Fig. 5. The three main histological subtypes of NSCLC are adenocarcinoma (AC), squamous cell carcinoma (SqCC), and large cell carcinoma (LCC).

9.1 Adenocarcinoma (AC)

AC is the most common histological subtype of lung cancer, accounting for approximately half of the cases of NSCLC (43.3%; SEER Cancer Statistics Review, 1975–2012), and usually appears in the glandular

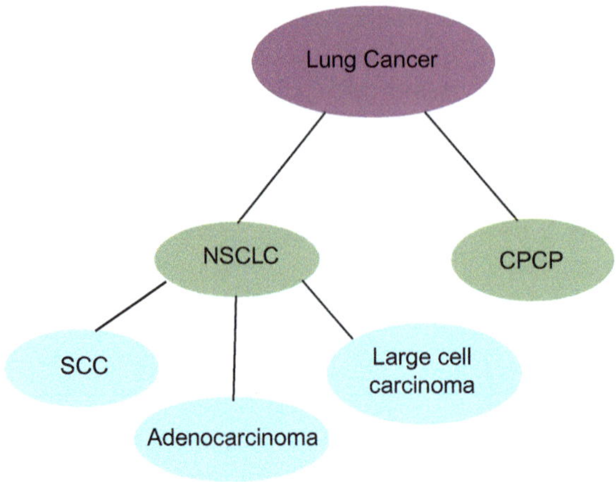

FIG. 5 Subdivision of lung cancer categories.

epithelium of the lung's periphery from cells alveolar bronchial stem, club (formerly Clara) type II cells or pneumocytes [38,39]. It is the predominant subtype that appears in patients who never smoked [40].

9.2 Squamous cell carcinoma (SqCC)

In contrast, SqCC is responsible for approximately 30% of NSCLC (22.6%; SEER Cancer Statistics Review, 1975–2012). It develops mainly in the central airways and segmental bronchi, and is strongly associated with a history of smoking [38,41].

9.3 Large cell carcinoma (LCC)

Regarding LCC, the cell of origin has not yet been defined, but it has been postulated that it originated from differentiated neuroendocrine cell lines, compromised neuroendocrine progenitor cells, or non-neuroendocrine cells that acquire neuroendocrine differentiation in the lungs [42,43].

10 Inflammation and lung cancer

Inflammation promotes the onset and progression of cancer [44]. Specifically, malignant transformation, cell proliferation and survival, angiogenesis, invasion, and metastasis [45], with a strong relationship between lung cancer risk factors and changes in inflammatory cytokine levels, oxidative stress markers, and composition of immune cells.

10.1 Risk factors for lung cancer, inflammatory cytokines, and oxidative stress

Exposure to tobacco smoke is the main risk factor associated with lung cancer—smokers are 14 times more likely to develop lung cancer than those who have never smoked [46]. Although lung cancer is often seen as a smoker's disease, if lung cancer in never smokers were their disease, it would be ranked as the seventh most common cause of cancer deaths worldwide and would account for 300,000 deaths each year [40]. Other factors known to influence lung cancer risk include environmental carcinogens, such as arsenic, radon, and asbestos, air pollution, viral infection, and genetic risk factors, including family history of lung cancer [40,47,48]. In addition, individuals with inflammatory lung disease, such as chronic obstructive pulmonary disease (COPD), are at increased risk of developing lung cancer [49].

Smoking is known to drive altered local and systemic levels of inflammatory cytokines and reactive oxygen species (ROS) in the development of smoking-related lung cancers. For example, inflammatory markers, including C-reactive protein (CRP) and chemokine ligand (motif CC) (CCL) 17 and CCL22, are elevated in the serum of ex-smokers or current smokers compared to those who never smoked [50]. In addition, elevated levels of circulating inflammatory molecules CRP, CCL22, CCL17, and chemokine ligand (motif CXC) 5, CXCL7, CXCL9, and CXCL13 are associated with increased risk of lung cancer in current smokers and ex-smokers [51]. Smokers with lung cancer increase their serum levels of CCL20, an inflammatory molecule that promotes tumor cell proliferation and migration, which is correlated with advanced disease and poor prognosis in lung cancer [52]. Surfactant protein D (SFTPD) is an important regulator of innate immunity, inflammation, and oxidative stress secreted by type II pneumocytes in the airways; the decrease in the expression of this protein in smokers' bronchoalveolar lavage (BAL) correlates with the progression of bronchial dysplasia [53]. In addition to inflammatory cytokines and chemokines, ROS is also an important mediator of smoke-induced damage, chronic inflammation, and cancer development, promoting oxidative DNA damage and genomic instability [54].

10.2 Immune response T helper 17 (Th17)

Respiratory diseases that involve chronic inflammation are associated with the risk of lung cancer, where patients with COPD inflammatory lung disease have a 10-fold increase in the risk of lung cancer, with COPD being linked to the activation of inflammatory signaling pathways. -tumorigenic in immune cells [55], which can communicate through the nuclear factor kappa-light-chain-enhancer of activated B cells (NF-κB) and signal transducer and activator of cytokine production dependent on transcription 3 (STAT3). The oncogenic activation of STAT3 and NF-κB induces the production of cytokines by tumor cells that can regulate the immunosuppressive and tumor-promoting functions of immune cells infiltrated by tumor via transactivation of these same transcription factors [56], being one mechanism for the development of COPD-related lung cancer [56,57].

10.3 Inflammatory cells and lung cancer

The biological process of inflammation depends on the recruitment of different types of immune cells, and the inflammatory response depends on the activities of innate and adaptive immune cells to maintain tissue homeostasis [58]. However, the infiltration of immune cells is observed in tumors, promoting the development, progression, and metastasis of cancer [59], which may contribute to immune evasion and the development of cancer. Innate and adaptive immune cells, including macrophages, neutrophils, natural killer (NK) cells, and B cells, have been implicated in antitumor and pro-tumor activities, as well as the antitumor and pro-tumor roles of T cells in the development of cancer. Therapeutic strategies aimed at this type of adaptive immune cell have been the focus of recent immunotherapy development and applications, including for the treatment of lung cancer [60].

T cells play several roles in the immune response and are highly relevant to the biology of lung cancer. CD8 + cytotoxic T lymphocytes (CTLs) facilitate immune vigilance by recognizing the T cell receptor (TCR) of antigens linked to the major histocompatibility complex (MHC) -I in antigen-presenting cells and, when activated, produce gamma interferon (IFN) -γ), perforin, and granzyme B that contribute to the cytolysis of tumor cells [61]. In addition to CD8 + CTLs, several CD4 + T cells, including subsets of regulatory CD4 + T cells (Treg) and auxiliary T cells (Th17), have emerged as major players in a variety of diseases involving inflammation, including cancer. Both CD4 + the subsets of T cells facilitate a pro-tumor environment by promoting and maintaining an environment of immunosuppressive and pro-tumor inflammation that can favor tumorigenesis, cancer progression, and metastasis. The Treg and Th17 subsets are generally considered to play opposite roles in the regulation of immunity, where the determination of cell fate arises from a balance of transcription factors that govern the differentiation of CD4 + T cells and the generation of Treg and Th17 cells [62]. For example, the expression of the forkhead box transcription factor P3 (FoxP3) represses the Th17 transcription factors related to the orphan receptor gamma t (RORγt) and ROR alpha (RORα) receptor to drive Treg differentiation [63]. Th17 differentiation can occur when pro-inflammatory cytokines, including IL-6, IL-21, and IL-23, inhibit FoxP3 expression and subsequent RORγt repression [63,64]. In addition to the generation of de novo Tregs from FoxP3—T cells, Tregs can also be generated in homeostatic or pathological conditions through the proliferation of FoxP3+cells derived from the thymus [65]. In addition, a new mechanism of Treg-dependent promotion of Th17 differentiation via IL-2 sequestration promotes IL-17-induced inflammation and tumorigenesis in colon cancer, highlighting the complex interaction between these two types of cells in the context of cancer [66].

10.4 Lung cancer and Tregs

Tregs represent a vital component of the adaptive immune system, which works to prevent autoimmunity and chronic inflammation [67], and represent a phenotypically diverse cell line, classified according to their differentiation site, either in the thymus or in different locations. Extrathymic Tregs [68] are generally characterized as CD4+CD25 high and express the transcriptional regulatory factor FoxP3, which can induce immunosuppression through contact-dependent mechanisms, such as the expression of protein 4 associated with cytotoxic T lymphocytes (CTLA-4), programmed cell death 1 (PD-1), programmed death-ligand 1 (PD-L1), lymphocytes-activation protein 3 (LAG-3), CD39/73 and neuropilin 1 (Nrp1), or through independent contact mechanisms, including the sequestration of IL-2 and the production of soluble immunosuppressive molecules IL-10, TGF-β, adenosine, prostaglandin E 2 (PGE 2), or galectin-1 [69] (Fig. 6). In carcinogenesis, systemic expansion and intratumoral accumulation of immunosuppressive Tregs can disrupt antitumor immunity, leading to the growth and metastasis of a variety of malignancies, including lung, breast, prostate, and ovary [68]. Certain cell surface molecules have been shown to have stabilizing effects on the population of Treg cells: CD39 (ectonucleoside triphosphate diphosphohydrolase 1; ENTPD1) increases the stability of CD4+FoxP3+Tregs, contributing to its immunosuppressive function [70]. By suppressing antitumor effector cells, Tregs emerge as active contributors to cancer progression [71].

Tregs are implicated in the early stages of tumor development, where exposure to tobacco carcinogen increased FoxP3+lung lymphocytes before tumor development, suggesting the potential for Tregs to generate a favorable niche for the development of Kras-driven lung tumors, mutations found mainly in smoker-related lung cancers [72].

Tregs can inhibit CD8+T cell-mediated antitumor immunity (Fig. 7), with depletion of Tregs resulting in tumor cell death and elevated levels of granzyme A, granzyme B, perforin, and IFN-γ in the infiltration of CD8 T cells + in early stages of tumorigenesis [73]. In addition, the development of SCLC influences the immunosuppressive activities of Tregs, where SCLC cell lines induce the generation of Treg from CD4+T cells through the production of IL-15 (Fig. 7). In lung tumors, Tregs are also associated with the

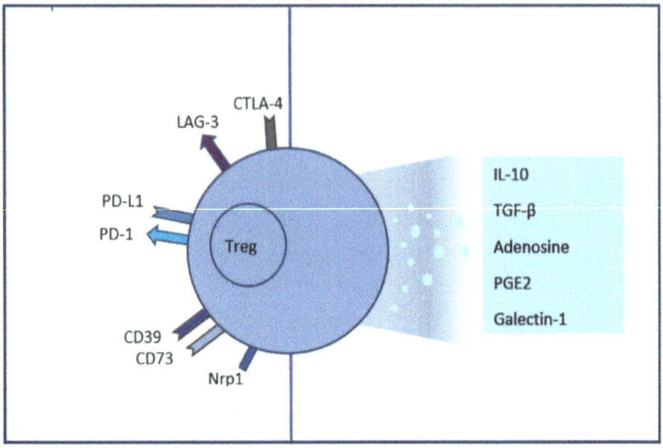

FIG. 6 Treg mechanisms dependent and independent of contacts in the mediation of tumorigenesis.

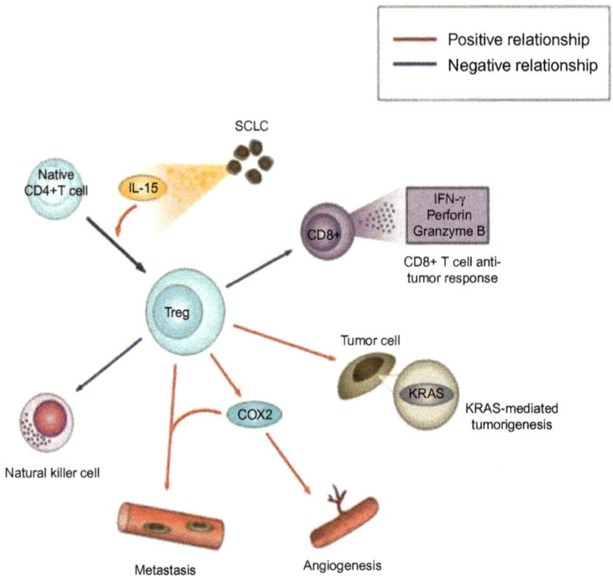

FIG. 7 The immunosuppressive and pro-tumorigenic processes in the development of lung cancer depend on the quantitative relationships of Treg populations. *Arrows* indicate Treg-dependent processes, with *red (dark gray in the printed version)* indicating positive relationships and *blue (light gray in the printed version)* indicating negative Treg-dependent relationships.

expression of the angiogenic and metastatic enhancer cyclooxygenase-2 (COX2), where high numbers of intratumoral FoxP3+lymphocytes are positively correlated with high intratumoral expression of COX2, and can be induced by the nicotine-carcinogenic nicotine of nitrosamine derivative (NNK) [74,75] (Fig. 7).

Tregs promote metastases and the development of metastatic tumor foci that can be seen in peripheral blood because they increase with the stage of the disease and are higher in patients with metastatic tumors [76], as well as in metastatic lymph nodes compared to non-metastatic lymph nodes in patients with CA [77]. In Lewis lung carcinoma, Tregs inhibit cytotoxicity mediated by NK cells in a TGF-β-dependent manner, and Tregs depletion contributes to increased antimetastatic activities of NK cells [78] (Fig. 7).

The prognosis, the relative accumulation of Tregs in NSCLC tumors, and peripheral blood of patients with SCLC (in relation to effector T cell populations) are associated with an increased risk of recurrence and a high proportion of FoxP3+lymphocytes in SCLC lung tumor biopsies being related to decreased survival, and in NSCLC, high levels of intratumoral FoxP3+lymphocytes are associated with reduced recurrence-free survival [74].

In a clinical setting, targeting the CTLA-4 and PD-1 immune control point molecules has recently received a lot of attention in a variety of cancers, including lung cancer [79]. Binding of CTLA-4 and PD-1 immune receptors to Tregs leads to inhibition of T cell activation, particularly CD8+effector T cells, where the anti-CTLA-4 immunotherapeutic agent ipilimumab improves the results of patients with SCLC and NSCLC, while nivolumab (anti-PD-1 therapy) is not remarkably effective in patients with lung CCSq [80].

Currently, clinical biomarkers are limited for these treatments, as patients with little or no expression of these molecules in their lung tumors may still have beneficial therapeutic responses. For example, patients harboring tumors with negative expression of the PD-1 ligand, PD-L1, may also benefit from anti-PD1/PD-L1 therapies, suggesting that the clinical use of PD-L1 positivity as a selection criterion can exclude patients who would potentially benefit from these treatments [69].

10.5 Th17 cells and lung cancer

Th17 cells are a group of CD4+helper T cells that are phenotypically distinct from Th1 and Th2 cells and have been characterized in many inflammatory lung diseases, including COPD [81,82]. Th17 cells express the transcription factors RORγt/RORC2 and RORα, which lead to Th17 differentiation and produce pro-inflammatory cytokines, including IL-17A, which modulate the tumor microenvironment [83] (Fig. 8). A family of cytokines IL-17 contributes to inflammation, production of cytokines and chemokines, recruitment of neutrophils in the context of inflammation, infection, and pulmonary antitumor immunity [84] (Fig. 8). Changes in IL-17 and its signaling pathways are relevant to the development of lung cancer with IL-17 polymorphisms and epigenetic changes in the IL-17 signaling pathway correlated with increased lung cancer predisposition [82,85].

Thus, Th17 cells are an integral component of the inflammatory medium in the tumor microenvironment and may be causally involved in the promotion of distinct lung tumor phenotypes.

The IL-17 cytokine family consists of six members:

- IL-17A (also known as IL-17)
- IL-17B
- IL-17C
- IL-17D
- IL-17E (also called IL-25)
- IL-17F

Th17 cells produce IL-17A and IL-17F. Serum levels of IL-17 are strongly associated with the development of lung cancer. Serum IL-17 levels are significantly higher in NSCLC compared to individuals without cancer, showing an additional diagnostic marker for NSCLC. IL-17 cytokines contribute to inflammation by facilitating the induction of pro-inflammatory cytokines and chemokines. Exogenous treatment of embryonic fibroblasts with IL-17A or IL-17F induces greater expression of IL-6, CXCL1, CCL2, and CCL7. High expression of chemokines CCL7, CCL22, CCL20, CCL11, and chemokine ligand 1 (motif C-X3-C) (CX3CL1) are found in the lungs with chronic overexpression of IL-17th. In addition to the positive regulation of these chemokines, the high expression of CCL1, CCL2, and CXCL1

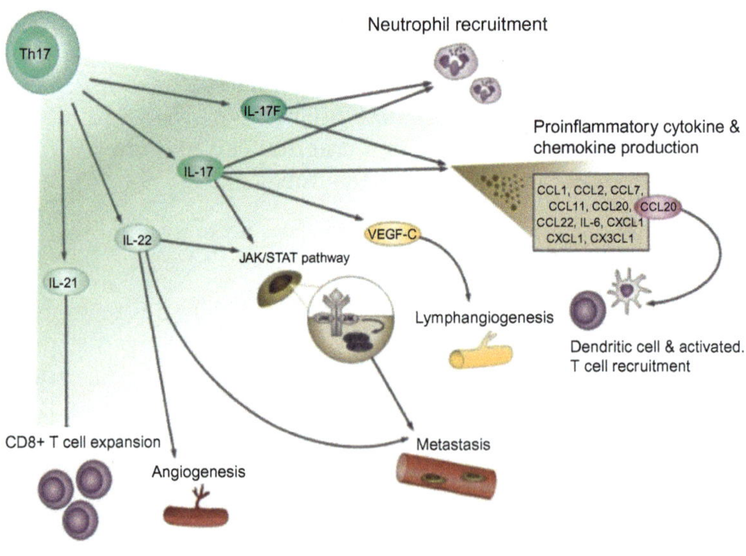

FIG. 8 The pro-inflammatory and pro-tumorigenic roles of Th17 cells in the pathogenesis of lung cancer depend on the production of IL-17, pro-inflammatory cytokines that promote lung tumorigenesis.

must also be present in the lung epithelial cell line after exogenous treatment with IL-17A [86] (Fig. 8).

Genetic variation and epigenetic changes in the IL-17F pathway can affect the development of lung cancer, especially the IL-17F inflammatory response pathway [82] in the small airways of COPD. The receptor IL-17C (IL17RC) and CXCL1 are identified as hypermethylated and under-expressed, while DNA hypomethylation and overexpression of colony-stimulating factor 2 (CSF2), a pro-inflammatory cytokine-induced by IL-17F, are observed [82].

Many other genes in the IL-17F inflammatory pathway are also altered by DNA methylation in the small airways of COPD, and changes in these genes are known to contribute to carcinogenesis. Single nucleotide polymorphisms (SNPs) in IL F genes – 17 are significantly associated with the development of lung cancer. For example, the IL-17F 7488G allele is associated with advanced or metastatic lung cancer in a Tunisian population [85]. COPD is characterized by enlarged Th17 cells (CD3+CD4+IL-17A+) in peripheral blood. Likewise, elevated levels of Th17 cell cytokines are present in Kras-induced lung cancer with simultaneous induction of COPD-like inflammation through exposure to *Haemophilus influenzae* lysate (NTHi).

In malignant disease, Th17 (IL-17A, RORα4, and RORγt) are more expressed when compared to normal tissues [87]. IL-17 is known to contribute to the induction of lung cancer prometastatic factor expression, its elevated expression in peripheral blood is related to the TNM stage, and the increased expression of IL-17 receptor (IL-17RC) in NSCLC tumor cells is associated with invasive potential. The treatment of NSCLC A549 and H520 cell lines with IL-17 results in increased STAT3 phosphorylation by positively regulating the production of prometastatic factor including vascular endothelial growth factor (VEGF). As a lymph node metastasis mechanism, IL-17 can promote lymphangiogenesis by positively regulating the expression of lymphangiogenic factor vascular endothelial growth factor C (VEGF-C) in lung cancer cells. Increased density of IL-17 positive cells in NSCLC tumors is related to lymphatic vessel density. In addition, the loss of IL-17 reduces metastases and increases survival [88,89].

Th17 cells produce IL-22, which contributes to pro-survival signaling, angiogenesis and metastasis, part of which may be associated with activation of the STAT3 signaling pathway in cancer cells [90]. High levels of IL-22 are detected locally in primary tumors and malignant pleural effusions (MSEs), and systemically in the serum of patients with NSCLC [91].

Th17 cells have complex biological functions and can, paradoxically, contribute to antitumor immunity [92]. Increased levels of IL-21 can induce tumor regression through the expansion of CD8+tumor infiltrating lymphocytes in NSCLC, as well as ovarian cancer and melanoma (Fig. 8).

In addition, high pleural Th17 cell counts are associated with increased survival of NSCLC (lung histologies AC and SqCC) [93].

10.6 Quantitative relationships between Th17s and Tregs in lung cancer prognosis

Although several populations of immune cells have been studied extensively in the context of cancer biology, the field has mainly focused on individual cell populations. Cancer immunology has shifted to a more integrated understanding of the potential interactions between immune cell populations within the tumor microenvironment, including a focus on cell proportions, crosstalk, and phenotypic plasticity in the context of cancer prognosis, where advances include the development of an Immunocore as a potential component of cancer classification with prognostic relevance in several different tumor types, including NSCLC. In addition to lung cancer, changes in Treg/Th17 levels occur in hematological tumors and other types of solid tumor, in addition to autoimmune diseases and viral and bacterial infections, indicating that the balance of these two subsets plays an important role in the regulation of inflammation and of cancer development [94]. The subsets of CD4+T cells may have opposite prognostic values in lung cancer, with a higher ratio of Tregs to Th17s being correlated with more aggressive malignancies in an advanced stage—that is, Th17 and Treg cells are widely considered to play roles in pro- and anti-inflammatory drugs, respectively, although it should be noted that the plasticity of T cells allows a functional continuum between these two subsets of CD4+T cells [95]. In summary, the balance of these two CD4+subsets of T cells in the tumor and systemic sites are associated with the development, progression, and prognosis of lung cancer.

The peripheral blood of patients with NSCLC has a Th17 (CD4+ IL-17 +) and Treg (CD4+CD25+FoxP3 +) percentage higher than in patients without cancer, but the cells of the CD4+T subsets are inversely correlated in the blood. Serum Th17 cells are known to correlate positively with IL-1β, IL-6, IL-23, while Tregs are known to correlate positively with TGF-β1 and IL-10. NSCLC patients with Stage IV disease have higher Treg/Th17 ratios compared to patients with Stage I–III disease (Fig. 6). On the other hand, the inversion occurs with serum levels of carcinoembryonic antigen (CEA), an elevated onco fetal marker in lung cancers with poor prognosis [96].

In lung cancer, a subset CD45RA − CD45RO+FoxP3 hi, enriched in NSCLC, is characterized by increased expression of RORγt and IL-17, showing the importance of elucidating specific molecular characteristics and functions, and indicating that IL-17+FoxP3+in the presence of T cells can influence inflammation in the tumor microenvironment and improve the understanding of the Treg/Th17 ratio in cancer prognosis. The balance between Tregs/Th17 and the lung cancer prognosis highlights the clinical relevance of these cell types as biomarkers and potential therapeutic targets [97].

10.7 DNA methylation and lung cancer

DNA methylation is important in the development of lung cancer during transformation, as the immortalization of normal bronchial epithelial cells is increased by exposure to low doses of carcinogens. The underlying mechanism is involved in the hypermethylation of 5–10 genes due to the high expression of DNA methyltransferase 1 (DNMT1). The ablation of DNMT1 reverses this process. In addition, the stable knockdown of DNMT1 before exposure to the carcinogen was sufficient to prevent cell transformation [98].

Aberrant post-translational modifications of histones also have predictive and prognostic significance in NSCLC [99,100], and the deregulation of some of these enzymes, among the oxidative stress response mechanism, critical regulators such as lysine acetyltransferases and histone deacetylases (HDACs), and the NF-kB pathway in the lungs can lead to the development of cancer cells, as shown in Fig. 9.

Elevated HDAC levels are observed in patients who have NSCLC in stages III or IV, where the HDAC3 protein can be found elevated in 92% of lung SCCs [101]. HDACs form large multiprotein complexes to regulate gene expression in NSCLC, as mSin3A, a member of a multicomponent co-repressor complex containing HDACs, is decreased in NSCLC [102].

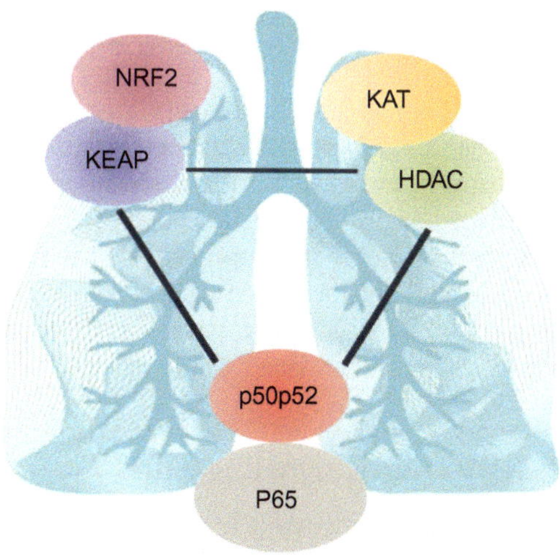

FIG. 9 Overview of the balance between oxidative stress response mechanisms.

10.8 ATP and NSCLC-dependent chromatin remodeling complexes SWI/SNF

In NSCLC, members of the ATP-dependent chromatin remodeling complex SWItch/sucrose (SWI)/non-fermentable (SNF) are also altered. SWI/SNF is found in the formation of a complex containing neuron-restricting silencing factor and its co-repressors, mSin3A and CoREST [104]. Although this complex regulates specific neuronal genes, its deregulation of genes regulated by neuron restrictive silencing factor may contribute to increasing tumorigenicity [103] through the expressions of the SWI/SNF ATPase, BRG1, and BRM subunits (BRG1/BRM) that are shown to be mutated or lost in approximately 30% of the sequences, and where BRM is shown to be regulated epigenetically, and where HDAC inhibitors are shown to reactivate BRM expression.

The expression of metastasis-associated protein 1 (MTA-1) is significantly elevated in NSCLC and is associated with tumor invasion and metastasis [104]. As MTA-1 is known to be associated with HDACs [105,106], this suggests that HDTA activity associated with MTA-1 is involved with NSCLC invasion and metastasis. Mutations within the lysine acetyltransferase KAT3A also occur in NSCLC, and polymorphisms are associated with an increased risk of lung cancer, including lysine methyltransferase KMT1B, and KMT8, methyl-CpG (MBD-1).

10.9 KATs/HDACs and oxidative stress pathways

Oxidative stress in adipose and renal cells increases the levels of HDAD SIRT2 dependent on NAD +. SIRT2 then binds to FOXO3a and reduces its level of acetylation, resulting in increased FOXO3a binding to DNA, increasing the expression of p27 (Kip1), manganese SOD, and Bim, and decreasing the cellular levels of ROS in these cells [107]. SIRT3 prevents oxidative stress-mediated cell death by deacetylation of the Ku70 protein, which results in increased Ku70 binding to the pro-apoptotic Bax protein preventing translocation to the mitochondria, thus protecting cardiomyocytes from stress-induced apoptosis oxidation [108].

The Bach1 transcription factor is a repressor of the response to oxidative stress, involved in cell senescence induced by oxidative stress by a complex containing p53, HDAC1, and the nuclear co-repressor N-CoR to suppress target genes.

10.10 KEAP1-NRF2-ARE signaling path

One of the most important pathways to protect against oxidative stress is Kelch-like ECH 1 protein (KEAP1) - related to NFE2.

Factor 2 (NRF2) - antioxidant response element (ARE), or KEAP1-NRF2-ARE, is a signaling pathway that plays critical roles in protecting cells from endogenous and exogenous stresses and plays important roles in the lungs.

The negative regulation of KEAP1 by CpG DNA methylation occurs in both NSCLC cell lines and tumor tissues [109]. In NSCLC cell lines, when the loss of KEAP1 function occurs, NRF2 activation and cell growth occur [110].

10.11 Epigenetics, NSCLC, and oxidative stress pathways/genes

10.11.1 MRF2

There is a strong relationship between direct chromatin remodeling and the cellular response to oxidative stress. The KEAP1-NRF2-ARE pathway is strongly implicated as a pathway for antioxidant stress in the lung. Under normal conditions, NRF2 is found associated with the cytoplasmic inhibitor, KEAP1, but after oxidative stress, NRF2 becomes phosphorylated and dissociates from KEAP1; it translocates to the nucleus to bind to a ROS inducing the expression of genes encoding detoxifying antioxidant proteins, to rescue cells from oxidative damage. Two transcriptional activation domains, Neh4 and Neh5, are associated with KAT3B and act synergistically in obtaining maximum gene expression by NRF2, while other proteins with histone modification capacity demonstrate association and regulation of the activity of the NRF2 transactivation domain and include the lysine acetyltransferases KAT3A/KAT3B/KAT2B and arginine methyltransferases CARM1 and PRMT1.

Under conditions of oxidative stress, the component of the SWI/SNF BRG1 complex interacts directly with NRF2 to mediate selectively the induction of responsive genes [111]. The NFκB p65 subunit represses the NRF2 antioxidant response element pathway (ARE) at the transcriptional level in cells where NFκB and NRF2 are activated simultaneously, p65 unidirectionally antagonizes the transcriptional activity of NRF2, and the ARE-dependent expression of hemoxigenase-1 (HO-1) is strongly suppressed. The mechanism to cause this effect is the selective prevention of the association of lysine acetyltransferase KAT3A with p65, resulting in the inactivation of p65, followed by the recruitment by p65 of HDAC3 to ARE-dependent promoters to suppress transcription [112].

10.12 Heme oxygenase 1

Heme oxygenase 1 (HO-1) is a cell stress response gene that has been shown to protect the lung from insults based on oxidative stress, such as cigarette smoke and cancer, by degrading heme to biliverdin-IXα, carbon monoxide, and iron, each with functions in cytoprotection. In low concentrations, carbon monoxide can confer cytological and tissue-protective effects like those of endogenous expression of HO-1, involving antioxidant, anti-inflammatory, antiproliferative, and anti-apoptotic effects. The regulation of HO-1 involves the remodeling of chromatin, and the exposure of cells to heme results in hyperacetylation and hypermethylation again of histone H3 in HO-1 intensifiers, resulting in its enhanced transcription.

The BRG1 catalytic subunit of SWI2/SNF2 chromatin remodeling complexes uses NRF2 to regulate HO-1 expression in a colon cell line in response to oxidative stress [113]. The deletion of a given region in NRF2 (the Neh5 domain) results in a reduction in the expression of HO-1. The loss of this domain reduces the ability of KAT3B and BRG1 (gene 1 related to Brahma) to associate with NRF2, thus reducing their cooperative increase in the activity of the HO-1 promoter [113].

10.13 Hypoxia-inducible factor 1α

Hypoxia is a common event in solid tumors, but it is also strongly associated with NSCLC, as well as levels of HIF-1α or carbonic anhydrase IX. HIF-1α can be expressed in epithelial cells associated with increased progression of structural changes, increased thickness of the reticular basement membrane (RBM), and a reduction in the number of blood vessels in the sub-epithelium of patients with NSCLC, while in hypoxia conditions it, induces oxidative stress through excessive ROS generation, related to positive regulation of HIF-1α, VEGF, and DNA oxidation, in addition to regulating epigenetic regulators, including lysine demethylases.

Epigenetic mechanisms affect HIF-1α mainly by stabilizing the protein itself, using HDACs. In addition, MTA-1 also stabilizes HIF-1α and increases tumor metastasis through angiogenesis due to direct associations with HDACs. In fact, in many cell culture models, inhibition of HDACs has been shown both to destabilize HIF-1α and to reduce angiogenesis [114].

HIF-1α is functionally associated with lysine acetyltransferases (KATs and HDACs) to regulate gene expression in response to hypoxia. These associations are known to be essential to lead to hypoxia-responsive transcription. The domains required for KAT3A and KAT3B interactions with HIF-1α are genetically non-redundant and indispensable for the HIF-1α transactivation function, while a similar and greater effect is necessary for HIF-1α domain and association to occur. In fact, for some target genes responsive to HIF-1α, both lysine acetyltransferases and HDACs are needed for more than 90% of the gene response, suggesting that a functional interaction between lysine acetylases and HDACs is essential for almost any transcription responsive to HIF-1α [115].

10.13.1 PGC-1α

The 1α co-activator of the γ receptor activated by peroxisome proliferator (PGC-1α) after being isolated from a brown fat cDNA library and linked to adaptive thermogenesis, has been associated with several pathways such as mitochondrial oxidative metabolism, the decision of the fate of the adipocyte cell, glucose homeostasis, and hepatic gluconeogenesis [116]. It is well established that PGC-1α acts as a transcriptional co-activator in association with both HDACs and lysine acetyltransferases to regulate gene expression. Under stress conditions, PGC-1α increases the expression of major antioxidant enzymes, including copper/zinc SOD (SOD1), manganese SOD (SOD2), catalase, and GPx1. Nitric oxide also regulates the protection of mitochondria by modulating the expression of PGC-1α, resulting in changes in the expression of members of the ROS mitochondrial detoxification system [117].

PGC-1 transcriptional co-activators are the main regulators of several crucial aspects of energy metabolism, and NRF1 and NRF2 are the main targets of PGC-1 s in mitochondrial biogenesis.

10.14 Oxidative stress, acetylation, and its role in NF-kB regulation

Chronic inflammation is an invasive element strongly associated with cancer. In fact, pro-inflammatory cascades are important elements in the responses mediated by oxidative stress in NSCLC [118]. One of the critical mediators of the pro-inflammatory cascade signaling pathways involves the activation or regulation of NFκB. The NFκB/Rel family consists of five subunits, but NFκB consists of a heterodimeric protein composed of a p50 and a p65 subunit (RelA) [119].

One of the indicators that oxidative stress affects NFκB activation within the lung shows that oxidative stress and TNF-α activating NFκB/AP-1 in lung cancer cells results in acetyl-induced histone and increased pro- inflammatory release of IL-8 from these cells [120]. The activation of NFκB itself depends on the activities of lysine acetyltransferases and HDACs. The lysine acetyltransferases KAT3B and KAT3A are key co-activators in the regulation of the expression of the gene driven by NFκB through associations with the RelA/p65 subunit, while the lysine acetyltransferase KAT13A potentiates NFκB transactivation through association with the p50 subunit [121]. In the context of the lungs, oxidative stress induces the expression of pro-inflammatory genes, such as IL-6 and IL-8, through the recruitment of NFκB from the activity of KAT3A or KAT3B lysine acetyltransferase to stimulate the transcription of these genes. Since lysine acetyltransferases have been found to associate with NFκB, interactions between HDACs and NFκB are necessary to suppress the expression of genes regulated by NFκB or to control the levels of gene expression induced by NFκB [122].

One of the critical regulators of transcriptional NFκB activation is I-Kappa-B Kinase-Alpha (IKKA). The cytoplasmic sequestration of p65 by IKBA is sufficient to translocate nuclear co-repressors, silencing mediator to retinoic acid and thyroid hormone receptor (SMRT) to the cytoplasm. Since SMRT normally acts by suppressing transcription through association with HDACs, this results in the suppressed transcription of its target genes. The association between the ability of p65 and IKKA to bind directly to SMRT can be inhibited in a dose-dependent manner by lysine acetyltransferase KAT3A, further implying a role for lysine acetylation in this process. IKKA activates NFκB-mediated transcription by phosphorylation (SMRT), resulting in the exchange of transcriptional co-repressor (HDAC3) by co-activating complexes (KAT3B), allowing KAT3B acetyl RelA/p65 [123].

Inhibition of NFκB activation occurs when the Daxx protein binds to the RelA/p65 region that contains the main acetylation sites mediated by KAT3B/KAT3A, preventing acetylation of RelA/p65 by direct interference, or when Daxx associates directly with HDAC2, representing a mechanism by which KATs and HDACs compete for critical lysines in the NFκB subunits, showing that a small modification similar to the KAT3A ubiquitin negatively modulates its transcription potential when recruiting a Daxx complex containing HDAC2 [124].

Cigarette smoke is associated with the activation of NFκB in part through the post-translation modification of proteins HDAC1, HDAC2, and HDAC3, by the formation of nitrotyrosine and aldehyde adduct [125]. SIRT1 also regulates NFκB transactivation by physically associating with one of its RelA/p65 subunits and removing acetylation in a critical lysine in RelA/p65, avoiding NFκB-dependent transcription, in addition to decreasing the regulation of SIRT1 in the lungs, resulting in increased acetylation of RelA/p65 with concomitant activation of NFκB-mediated proinflammatory cytokine release [125,126].

11 Oxidative stress in breast cancer

Breast cancer is the second most diagnosed cancer worldwide and the leading cause of cancer death in women. The severity and aggressiveness of breast cancer are assessed by examining the physical and anatomical properties of the disease, in particular using histological grading and TNM staging, where T (0–4) is used to describe the size and location of the tumor, N (0–3) is responsible for the invasion of the lymph nodes, and M measures the spread of the tumor as distant metastasis. With the advent of genomics, the disease could be classified into six specific subtypes according to the intrinsic gene expression: Luminal-A, Luminal-B, HER2-positive, Basal-like, Normal-like, and the so-called claudin-low, which is responsible for 7%–14% of all breast cancers [127–129].

Positive breast cancer malignancy for the estrogen receptor at high and significant levels of 8-OHdG and with detection in blood serum showing prognostic value [130]. Mitochondrial ROS are mainly activated by estrogen, which activates redox-sensitive proteins involved in cell proliferation and anti-apoptotic pathways, to support such signaling without risking interruption of the cell cycle and apoptosis. Estrogen also increases an antioxidant response inducing, for example, the nuclear-erythroid-2-related factor 2 (NRF2) transcription factor. This enzyme is the main redox regulator; under oxidative stress, its Kelch-like ECH-associated protein 1 inhibitor (Keap1) undergoes a conformational change that allows the dissociation of NRF2 and the consequent translocation to the nucleus, where it increases the transcription of different agents contrary to ROS. NRF2 is overexpressed in breast cancer, where it promotes cell survival, proliferation, migration, and metastasis [131]. In addition, the interaction between NRF2 and BRCA1 increases and stabilizes the expression of NRF2, and estrogen can partially mimic this action in BRCA1 null cells [20].

11.1 Controversy over the effect of antioxidants on breast cancer

Suppression of HER2 in the oxidative systemic profile in patients with breast cancer occurs by increased oxidative stress, attenuated by increased levels of stabilized SOD e glutathione (GSH), which are indicative of active antioxidant response. In addition, HER2 interacts with NRF2 to promote the transcription of antioxidant and detoxification genes, and this partnership gives drug resistance to breast cancer cells, causing antioxidants to favor the neoplastic transformation of the breast: by reducing ROS concentrations, they can prevent cell death dependent on ROS, which makes the role of antioxidants controversial in breast cancer [132,133].

For example, the antioxidant superoxide dismutase 2 (SOD2), which converts the highly toxic superoxide radical to more stable hydrogen peroxide in the mitochondria, can act both as an oncogene and as a tumor suppressor. It is found modulated downward in early-stage breast cancer, while it is regulated upward in advanced tumors. Despite these results, SOD mimics are proposed for therapeutic purposes. Catalase, glutathione peroxidases, and peroxiredoxins are among the other antioxidant enzymes that balance the production of ROS [134,135].

The addition of a redesigned protein form of the catalase enzyme for treatment with EGFR inhibitor erlotinib helps overcome resistance by targeting the stem-like portion of TNBC cells specifically. Downward modulation of peroxiredoxin-1 (PRDX1) is beneficial for breast cancer therapy, especially in conjunction with pro-oxidant agents, and specific aquaporins allow H_2O_2 to cross cell membranes more quickly than by single diffusion. Aquaporins in breast cancer represent therapy as a target, due to its role in the migration of CXCL12/CXCR4-dependent cancer cells [136].

11.2 miRNA

miRNAs are small molecules of single-stranded non-coding RNAs (~ 18–25 nucleotides), capable of controlling gene expression at the post-transcriptional level. miRNA biogenesis begins when RNA polymerase II/III transcribes into a long primary transcript with a clamp structure, called pre-miRNA [137]. Pri-miRNA is the substrate of Drosha and Dicer, two members of the RNase III enzyme family. First, Drosha cleaves the pri-miRNA into a ~ 70 nucleotide pre-miRNA in the nucleus, which is then exported to the cytoplasm by Exportin-5 Ran-GTPase, where Dicer catalyzes its conversion to a short miRNA/miRNA * duplex (~ 20 bp). To complete miRNA biogenesis, the transactivation-responsive RNA-binding protein (TRBP) leads to the assembly of the miRNA-induced silencing complex (miRISC), mediating the interaction between DICER and Argonaut protein (AGO1, AGO2, AGO3, or AGO4). Finally, the miRISC complex selects a single strand of the duplex (mature miRNA), which recognizes the "seed" region in the target miRNA, usually placed in the 5′ RTU, inducing translational repression or deadenylation and degradation. *Caenorhabditis elegans*, involved in larval development [138].

miRNAs are involved in almost all biological processes, including oxidative stress and cancer [139], and can act as oncosuppressants or oncogenes, which are usually found, respectively, regulated downward and upward in tumor cells (for example, miR-205 and miR-21, respectively). They have a relevant role in the response to DNA damage, occurring after exogenous oxidative stress, such as chemotherapy, and have the ability to target several genes involved in DNA repair, regulating the responsiveness of therapy [140,141].

11.3 miRNA modulates oxidative stress master regulators: NRF2 and NF-kB

NRF2 is an important transcription factor whose induction, or derepression, depends on the redox state of the cell. It is normally found inactive in the cytoplasm linked to its homodimeric repressor Keap1, which anchors the Cullin-3 protein (CUL3) to form an E3 ubiquitin ligase complex; the complex is responsible for the ubiquitination of NRF2 and consequent proteasomal degradation [142]. When cell ROS concentrations increase, specific Keap1 cysteinyl residues are modified and NRF2 is released and free to translocate to the nucleus, where it recognizes the so-called "Antioxidant Response Elements" (ARE) sequences in the target gene's promoters and increases the transcription process [143].

NRF2 promotes the expression of antioxidants and detoxifying enzymes, and the pathogenic activation and accumulation of NRF2 can be triggered by different events. One of the most frequent changes concerns the expression of Keap1 or its ability to bind stably and degrade NRF2, which can be induced to perform the oncogenic activity by miRNAs in different diseases: miR-200a targets the Keap1 mRNA and induces its degradation. Interestingly, there is inhibition of NRF2 by miR-28 in the breast cancer cell line MCF7; miR-93 and miR-153 are targets of NRF2, and their overexpression is associated with breast carcinogenesis [144,145] (Fig. 10).

NF-κB can be found as a homo- and heterodimer of five different proteins, RelA, RelB, c-Rel, p50, and p52. It is inhibited in the cytoplasm by IκB families, which interfere in the target activity when interacting with its important domain of homology (RHD), and is involved in the formation of dimers, and binding to DNA [146]. IκB proteins are usually degraded in response to inflammatory stimuli like TNFα and lipopolysaccharide (LPS). The consequent NF-κB signaling, modulated by ROS, is specific to the cell type and context. This is probably due to the wide range of action of the transcription factor: cell growth, proliferation, migration, and apoptosis are among the pathways it influences [147,148].

NF-κB signaling is often found upregulated in cancers and breast cancer. The protein is reported to be constitutively activated and associated with aggressive and chemoresistant malignancies [149].

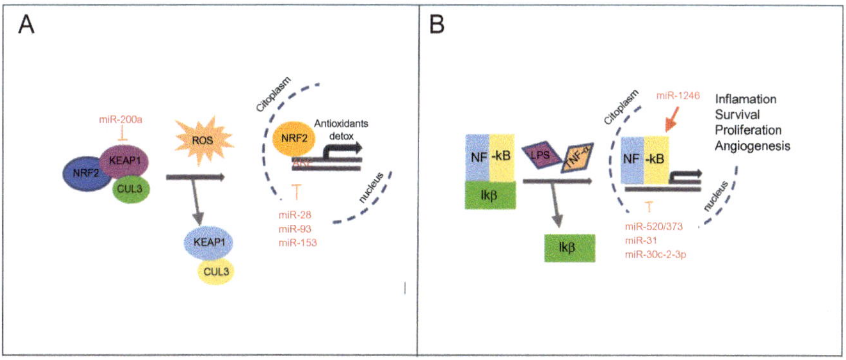

FIG. 10 miRNA that modulates the main regulators of oxidative stress NRF2 (A) and NF-kB (B) in breast cancer (the *red (dark gray in the printed version) arrow* indicates modulation, the *red (dark gray in the printed version)* "T" indicates inhibition).

MiRNAs also play an important role in this scenario. First, NF-κB favors the invasion of breast cancer cells by inducing the expression of oncomiR miR-21 in response to DNA damage [150]. The miR-520/373 family plays a tumor-suppressing role in ER-negative breast cancer, by targeting the NF-κB and TGF-β signaling pathways; miR-31 sensitizes cancer cells to apoptosis by impairing the pathway of NF-κB; miR-30c-2-3p reduces proliferation and invasion of MDA-MB-231 cells through negative modulation of TNFR/NF-κB signaling proteins and the cell cycle; miR-1246 induces signaling pro-inflammatory NF-κB in breast cancer cells; and miR-221/222 promotes stem-like properties and tumor growth of breast cancer through PTEN and sustained activation of Akt/NF-κB/COX-2 [151].

Despite having an oscillatory expression in a physiological context, NF-κB emerges as an adequate oncogene in breast cancer. In addition, due to its wide spectrum of interactions, there are countless miRNAs involved in the regulation of the signaling cascade; consequently, many are the tips for therapeutic interventions.

11.4 miRNA modulates pathways altered by oxidative stress

11.4.1 Metabolism

The main objective of cancer cells is proliferation and survival, which require a large amount of energy in a short period. Therefore, cancer cells tend to modify their metabolism to respond to this demand, and cancers thus prefer rapid glycolysis to phosphorylation mitochondrial oxidation more efficient, allowing to avoid an excessive production of ROS related to the mitochondria [152].

In breast cancers, the metabolic state appears to be linked to the molecular subtype. The most aggressive TNBCs are characterized by a glycolytic phenotype, while luminal malignancies retain oxidative phosphorylation as the main source of energy [153]. It is important to note that it is not uncommon to find heterogeneity also between cells of the same mass tumor, a scenario that can be very harmful.

Through a mixed population of TNBC cells, if all of them migrate towards a glycolytic phenotype, this event would be counterproductive to the tumor, but through the down-modulation of the lactate transporter MCT1, miR-342-3p is able to interrupt the energy flows between neighboring glycolytic and oxidative cells, promoting displacement and, finally, triggering competition for glucose [154], inducing oxidative stress in cancer cells.

One of the most cited mechanisms for metabolic reprogramming of breast cancer cells involving miRNAs is miR-155, which promotes the expression of hexokinase II (HKII), necessary to initiate glycolysis. This miRNA modulates multiple pathways that control HKII:

- *First:* miR-155, through direct down-modulation of C/EBPβ, reduces miR-143, an HKII inhibitor.
- *Second:* miRNA releases STAT3 from its SOCS1 suppressor to increase HKII.
- *Third:* miR-155 positively regulates HKII by interfering with the PIK3R1-FOXO3a-cMYC axis.

miR-27b promotes the progression of breast cancer by targeting Pyruvate Dehydrogenase Protein X (PDHX), thereby altering the cell's metabolic configuration [155]. The PI3K/Akt pathway, whose cells are often mutated in cancers of the breast, has a profound impact on the metabolism and production of ROS, directly regulating mitochondrial bioenergetics and NOX enzymes. Vice versa, oxidative stress activates PI3K and suppresses the activity of PTEN, an inhibitor of PI3K/Akt signaling [156–158]. Due to the relevance of the pathway, there are numerous miRNAs found involved in its regulation in breast cancer, such as tumor suppressor miR-204-5p, which targets PIK3CB, and PTEN inhibitor oncomiR miR-1297 and miR-498 [159–161] (Fig. 11).

11.4.2 Hypoxia

Hypoxia refers to a pathological level of oxygen tension, caused by high proliferative rates of cancer cells and insufficient vasculature. The lack of oxygen supply therefore induces cancer cells to undergo the epithelial-mesenchymal transition (EMT), which corresponds to the acquisition of migratory and invasive properties, stem characteristics, and resistance to apoptosis. The transcription factor HIF-1α stimulates angiogenesis and triggers a positive feedback cycle in the proliferation

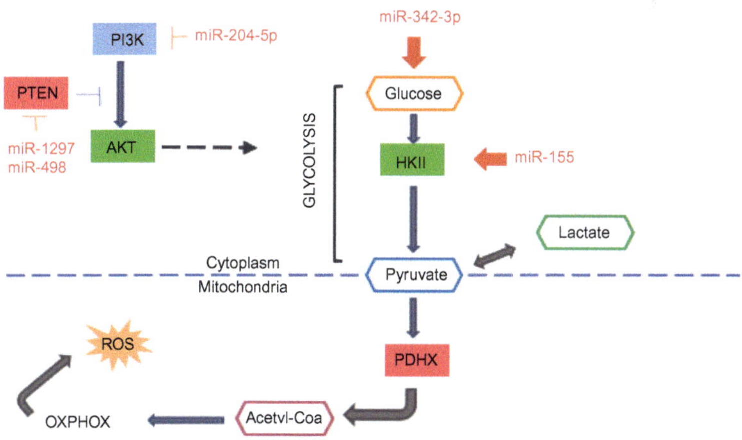

FIG. 11 miRNAs involved in the metabolism and regulation of cancer characteristics influenced by oxidative stress in breast cancer (where the *red (dark gray in the printed version) arrow* means up-modulation and the red *(dark gray in the printed version)* "T" means inhibition).

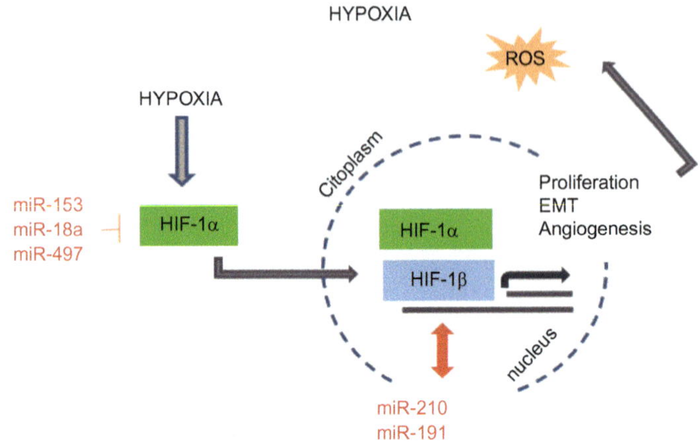

FIG. 12 miRNAs involved in the regulation of cancer characteristics influenced by oxidative stress in breast cancer and hypoxia (where the *red (dark gray in the printed version) arrow* means up-modulation and the *red (dark gray in the printed version)* "T" means inhibition).

pathways. Consequently, oxidative stress increases with reoxygenation and leakage of mitochondrial electrons, and miRNA are causally linked to hypoxia at initiation progression of breast cancer, such as miRNA-210.

In addition, miRNA-153 acts as a tumor suppressor in breast cancer, targeting HIF-1α [162], inhibiting migration, proliferation, and tube formation in HUVEC cells and angiogenesis in the MDA- in vivo model MB-231 by inhibiting the HIF-1α/VEGFA axis. The high expression of miR-191 in breast cancer cell lines induces a more aggressive tumor under hypoxia [163], and consequently its inhibition can be explored as a therapeutic option for hypoxic breast cancer. miR-18a targets HIF-1α, whose high expression is associated with shorter DMFS (distant metastasis-free survival) in patients with breast tumors like baseline [164]. In MDA-MB-231 metastatic cells, ectopic expression of miR-18a reduces the primary tumor and lung metastasis. miR-497 suppresses hypoxia conditions and for this reason is generally down-regulated in breast cancer cells [165]. miR-497 also targets a pro-angiogenic molecule, vascular endothelial growth factor (VEGF), and its ectopic expression reduces tumor growth and angiogenesis in the breast cancer tumor model (Fig. 12).

It is important to note that miRNAs are important in tumor progression and in hypoxia and metastasis, and that they could be used as therapeutic tools to reduce tumor aggressiveness and spread.

12 Response to therapy

It is well known that miRNAs can influence the response to therapy in breast cancer and are used as potential therapeutic tools, either alone or in combination with standard therapy to prevent cancer progression. Chemotherapy and radiation therapy still represent the standard therapy for breast cancer; miRNAs can target different genes, reducing drug resistance and promoting therapeutic response.

miR-302b, targeting E2F1 and DNA repair, increases the cisplatin response in breast cancer cells, chemotherapy drugs such as platinum and anthracycline compounds and ionizing radiation induce oxidative stress generating high levels of ROS [166]. The induction of oxidative stress can lead to preferential death of cancer cells.

Currently, the main problem of chemotherapy and radiotherapy is the development of resistance mechanisms, and miRNAs play an important role in the response to these therapies, targeting oxidative stress molecules. miR-125b is involved in resistance to chemotherapy, affecting oxidative stress pathways in breast cancer [167]; miR-125b, by targeting HAX-1, an anti-apoptotic gene, has an impact on resistance to doxorubicin. The mechanism behind this phenomenon is a decrease in MMP levels after downregulation of HAX-1 and the release of ROS from the mitochondria into the cytoplasm. Thus, miR-125b is capable of resensitizing breast cancer cells for treatment with doxorubicin using the ROS pathway (Fig. 13). Regarding chemoresistance, miR-24, which is positively regulated in breast cancer stem cells, induces resistance to cisplatin by targeting the pro-apoptotic factor BimL [168] and targets FIH1, which induces

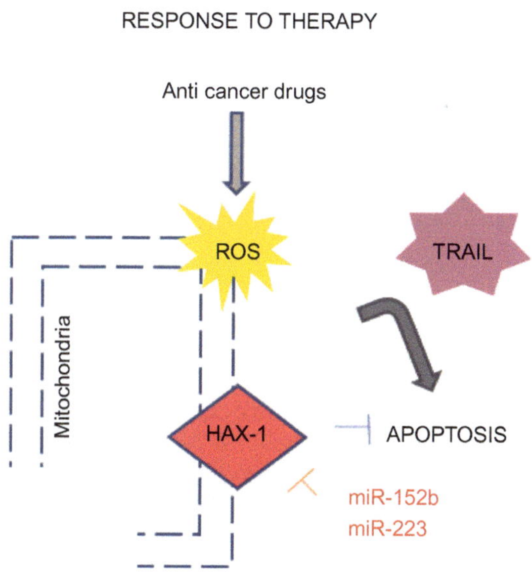

FIG. 13 miRNAs involved in the regulation of cancer characteristics influenced by oxidative stress in breast cancer and response to therapy (where the *red (dark gray in the printed version) arrow* means up-modulation and the *red (dark gray in the printed version)* "T" means inhibition).

repression of HIF-1α, induced in hypoxia conditions, leading to the growth of cancer stem cells, consequently inducing resistance to chemotherapy.

Breast cancer patients generally respond poorly to radiotherapy and the radio resistance mechanisms have not yet been elucidated. miR-668 is increased in radiation-resistant breast cancer cells; this phenomenon occurs because IκBα is a direct target of miR-668, leading to the activation of NF-κB. Generally, drug resistance is an important challenge in the treatment of breast cancer, especially for TNBC, which has no target therapy yet. miR-223 is related to resistance to apoptosis induced by TRAIL in TNBC cancer stem cells [169]. The reintroduction of miR-223 and TRAIL treatment in the MDA-MB-231 cell line induces a strong generation of ROS, by targeting HAX-1 to the mitochondria, and TNBC stem cells are more sensitive to TRAIL treatment. In addition, the miR-223/HAX-1 axis increases sensitivity to doxorubicin and cisplatin in TNBC stem cells.

The induction of oxidative stress is a crucial mechanism of action for many anticancer agents, including radiation, and when it is mediated it induces premature senescence, but not apoptosis in BCSCs [170,171]. BCSC senescence, when induced by H_2O_2, is associated with p53 activation, as well as with increased expression of p53 and p21, implying a role for the p53/p21 signaling pathway in senescence induced by oxidative stress in BCSCs.

12.1 BCSCs exhibit lower levels of ROS

Hematopoietic stem cells (HSCs) are susceptible to oxidative stress induced by ionizing radiation and [172,173] when presenting with low levels of ROS, they exhibit greater potential for self-renewal and the ability to hematopoiesis reconstitution than cells with high levels of ROS [174], providing long-term blood cell restocking [175,176].

12.2 Oxidative stress induces premature senescence in BCSCs

Senescence induction is an important tumor prevention mechanism; by limiting the reproductive capacity of cells and oxidative stress mediated by H_2O_2, it can easily induce premature senescence in a variety of cells [177,178], including in breast cancer cells and BCSCs, causing a dose-dependent increase in senescent cells in MCF-7 cells, therefore causing induction of premature senescence in BCSCs and breast cancer cells.

Activation of the p53/p21 signaling pathway plays a key role in mediating senescence induced by ionizing radiation in lung cancer cells [179], and treatment with H_2O_2 leads to a dose-dependent increase in the levels of phosphorylated p53 expression (p-p53), activating the p53 signaling pathway. This increases the levels of p53 and p21 expression, which induces senescence in BCSCs, in part by activating the p53/signaling pathway p21.

Oxygen metabolism in the mitochondria is a crucial process to produce ATP to provide energy for cell survival and functional activity needs, but it also leads to the generation of ROS due to electron leakage during oxidative phosphorylation and reduction of molecules of O_2 [180]. Mitochondrial ROS include superoxide anion, hydroxyl radical, H_2O_2, and singlet oxygen [180]. If the accumulation of ROS exceeds the elimination capacity of the antioxidant system, it can disturb the redox balance, resulting in oxidative stress.

References

[1] V. Sosa, T. Moliné, R. Somoza, R. Paciucci, H. Kondoh, M.E. LLeonart, Oxidative stress and cancer: an overview, Ageing Res. Rev. 12 (1) (2013) 376–390.

[2] S. Prasad, S.C. Gupta, A.K. Tyagi, Reactive oxygen species (ROS) and cancer: role of antioxidative nutraceuticals, Cancer Lett. 28 (2017) 95–105.

[3] J.C. Reed, J.M. Jurgensmeier, S. Matsuyama, Bcl-2 family proteins and mitochondria, Biochim. Biophys. Acta 1366 (1–2) (1998) 127–137.

[4] J.K. Brunelle, A. Letai, Control of mitochondrial apoptosis by the Bcl-2 family, J. Cell Sci. 122 (Pt 4) (2009) 437–441.

[5] P. Li, D. Nijhawan, I. Budihardjo, S.M. Srinivasul, M. Ahmad, E.S. Alnemri, et al., Cytochrome c and dATP-dependent formation of Apaf-1/caspase-9 complex initiates an apoptotic protease cascade, Cell 14 (4) (1997) 479–489.

[6] H. Zhong, H. Yin, Role of lipid peroxidation derived 4-hydroxynonenal (4-HNE) in cancer: focusing on mitochondria, Redox Biol. 4 (2015) 193–199.

[7] J. Kaur, C. Politis, R. Jacobs, Salivary 8-hydroxy-2-deoxyguanosine, malondialdehyde, vitamin C, and vitamin E in oral pre-cancer and cancer: diagnostic value and free radical mechanism of action, Clin. Oral Investig. 20 (2) (2016) 315–319.

[8] K. Le Gal, M.X. Ibrahim, C. Wiel, V.I. Sayin, M.K. Akula, C. Karlsson, et al., Antioxidants can increase melanoma metastasis in mice, Sci. Transl. Med. 7 (308) (2015) 308re8, https://doi.org/10.1126/scitranslmed.aad3740. 26446958.

[9] I.I.C. Chio, D.A. Tuveson, ROS in cancer: the burning question, Trends Mol. Med. 23 (5) (2017) 411–429.

[10] G. Bauer, N. Zarkovic, Revealing mechanisms of selective, concentration-dependent potentials of 4-hydroxy-2-nonenal to induce apoptosis in cancer cells through inactivation of membrane-associated catalase, Free Radic. Biol. Med. 81 (2015) 128–144.

[11] Y. Xia, S. Shen, I.M. Verma, NF-κB, an active player in human cancers, Cancer Immunol. Res. 2 (9) (2014) 823–830.

[12] A. Oeckinghaus, S. Ghosh, The NF-kappaB family of transcription factors and its regulation, Cold Spring Harb. Perspect. Biol. 1 (4) (2009) a000034, https://doi.org/10.1101/cshperspect.a000034.

[13] O. Kutuk, H. Basaga, Apoptosis signalling by 4-hydroxynonenal: a role for JNK-c-Jun/AP-1 pathway, Redox Rep. 12 (1) (2007) 30–34.

[14] Y.G. Ko, Y.S. Kang, H. Park, W. Seol, J. Kim, T. Kim, et al., Apoptosis signal-regulating kinase 1 controls the proapoptotic function of death-associated protein (Daxx) in the cytoplasm, J. Biol. Chem. 19 (42) (2001) 39103–39106.

[15] P. Chaudhary, R. Sharma, A. Sharma, R. Vatsyayan, S. Yadav, S.S. Singhal, et al., Mechanisms of 4-hydroxy-2-nonenal induced pro- and anti-apoptotic signaling, Biochemistry 27 (29) (2010) 6263–6275.

[16] S.O. Abarikwu, A.B. Pant, E.O. Farombi, 4-Hydroxynonenal induces mitochondrial-mediated apoptosis and oxidative stress in SH-SY5Y human neuronal cells, Basic Clin. Pharmacol. Toxicol. 110 (5) (2012) 441–448.

[17] N.A. Tchurikov, D.M. Fedoseeva, D.V. Sosin, A.V. Snezhkina, N.V. Melnikova, A.V. Kudryavtseva, et al., Hot spots of DNA double-strand breaks, and genomic contacts of human rDNA units are involved in epigenetic regulation, J. Mol. Cell Biol. 7 (4) (2015) 366–382.
[18] R. Olinski, D. Gackowski, R. Rozalski, M. Foksinski, K. Bialkowski, Oxidative DNA damage in cancer patients: a cause or a consequence of the disease development? Mutat. Res. 29 (1-2) (2003) 177–190.
[19] M. Lund, M. Melbye, L.J. Diaz, M. Duno, J. Wohlfahrt, J. Vissing, Mitochondrial dysfunction and risk of cancer, Br. J. Cancer 17 (6) (2015) 1134–1140.
[20] C. Gorrini, P.S. Baniasadi, I.S. Harris, J. Silvester, S. Inoue, B. Snow, et al., BRCA1 interacts with Nrf2 to regulate antioxidant signaling and cell survival, J. Exp. Med. 29 (8) (2013) 1529–1544.
[21] M. Ogrunc, R. Di Micco, M. Liontos, L. Bombardelli, M. Mione, M. Fumagalli, et al., Oncogene-induced reactive oxygen species fuel hyperproliferation and DNA damage response activation, Cell Death Differ. 21 (6) (2014) 998–1012.
[22] I.I.C. Chio, S.M. Jafarnejad, M. Ponz-Sarvise, Y. Park, K. Rivera, W. Palm, et al., NRF2 promotes tumour maintenance by modulating mRNA translation in pancreatic cancer, Cell 11 (4) (2016) 963–976.
[23] L. Jiang, A.A. Shestov, P. Swain, C. Yang, S.J. Parker, Q.A. Wang, et al., Reductive carboxylation supports redox homeostasis during anchorage-independent growth, Nature 14 (7598) (2016) 255–258.
[24] E. Piskounova, M. Agathocleous, M.M. Murphy, Z. Hu, S.E. Huddlestun, Z. Zhao, et al., Oxidative stress inhibits distant metastasis by human melanoma cells, Nature 12 (7577) (2015) 186–191.
[25] O.I. Aruoma, B. Halliwell, B.M. Hoey, J. Butler, The antioxidant action of N-acetylcysteine: its reaction with hydrogen peroxide, hydroxyl radical, superoxide, and hypochlorous acid, Free Radic. Biol. Med. 6 (6) (1989) 593–597.
[26] A.C. White, E.K. Maloney, S.L. Lee, J.J. Lanzillo, B.L. Fanburg, Reduction of endothelial cell related TGFbeta activity by thiols, Endothelium 6 (3) (1999) 231–239.
[27] J. Ye, J. Fan, S. Venneti, Y.W. Wan, B.R. Pawel, J. Zhang, et al., Serine catabolism regulates mitochondrial redox control during hypoxia, Cancer Discov. 4 (12) (2014) 1406–1417.
[28] A.L. Orr, L. Vargas, C.N. Turk, J.E. Baaten, J.T. Matzen, V.J. Dardov, et al., Suppressors of superoxide production from mitochondrial complex III, Nat. Chem. Biol. 11 (1) (2015) 834–836.
[29] A.Y. Abramov, A. Scorziello, M.R. Duchen, Three distinct mechanisms generate oxygen free radicals in neurons and contribute to cell death during anoxia and reoxygenation, J. Neurosci. 27 (5) (2007) 1129–1138.
[30] M.H. Choi, I.K. Lee, G.W. Kim, B.U. Kim, Y.H. Han, D.Y. Yu, et al., Regulation of PDGF signalling and vascular remodelling by peroxiredoxin II, Nature 19 (7040) (2005) 347–353.
[31] E.C. Cheung, P. Lee, F. Ceteci, C. Nixon, K. Blyth, O.J. Sansom, et al., Opposing effects of TIGAR- and RAC1-derived ROS on Wnt-driven proliferation in the mouse intestine, Genes Dev. 1 (1) (2016) 52–63.
[32] P.L. Tan, T. Shavlakadze, M.D. Grounds, P.G. Arthur, Differential thiol oxidation of the signaling proteins Akt, PTEN or PP2A determines whether Akt phosphorylation is enhanced or inhibited by oxidative stress in C2C12 myotubes derived from skeletal muscle, Int. J. Biochem. Cell Biol. (2015) 72–79.
[33] S.E. Artandi, R.A. DePinho, A critical role for telomeres in suppressing and facilitating carcinogenesis, Curr. Opin. Genet. Dev. 10 (1) (2000) 39–46.
[34] M.V. Blagosklonny, Molecular damage in cancer: an argument for mTOR-driven aging, Aging (Albany NY) 3 (12) (2011) 1130–1141.
[35] M.B. Morelli, C. Amantini, M. Santoni, A. Soriani, M. Nabissi, C. Cardinali, et al., Axitinib induces DNA damage response leading to senescence, mitotic catastrophe, and increased NK cell recognition in human renal carcinoma cells, Oncotarget 3 (34) (2015) 36245–36259.

[36] A. Macieira-Coelho, Putative mechanisms responsible for the decline in cancer prevalence during organism senescence, Biogerontology 16 (4) (2015) 559–565.
[37] A. Jemal, F. Bray, M.M. Center, J. Ferlay, E. Ward, D. Forman, Global cancer statistics, CA Cancer J. Clin. 61 (2) (2011) 69–90.
[38] L.A. Pikor, V.R. Ramnarine, S. Lam, W.L. Lam, Genetic alterations defining NSCLC subtypes and their therapeutic implications, Lung Cancer 82 (2) (2013) 179–189.
[39] C.F. Kim, E.L. Jackson, A.E. Woolfenden, S. Lawrence, I. Babar, S. Vogel, et al., Identification of bronchioalveolar stem cells in normal lung and lung cancer, Cell 17 (6) (2005) 823–835.
[40] S. Sun, J.H. Schiller, A.F. Gazdar, Lung cancer in never smokers—a different disease, Nat. Rev. Cancer 7 (10) (2007) 778–790.
[41] A.F. Gazdar, E. Brambilla, Preneoplasia of lung cancer, Cancer Biomark. 9 (1–6) (2010) 385–396.
[42] K.S. Park, M.C. Liang, D.M. Raiser, R. Zamponi, R.R. Roach, S.J. Curtis, et al., Characterization of the cell of origin for small cell lung cancer, Cell Cycle 15 (16) (2011) 2806–2815.
[43] E.A. Semenova, R. Nagel, A. Berns, Origins, genetic landscape, and emerging therapies of small cell lung cancer, Genes Dev. 15 (14) (2015) 1447–1462.
[44] D. Hanahan, R.A. Weinberg, Hallmarks of cancer: the next generation, Cell 4 (5) (2011) 646–674.
[45] M. Gomes, A.L. Teixeira, A. Coelho, A. Araújo, R. Medeiros, The role of inflammation in lung cancer, Adv. Exp. Med. Biol. (2014) 1–23, https://doi.org/10.1007/978-3-0348-0837-8_1.
[46] M. Sato, D.S. Shames, A.F. Gazdar, J.D. Minna, A translational view of the molecular pathogenesis of lung cancer, J. Thorac. Oncol. 2 (4) (2007) 327–343.
[47] S.I. Hajdu, Much overlooked causes of lung cancer, Ann. Clin. Lab. Sci. 41 (1) (2011) 97–101.
[48] R. Hubaux, D.D. Becker-Santos, K.S. Enfield, D. Rowbotham, S. Lam, W.L. Lam, et al., Molecular features in arsenic-induced lung tumors, Mol. Cancer 12 (2013) 20, https://doi.org/10.1186/1476-4598-12-20.
[49] A.M. Houghton, Mechanistic links between COPD and lung cancer, Nat. Rev. Cancer 13 (4) (2013) 233–245.
[50] M.S. Shiels, H.A. Katki, N.D. Freedman, M.P. Purdue, N. Wentzensen, B. Trabert, et al., Cigarette smoking and variations in systemic immune and inflammation markers, J. Natl. Cancer Inst. 1 (11) (2014) dju294.
[51] M.S. Shiels, R.M. Pfeiffer, A. Hildesheim, E.A. Engels, T.J. Kemp, J.H. Park, et al., Circulating inflammation markers and prospective risk for lung cancer, J. Natl. Cancer Inst. 18 (24) (2013) 1871–1880.
[52] G.Z. Wang, X. Cheng, X.C. Li, Y.Q. Liu, X.Q. Wang, X. Shi, Tobacco smoke induces production of chemokine CCL20 to promote lung cancer, Cancer Lett. 10 (1) (2015) 60–70.
[53] D.D. Sin, S.F.P. Man, A. McWilliams, S. Lam, Surfactant protein D and bronchial dysplasia in smokers at high risk of lung cancer, Chest 134 (3) (2008) 582–588.
[54] E. Filaire, C. Dupuis, G. Galvaing, S. Aubreton, H. Laurent, R. Richard, Lung cancer: what are the links with oxidative stress, physical activity, and nutrition, Lung Cancer 82 (3) (2013) 383–389.
[55] R.A. El-Zein, R.P. Young, R.J. Hopkins, C.J. Etzel, Genetic predisposition to chronic obstructive pulmonary disease and/or lung cancer: important considerations when evaluating risk, Cancer Prev. Res. (Phila.) 5 (4) (2012) 522–527.
[56] K. Vermaelen, G. Brusselle, Exposing a deadly alliance: novel insights into the biological links between COPD and lung cancer, Pulm. Pharmacol. Ther. 26 (5) (2013) 544–554.
[57] Y. Sekine, A. Hata, E. Koh, K. Hiroshima, Lung carcinogenesis from chronic obstructive pulmonary disease: characteristics of lung cancer from COPD and contribution of signal transducers and lung stem cells in the inflammatory microenvironment, Gen. Thorac. Cardiovasc. Surg. 62 (7) (2014) 415–421.

[58] K.E. de Visser, A. Eichten, L.M. Coussens, Paradoxical roles of the immune system during cancer development, Nat. Rev. Cancer 6 (1) (2006) 24–37.
[59] T. Kitamura, B.Z. Qian, J.W. Pollard, Immune cell promotion of metastasis, Nat. Rev. Immunol. 15 (2) (2015) 73–86.
[60] H.J. Kim, H. Cantor, CD4 T-cell subsets and tumour immunity: the helpful and the not-so-helpful, Cancer Immunol. Res. 2 (2) (2014) 91–98.
[61] B. Lakshmi Narendra, K. Eshvendar Reddy, S. Shantikumar, S. Ramakrishna, Immune system: a double-edged sword in cancer, Inflamm. Res. 62 (9) (2013) 823–834.
[62] E.M. Eisenstein, C.B. Williams, The T(reg)/Th17 cell balance: a new paradigm for autoimmunity, Pediatr. Res. 65 (5Pt2) (2009) 26R–31R, https://doi.org/10.1203/PDR.0b013e31819e76c7.
[63] L. Zhou, B.E. Lopes, M.M. Chong, I.I. Ivanov, R. Min, G.D. Victora, et al., TGF-beta-induced Foxp3 inhibits T(H)17 cell differentiation by antagonizing RORgammat function, Nature 8 (7192) (2008) 236–240.
[64] S.F. Ziegler, J.H. Buckner, FOXP3 and the regulation of Treg/Th17 differentiation, Microbes Infect. 11 (5) (2009) 594–598.
[65] H.M. Lee, J.L. Bautista, C.S. Hsieh, Thymic and peripheral differentiation of regulatory T cells, Adv. Immunol. (2011) 25–71, https://doi.org/10.1016/B978-0-12-387827-4.00002-4.
[66] A.L. Geis, H. Fan, X. Wu, S. Wu, D.L. Huso, J.L. Wolfe, Regulatory T-cell response to enterotoxigenic bacteroides fragilis colonization triggers IL17-dependent colon carcinogenesis, Cancer Discov. 5 (10) (2015) 1098–1109.
[67] M. Beyer, J.L. Schultze, Regulatory T cells in cancer, Blood 1 (3) (2006) 804–811.
[68] P.A. Savage, S. Malchow, D.S. Leventhal, Basic principles of tumour-associated regulatory T cell biology, Trends Immunol. 34 (1) (2013) 33–40.
[69] J. He, Y. Hu, M. Hu, B. Li, Development of PD-1/PD-L1 pathway in tumour immune microenvironment and treatment for non-small cell lung cancer, Sci. Rep. 17 (2015) 13110.
[70] M.C. Takenaka, S. Robson, F.J. Quintana, Regulation of the T cell response by CD39, Trends Immunol. 37 (7) (2016) 427–439.
[71] D.S. Vinay, E.P. Ryan, G. Pawelec, W.H. Talib, J. Stagg, E. Elkord, et al., Immune evasion in cancer: mechanistic basis and therapeutic strategies, Semin. Cancer Biol. 35 Suppl (2015) S185–S198.
[72] G.A. Granville, R.M. Memmott, A. Balogh, J. Mariotti, S. Kawabata, W. Han, et al., A central role for Foxp3+ regulatory T cells in K-Ras-driven lung tumorigenesis, PLoS One 4 (3) (2009), e5061.
[73] A.P. Ganesan, M. Johansson, B. Ruffell, A. Yagui-Beltrán, J. Lau, D.M. Jablons, et al., Tumour-infiltrating regulatory T cells inhibit endogenous cytotoxic T cell responses to lung adenocarcinoma, J. Immunol. 15 (4) (2013) 2009–2017.
[74] K. Shimizu, M. Nakata, Y. Hirami, T. Yukawa, A. Maeda, K. Tanemoto, Tumour-infiltrating Foxp3+ regulatory T cells are correlated with cyclooxygenase-2 expression and are associated with recurrence in resected non-small cell lung cancer, J. Thorac. Oncol. 5 (5) (2010) 585–590.
[75] S. Razani-Boroujerdi, M.L. Sopori, Early manifestations of NNK-induced lung cancer: role of lung immunity in tumour susceptibility, Am. J. Respir. Cell Mol. Biol. 36 (1) (2007) 13–19.
[76] N. Erfani, S.M. Mehrabadi, M.A. Ghayumi, M.R. Haghshenas, Z. Mojtahedi, A. Ghaderi, et al., Increase of regulatory T cells in metastatic stage and CTLA-4 over expression in lymphocytes of patients with non-small cell lung cancer (NSCLC), Lung Cancer 77 (2) (2012) 306–311.
[77] T. Schneider, S. Kimpfler, A. Warth, P.A. Schnabel, H. Dienemann, D. Schadendorf, et al., Foxp3(+) regulatory T cells and natural killer cells distinctly infiltrate primary tumors and draining lymph nodes in pulmonary adenocarcinoma, J. Thorac. Oncol. 6 (3) (2011) 432–438.

[78] M.J. Smyth, M.W. Teng, J. Swann, K. Kyparissoudis, D.I. Godfrey, Y. Hayakawa, CD4+CD25+ T regulatory cells suppress NK cell-mediated immunotherapy of cancer, J. Immunol. 1 (3) (2006) 1582–1587.
[79] H. Borghaei, L. Paz-Ares, L. Horn, D.R. Spigel, M. Steins, N.E. Ready, et al., Nivolumab versus docetaxel in advanced nonsquamous non-small-cell lung cancer, N. Engl. J. Med. 22 (17) (2015) 1627–1639.
[80] N. Guibert, J. Mazières, Nivolumab for treating non-small cell lung cancer, Expert. Opin. Biol. Ther. 15 (12) (2015) 1789–1797.
[81] A. Wang, Z. Wang, Y. Cao, S. Cheng, H. Chen, H. Bunjhoo, et al., CCL2/CCR2-dependent recruitment of Th17 cells but not Tc17 cells to the lung in a murine asthma model, Int. Arch. Allergy Immunol. 166 (1) (2015) 52–62.
[82] E.A. Vucic, R. Chari, K.L. Thu, I.M. Wilson, A.M. Cotton, J.Y. Kennett, et al., DNA methylation is globally disrupted and associated with expression changes in chronic obstructive pulmonary disease small airways, Am. J. Respir. Cell Mol. Biol. 50 (5) (2014) 912–922.
[83] S.L. Gaffen, R. Jain, A.V. Garg, D.J. Cua, The IL-23-IL-17 immune axis: from mechanisms to therapeutic testing, Nat. Rev. Immunol. 14 (9) (2014) 585–600.
[84] Y. Iwakura, H. Ishigame, S. Saijo, S. Nakae, Functional specialization of interleukin-17 family members, Immunity 25 (2) (2011) 149–162.
[85] W. Kaabachi, A.B. Amor, S. Kaabachi, A. Rafrafi, K. Tizaoui, K. Hamzaoui, Interleukin-17A and -17F genes polymorphisms in lung cancer, Cytokine 66 (1) (2014) 23–29.
[86] H. Park, Z. Li, X.D. Yang, S.H. Chang, R. Nurieva, Y.H. Wang, et al., A distinct lineage of CD4 T cells regulates tissue inflammation by producing interleukin 17, Nat. Immunol. 6 (11) (2005) 1133–1141.
[87] S. Reppert, I. Boross, M. Koslowski, Ö. Türeci, S. Koch, H.A. Lehr, et al., A role for T-bet-mediated tumour immune surveillance in anti-IL-17A treatment of lung cancer, Nat. Commun. 20 (2011) 600, https://doi.org/10.1038/ncomms1609.
[88] Q. Li, Y. Han, G. Fei, Z. Guo, T. Ren, Z. Liu, IL-17 promoted metastasis of non-small-cell lung cancer cells, Immunol. Lett. 17 (2) (2012) 144–150.
[89] Y. Carmi, G. Rinott, S. Dotan, M. Elkabets, P. Rider, E. Voronov, R.N. Apte, Microenvironment-derived IL-1 and IL-17 interact in the control of lung metastasis, J. Immunol. 15 (6) (2011) 3462–3471.
[90] C. Lim, R. Savan, The role of the IL-22/IL-22R1 axis in cancer, Cytokine Growth Factor Rev. 25 (3) (2014) 257–271.
[91] W. Zhang, Y. Chen, H. Wei, C. Zheng, R. Sun, J. Zhang, et al., Antiapoptotic activity of autocrine interleukin-22 and therapeutic effects of interleukin-22-small interfering RNA on human lung cancer xenografts, Clin. Cancer Res. 15 (20) (2008) 6432–6439.
[92] N. Martin-Orozco, P. Muranski, Y. Chung, X.O. Yang, T. Yamazaki, S. Lu, et al., T helper 17 cells promote cytotoxic T cell activation in tumour immunity, Immunity 20 (5) (2009) 787–798.
[93] Z.J. Ye, Q. Zhou, Y.Y. Gu, S.M. Qin, W.L. Ma, J.B. Xin, et al., Generation and differentiation of IL-17-producing CD4+ T cells in malignant pleural effusion, J. Immunol. 15 (10) (2010) 6348–6354.
[94] C. Bryant, H. Suen, R. Brown, S. Yang, J. Favaloro, E. Aklilu, et al., Long-term survival in multiple myeloma is associated with a distinct immunological profile, which includes proliferative cytotoxic T-cell clones and a favourable Treg/Th17 balance, Blood Cancer J. 13 (9) (2013), https://doi.org/10.1038/bcj.2013.34, e148.
[95] L. Guéry, S. Hugues, Th17 cell plasticity and functions in cancer immunity, Biomed. Res. Int. 2015 (2015) 314620, https://doi.org/10.1155/2015/314620. Epub 2015 Oct 25.
[96] H. Prado-Garcia, S. Romero-Garcia, U. Rumbo-Nava, J.S. Lopez-Gonzalez, Predominance of Th17 over regulatory T-cells in pleural effusions of patients with lung cancer implicates a proinflammatory profile, Anticancer Res. 35 (3) (2015) 1529–1535.

[97] H. Onishi, T. Morisaki, M. Katano, Immunotherapy approaches targeting regulatory T-cells, Anticancer Res. 32 (3) (2012) 997–1003.
[98] L.A. Damiani, C.M. Yingling, S. Leng, P.E. Romo, J. Nakamura, S.A. Belinsky, Carcinogen-induced gene promoter hypermethylation is mediated by DNMT1 and causal for transformation of immortalized bronchial epithelial cells, Cancer Res. 1 (21) (2008) 9005–9014.
[99] F. Barlési, G. Giaccone, M.I. Gallegos-Ruiz, A. Loundou, S.W. Span, P. Lefesvre, et al., Global histone modifications predict prognosis of resected non-small-cell lung cancer, J. Clin. Oncol. 1 (28) (2007) 4358–4364.
[100] A. Van Den Broeck, E. Brambilla, D. Moro-Sibilot, S. Lantuejoul, C. Brambilla, B. Eymin, S. Gazzeri, Loss of histone H4K20 trimethylation occurs in preneoplasia and influences prognosis of non-small cell lung cancer, Clin. Cancer Res. 15 (22) (2008) 7237–7245.
[101] B. Bartling, H.S. Hofmann, T. Boettger, G. Hansen, S. Burdach, R.E. Silber, et al., Comparative application of antibody and gene array for expression profiling in human squamous cell lung carcinoma, Lung Cancer 49 (2) (2005) 145–154.
[102] H. Suzuki, M. Ouchida, H. Yamamoto, M. Yano, S. Toyooka, M. Aoe, et al., Decreased expression of the SIN3A gene, a candidate tumour suppressor located at the prevalent allelic loss region 15q23 in non-small cell lung cancer, Lung Cancer 59 (1) (2008) 24–31.
[103] H. Watanabe, T. Mizutani, T. Haraguchi, N. Yamamichi, S. Minoguchi, M. Yamamichi-Nishina, et al., SWI/SNF complex is essential for NRSF-mediated suppression of neuronal genes in human nonsmall cell lung carcinoma cell lines, Oncogene 19 (3) (2006) 470–479.
[104] H. Sasaki, S. Moriyama, Y. Nakashima, Y. Kobayashi, H. Yukiue, M. Kaji, I. Fukai, et al., Expression of the MTA1 mRNA in advanced lung cancer, Lung Cancer 35 (2) (2002) 149–154.
[105] H. Iguchi, G. Imura, Y. Toh, Y. Ogata, Expression of MTA1, a metastasis-associated gene with histone deacetylase activity in pancreatic cancer, Int. J. Oncol. 16 (6) (2000) 1211–1214.
[106] A. You, J.K. Tong, C.M. Grozinger, S.L. Schreiber, CoREST is an integral component of the CoREST- human histone deacetylase complex, Proc. Natl. Acad. Sci. U. S. A. 13 (4) (2001) 1454–1458.
[107] M. Kishimoto, T. Kohno, K. Okudela, A. Otsuka, H. Sasaki, C. Tanabe, et al., Mutations and deletions of the CBP gene in human lung cancer, Clin. Cancer Res. 15 (2 Pt 1) (2005) 512–519.
[108] N.R. Sundaresan, S.A. Samant, V.B. Pillai, S.B. Rajamohan, M.P. Gupta, SIRT3 is a stress-responsive deacetylase in cardiomyocytes that protects cells from stress-mediated cell death by deacetylation of Ku70, Mol. Cell. Biol. 28 (20) (2008) 6384–6401.
[109] R. Wang, J. An, F. Ji, H. Jiao, H. Sun, D. Zhou, Hypermethylation of the Keap1 gene in human lung cancer cell lines and lung cancer tissues, Biochem. Biophys. Res. Commun. 15 (1) (2008) 151–154.
[110] T. Ohta, K. Iijima, M. Miyamoto, I. Nakahara, H. Tanaka, M. Ohtsuji, et al., Loss of Keap1 function activates Nrf2 and provides advantages for lung cancer cell growth, Cancer Res. 1 (5) (2008) 1303–1309.
[111] J. Zhang, T. Ohta, A. Maruyama, T. Hosoya, K. Nishikawa, J.M. Maher, et al., BRG1 interacts with Nrf2 to selectively mediate HO-1 induction in response to oxidative stress, Mol. Cell. Biol. 26 (21) (2006) 7942–7952.
[112] G.H. Liu, J. Qu, X. Shen, NF-kappaB/p65 antagonizes Nrf2-ARE pathway by depriving CBP from Nrf2 and facilitating recruitment of HDAC3 to MafK, Biochim. Biophys. Acta, Mol. Cell Res. 1783 (2008) 713–727.
[113] J. Zhang, T. Hosoya, A. Maruyama, K. Nishikawa, J.M. Maher, T. Ohta, et al., Nrf2 Neh5 domain is differentially utilized in the transactivation of cytoprotective genes, Biochem. J. 15 (3) (2007) 459–466.

[114] S.H. Kim, J.W. Jeong, J.A. Park, J.W. Lee, J.H. Seo, B.K. Jung, et al., Regulation of the HIF-1alpha stability by histone deacetylases, Oncol. Rep. 17 (3) (2007) 647–651.

[115] L.H. Kasper, F. Boussouar, K. Boyd, W. Xu, M. Biesen, J. Rehg, et al., Two transactivation mechanisms cooperate for the bulk of HIF-1-responsive gene expression, EMBO J. 16 (22) (2005) 3846–3858.

[116] B.N. Finck, D.P. Kelly, PGC-1 coactivators: inducible regulators of energy metabolism in health and disease, J. Clin. Invest. 116 (3) (2006) 615–622.

[117] S. Borniquel, I. Valle, S. Cadenas, S. Lamas, M. Monsalve, Nitric oxide regulates mitochondrial oxidative stress protection via the transcriptional coactivator PGC-1alpha, FASEB J. 20 (11) (2006) 1889–1891.

[118] N. Azad, Y. Rojanasakul, V. Vallyathan, Inflammation and lung cancer: roles of reactive oxygen/nitrogen species, J. Toxicol. Environ. Health B Crit. Rev. 11 (1) (2008) 1–15.

[119] W.E. Naugler, M. Karin, NF-kappaB and cancer-identifying targets and mechanisms, Curr. Opin. Genet. Dev. 18 (1) (2008) 19–26.

[120] I. Rahman, P.S. Gilmour, J.A. Jimenez, W. MacNee, Oxidative stress and TNF-alpha induce histone acetylation and NF-kappaB/AP-1 activation in alveolar epithelial cells: potential mechanism in gene transcription in lung inflammation, Mol. Cell. Biochem. 234 (1-2) (2002) 239–248.

[121] S.Y. Na, S.K. Lee, S.J. Han, H.S. Choi, S.Y. Im, J.W. Lee, Steroid receptor coactivator-1 interacts with the p50 subunit and coactivates nuclear factor kappaB-mediated transactivations, J. Biol. Chem. 1 (18) (1998) 10831–10834.

[122] B.P. Ashburner, S.D. Westerheide, A.S. Baldwin Jr., The p65 (RelA) subunit of NF-kappaB interacts with the histone deacetylase (HDAC) corepressors HDAC1 and HDAC2 to negatively regulate gene expression, Mol. Cell. Biol. 21 (20) (2001) 7065–7077.

[123] J.E. Hoberg, F. Yeung, M.W. Mayo, SMRT derepression by the IkappaB kinase alpha: a prerequisite to NF-kappaB transcription and survival, Mol. Cell 22 (2) (2004) 245–255.

[124] H.Y. Kuo, C.C. Chang, J.C. Jeng, H.M. Hu, D.Y. Lin, G.G. Maul, et al., SUMO modification negatively modulates the transcriptional activity of CREB-binding protein via the recruitment of Daxx, Proc. Natl. Acad. Sci. U. S. A. 22 (47) (2005) 16973–16978.

[125] S.R. Yang, A.S. Chida, M.R. Bauter, N. Shafiq, K. Seweryniak, S.B. Maggirwar, et al., Cigarette smoke induces proinflammatory cytokine release by activation of NF-kappaB and posttranslational modifications of histone deacetylase in macrophages, Am. J. Physiol. Lung Cell. Mol. Physiol. 291 (1) (2006) L46–L57.

[126] S.R. Yang, J. Wright, M. Bauter, K. Seweryniak, A. Kode, I. Rahman, Sirtuin regulates cigarette smoke-induced proinflammatory mediator release via RelA/p65 NF-kappaB in macrophages in vitro and in rat lungs in vivo: implications for chronic inflammation and aging, Am. J. Phys. Lung Cell. Mol. Phys. 292 (2) (2007) L567–L576.

[127] T. Sørlie, C.M. Perou, R. Tibshirani, T. Aas, S. Geisler, H. Johnsen, et al., Gene expression patterns of breast carcinomas distinguish tumour subclasses with clinical implications, Proc. Natl. Acad. Sci. U. S. A. 11 (19) (2001) 10869–10874.

[128] R. Rouzier, C.M. Perou, W.F. Symmans, N. Ibrahim, M. Cristofanilli, K. Anderson, et al., Breast cancer molecular subtypes respond differently to preoperative chemotherapy, Clin. Cancer Res. 15 (16) (2005) 5678–5685.

[129] Z. Hu, C. Fan, D.S. Oh, J.S. Marron, X. He, B.F. Qaqish, et al., The molecular portraits of breast tumors are conserved across microarray platforms, BMC Genomics 7 (2006) 96, https://doi.org/10.1186/1471-2164-7-96.

[130] J. Musarrat, J. Arezina-Wilson, A.A. Wani, Prognostic and aetiological relevance of 8-hydroxyguanosine in human breast carcinogenesis, Eur. J. Cancer 32A (7) (1996) 1209–1214.

[131] C. Zhang, H.J. Wang, Q.C. Bao, L. Wang, T.K. Guo, W.L. Chen, et al., NRF2 promotes breast cancer cell proliferation and metastasis by increasing RhoA/ROCK pathway signal transduction, Oncotarget 8 (45) (2016) 73593–73606.

[132] I.S. Harris, A.E. Treloar, S. Inoue, M. Sasaki, C. Gorrini, K.C. Lee, et al., Glutathione and thioredoxin antioxidant pathways synergize to drive cancer initiation and progression, Cancer Cell 9 (2) (2015) 211–222.
[133] H.P. Monteiro, F.T. Ogata, A. Stern, Thioredoxin promotes survival signaling events under nitrosative/oxidative stress associated with cancer development, Biom. J. 40 (4) (2017) 189–199.
[134] I. Batinić-Haberle, J.S. Rebouças, I. Spasojević, Superoxide dismutase mimics: chemistry, pharmacology, and therapeutic potential, Antioxid. Redox Signal. 15 (6) (2010) 877–918.
[135] H. Satooka, M. Hara-Chikuma, Aquaporin-3 controls breast cancer cell migration by regulating hydrogen peroxide transport and its downstream cell signaling, Mol. Cell. Biol. 1 (7) (2016) 1206–1218.
[136] J. Krol, I. Loedige, W. Filipowicz, The widespread regulation of microRNA biogenesis, function, and decay, Nat. Rev. Genet. 11 (9) (2010) 597–610.
[137] R.C. Lee, R.L. Feinbaum, V. Ambros, The C. elegans heterochronic gene lin-4 encodes small RNAs with antisense complementarity to lin-14, Cell 3 (5) (1993) 843–854.
[138] M.V. Iorio, P. Casalini, C. Piovan, L. Braccioli, E. Tagliabue, Breast cancer and microRNAs: therapeutic impact, Breast Suppl 3 (2011) S63–S70, https://doi.org/10.1016/S0960-9776(11)70297-1.
[139] A. Cataldo, D.G. Cheung, A. Balsari, E. Tagliabue, V. Coppola, M.V. Iorio, et al., miR-302b enhances breast cancer cell sensitivity to cisplatin by regulating E2F1 and the cellular DNA damage response, Oncotarget 5 (1) (2016) 786–797.
[140] I. Plantamura, G. Cosentino, A. Cataldo, MicroRNAs and DNA-damaging drugs in breast cancer: strength in numbers, Front. Oncol. 8 (2018) 352, https://doi.org/10.3389/fonc.2018.00352.
[141] M. McMahon, K. Itoh, M. Yamamoto, J.D. Hayes, Keap1-dependent proteasomal degradation of transcription factor Nrf2 contributes to the negative regulation of antioxidant response element-driven gene expression, J. Biol. Chem. 13 (24) (2003) 21592–21600.
[142] T. Suzuki, M. Yamamoto, Molecular basis of the Keap1-Nrf2 system, Free Radic. Biol. Med. 88 (Pt B) (2015) 93–100, https://doi.org/10.1016/j.freeradbiomed.2015.06.006.
[143] B. Singh, A.M. Ronghe, A. Chatterjee, N.K. Bhat, H.K. Bhat, MicroRNA-93 regulates NRF2 expression and is associated with breast carcinogenesis, Carcinogenesis 34 (5) (2013) 1165–1172.
[144] B. Wang, Y. Teng, Q. Liu, MicroRNA-153 regulates NRF2 expression and is associated with breast carcinogenesis, Clin. Lab. 62 (1–2) (2016) 39–47, https://doi.org/10.7754/clin.lab.2015.150518.
[145] M.S. Hayden, S. Ghosh, Shared principles in NF-kappaB signaling, Cell 8 (3) (2008) 344–362.
[146] M.J. Morgan, Z.G. Liu, Crosstalk of reactive oxygen species and NF-κB signaling, Cell Res. 21 (1) (2011) 103–115.
[147] K. Lingappan, NF-κB in oxidative stress, Curr. Opin. Toxicol. 7 (2018) 81–86.
[148] S.M. Smith, Y.L. Lyu, L. Cai, NF-κB affects proliferation and invasiveness of breast cancer cells by regulating CD44 expression, PLoS One 3 (9) (2014), e106966.
[149] J. Niu, Y. Shi, G. Tan, C.H. Yang, M. Fan, L.M. Pfeffer, et al., DNA damage induces NF-κB-dependent microRNA-21 up-regulation and promotes breast cancer cell invasion, J. Biol. Chem. 22 (26) (2012) 21783–21795.
[150] B. Li, Y. Lu, L. Yu, X. Han, H. Wang, J. Mao, et al., miR-221/222 promote cancer stem-like cell properties and tumour growth of breast cancer via targeting PTEN and sustained Akt/NF-κB/COX-2 activation, Chem. Biol. Interact. 277 (2017) 33–42.
[151] S. Rodic, M.D. Vincent, Reactive oxygen species (ROS) are a key determinant of cancer's metabolic phenotype, Int. J. Cancer 1 (3) (2018) 440–448.
[152] J. Choi, D.H. Kim, W.H. Jung, J.S. Koo, Metabolic interaction between cancer cells and stromal cells according to breast cancer molecular subtype, Breast Cancer Res. 15 (3) (2013) R78.

[153] S.L. Romero-Cordoba, S. Rodriguez-Cuevas, V. Bautista-Pina, A. Maffuz-Aziz, E. D'Ippolito, G. Cosentino, et al., Loss of function of miR-342-3p results in MCT1 over-expression and contributes to oncogenic metabolic reprogramming in triple negative breast cancer, Sci. Rep. 16 (1) (2018) 12252.

[154] S.C. Eastlack, S. Dong, C. Ivan, S.K. Alahari, Suppression of PDHX by microRNA-27b deregulates cell metabolism and promotes growth in breast cancer, Mol. Cancer 16 (1) (2018) 100.

[155] T.L. Yuan, L.C. Cantley, PI3K pathway alterations in cancer: variations on a theme, Oncogene 18 (41) (2008) 5497–5510.

[156] N.R. Leslie, D. Bennett, Y.E. Lindsay, H. Stewart, A. Gray, C.P. Downes, Redox regulation of PI 3-kinase signalling via inactivation of PTEN, EMBO J. 15 (20) (2003) 5501–5510.

[157] N. Koundouros, G. Poulogiannis, Phosphoinositide 3-kinase/Akt signaling and redox metabolism in cancer, Front. Oncol. 8 (2018) 160.

[158] B.S. Hong, H.S. Ryu, N. Kim, J. Kim, E. Lee, H. Moon, et al., Tumour suppressor miRNA-204-5p regulates growth, metastasis, and immune microenvironment remodeling in breast cancer, Cancer Res. 1 (7) (2019) 1520–1534.

[159] C. Liu, Z. Liu, X. Li, X. Tang, J. He, S. Lu, MicroRNA-1297 contributes to tumour growth of human breast cancer by targeting PTEN/PI3K/AKT signaling, Oncol. Rep. 38 (4) (2017) 2435–2443.

[160] R. Kulshreshtha, M. Ferracin, S.E. Wojcik, R. Garzon, H. Alder, F.J. Agosto-Perez, et al., A microRNA signature of hypoxia, Mol. Cell. Biol. 27 (5) (2007) 1859–1867.

[161] H. Liang, J. Xiao, Z. Zhou, J. Wu, F. Ge, Z. Li, et al., Hypoxia induces miR-153 through the IRE1α-XBP1 pathway to fine tune the HIF1α/VEGFA axis in breast cancer angiogenesis, Oncogene 37 (15) (2018) 1961–1975.

[162] N. Nagpal, H.M. Ahmad, S. Chameettachal, D. Sundar, S. Ghosh, R. Kulshreshtha, HIF-inducible miR-191 promotes migration in breast cancer through complex regulation of TGFβ-signaling in hypoxic microenvironment, Sci. Rep. 5 (2015) 9650.

[163] R. Krutilina, W. Sun, A. Sethuraman, M. Brown, T.N. Seagroves, L.M. Pfeffer, et al., MicroRNA-18a inhibits hypoxia-inducible factor 1α activity and lung metastasis in basal breast cancers, Breast Cancer Res. 28 (4) (2014) R78.

[164] Z. Wu, X. Ca, C. Huang, J. Xu, A. Liu, miR-497 suppresses angiogenesis in breast carcinoma by targeting HIF-1α, Oncol. Rep. 35 (3) (2016) 1696–1702.

[165] C. Gorrini, I.S. Harris, T.W. Mak, Modulation of oxidative stress as an anticancer strategy, Nat. Rev. Drug Discov. 12 (12) (2013) 931–947.

[166] G. Hu, X. Zhao, J. Wang, L. Lv, C. Wang, L. Feng, et al., miR-125b regulates the drug-resistance of breast cancer cells to doxorubicin by targeting HAX-1, Oncol. Lett. 15 (2) (2018) 1621–1629.

[167] G. Roscigno, I. Puoti, I. Giordano, E. Donnarumma, V. Russo, A. Affinito, et al., MiR-24 induces chemotherapy resistance and hypoxic advantage in breast cancer, Oncotarget 21 (12) (2017) 19507–19521.

[168] M. Luo, L. Ding, Q. Li, H. Yao, miR-668 enhances the radioresistance of human breast cancer cell by targeting IκBα, Breast Cancer 24 (5) (2017) 673–682.

[169] X. Sun, Y. Li, M. Zheng, W. Zuo, W. Zheng, MicroRNA-223 increases the sensitivity of triple-negative breast cancer stem cells to TRAIL-induced apoptosis by targeting HAX-1, PLoS One 12 (9) (2016), e0162754.

[170] B.G. Allen, S.K. Bhatia, J.M. Buatti, K.E. Brandt, K.E. Lindholm, A.M. Button, et al., Ketogenic diets enhance oxidative stress and radio-chemo-therapy responses in lung cancer xenografts, Clin. Cancer Res. 15 (14) (2013) 3905–3913.

[171] D. Trachootham, J. Alexandre, P. Huang, Targeting cancer cells by ROS-mediated mechanisms: a radical therapeutic approach? Nat. Rev. Drug Discov. 8 (7) (2009) 579–591.

[172] Y. Wang, L. Liu, S.K. Pazhanisamy, H. Li, A. Meng, D. Zhou, Total body irradiation causes residual bone marrow injury by induction of persistent oxidative stress in murine hematopoietic stem cells, Free Radic. Biol. Med. 15 (2) (2010) 348–356.

[173] H. Li, Y. Wang, S.K. Pazhanisamy, L. Shao, I. Batinic-Haberle, A. Meng, et al., Mn (III) meso-tetrakis-(N-ethylpyridinium-2-yl) porphyrin mitigates total body irradiation-induced long-term bone marrow suppression, Free Radic. Biol. Med. 1 (1) (2011) 30–37.

[174] Y.Y. Jang, S.J. Sharkis, A low level of reactive oxygen species selects for primitive hematopoietic stem cells that may reside in the low-oxygenic niche, Blood 15 (8) (2007) 3056–3063.

[175] K. Ito, A. Hirao, F. Arai, K. Takubo, S. Matsuoka, K. Miyamoto, et al., Reactive oxygen species act through p38 MAPK to limit the lifespan of hematopoietic stem cells, Nat. Med. 12 (4) (2006) 446–451.

[176] T. Yahata, T. Takanashi, Y. Muguruma, A.A. Ibrahim, H. Matsuzawa, T. Uno, et al., Accumulation of oxidative DNA damage restricts the self-renewal capacity of human hematopoietic stem cells, Blood 15 (11) (2011) 2941–2950.

[177] E. Burova, A. Borodkina, A. Shatrova, N. Nikolsky, Sublethal oxidative stress induces the premature senescence of human mesenchymal stem cells derived from endometrium, Oxidative Med. Cell. Longev. 25 (2013) 474931.

[178] Y. Wang, A. Meng, D. Zhou, Inhibition of phosphatidylinostol 3-kinase uncouples H2O2-induced senescent phenotype and cell cycle arrest in normal human diploid fibroblasts, Exp. Cell Res. 1 (1) (2004) 188–196.

[179] H. Luo, C. Yount, H. Lang, A. Yang, E.C. Riemer, K. Lyons, et al., Activation of p53 with Nutlin-3a radiosensitizes lung cancer cells via enhancing radiation-induced premature senescence, Lung Cancer 81 (2) (2013) 167–173.

[180] J.N. Moloney, T.G. Cotter, ROS signalling in the biology of cancer, Semin. Cell Dev. Biol. 80 (2018) 50–64, https://doi.org/10.1016/j.semcdb.2017.05.023.

CHAPTER

4

Novel therapeutic approaches targeting oxidative stress in aging

Akshay Kumar and Pawan Kumar Maurya

Department of Biochemistry, Central University of Haryana, Mahendergarh, India

1 Introduction

Aging is the accretion process of different damaging changes in the organs, resulting in progressive decline in function and increased risks of diseases and death [1]. During the last two decades, average human life expectancy has increased worldwide because of improved medical facilities and reduced mortality in early life. Recent increases in human life expectancy have led to age-associated diseases in populations. The health of people of all age groups people has been improved because of better lifestyles, safe drinking water, food, and immunization against various deadly diseases. A total of 6.5years global human life expectancy has increased during 2000–2019 (source: https://apps.who.int/gho/data/view.main.SDG2016LEXREGv?lang=en). An average 16%–20% of life is inhabit in late life morbidity, lengthier in women's as compared to men, and people having lower social economic position and obesity. Aging impairs sensory, motor, and cognitive function, thus lowering the quality of life. A major goal of civilized society in the future is to reduce late-life morbidity [2]. The foremost feature of aged organisms is the accumulation of cellular senescence, a state of cell cycle permanently arrested in response to diverse damaging stimuli. Unnecessary and excessive accretion of senescent cells in the tissue affects the regenerative potential of the tissue negatively and creates a pro-inflammatory milieu favorable for various age-associated diseases, including neurodegenerative disease, cardiovascular disease, cancer, etc. [3].

Novel Therapeutic Approaches Targeting Oxidative Stress
https://doi.org/10.1016/B978-0-323-90905-1.00004-3

2 Hallmarks of aging

The aging process starts from several basic molecular changes like genomic instability, epigenetic changes, telomere attrition, and loss of proteostasis, which are known as hallmarks of aging. Three other antagonistic hallmarks are deregulated nutrient sensing, cellular senescence, and mitochondrial dysfunction, and two integrative hallmarks are stem cell exhaustion and altered intercellular communication. All these hallmarks that contribute to the aging process might be triggered by oxidative stress. Telomeres present on chromosomes are very sensitive to oxidative stress and have low repair capacity compared to other parts of the chromosome. Hence, oxidative damage may result in telomere attrition that accelerates the aging process and increases the risk of age-associated disease [4]. Activation of chronic DNA damage distinguishes as cellular senescence as well as committed towards different cyclin dependent kinase inhibitors (CDKi); these CDKi increase the release of proinflammatory and tissue remodeling factors. The introduction of antiapoptotic genes leads to changes in metabolic rates as well as a change in endoplasmic reticulum stress. These changes in cell signaling pathways lead to abnormalities in the morphology of cells, and a change in constituents of plasma membrane and other cytoplasmic organelles [3].

3 Causes of aging

Generation of reactive oxygen species (ROS) and reactive nitrogen species (RNS) are the foremost cause of aging and shorter life span. According to the "free radical" theory of aging proposed by Harman in 1956 and revised in 1972, free radicals have minimum one unpaired electron, and are the primary reactive oxygen species molecules such as superoxides (O_2^-), hydroxyl radicals (OH), and hydrogen peroxide (H_2O_2), which contribute to endogenous oxidative damage [4]. The cause of oxidative stress is the amount of reactive oxygen species (ROS) production greater than the cellular antioxidant defense capacity to excrete toxic species [5]. ROS are created as a by-product of the metabolic process in all aerobic cells. When the levels of ROS are higher than the antioxidant production capacity of the cell or tissue, oxidative damage occurs in the cell or tissue. The production of ROS is a natural physiological process. However, the damaging rate of oxidative stress increases during the aging process, and the levels of antioxidant and the repairing capacity of the cell decrease. The relationship between the oxidative damage and antioxidant production capacity during the aging process has been reported in various tissues of different species [6]. Lipids, protein, and nucleic acid are damaged by oxidative stress caused by ROS/RNS [7]. Production and aggregation of ROS and RNS that are higher than the antioxidant capacity of cells or tissues have been seen in many diseased conditions in all aerobic organisms. Environmental conditions, exposure to UV radiation, air pollution, heavy metals, toxins, etc. are the major sources of ROS/RNS production. Plenty of free radicals can be generated by injured cells and tissues. The mitochondrial respiratory chain is the main site of ROS production during aerobic respiration. Massive production of ROS and RNS increases inflammation and cellular proliferation, and plays a key role in aging and age-associated diseases [8].

4 Age-associated diseases

4.1 Alzheimer's disease

Alzheimer's disease is the most common disease found worldwide in older people. There are 36 million people worldwide affected by Alzheimer's, and this number is expected to increase to more than 115 million by 2050 [9]. Alzheimer's disease is referred to decrease the ability of memory recognition and cognitive impairment and presence of intracellular neurofibrillary tangles and extracellular amyloid plaques in the brain with age. The name Alzheimer's was given by its discoverer Alois Alzheimer in 1906. A 51-year-old female named Auguste Deter with behavioral changes, cognitive disturbance, disorientation, and delusions had been investigated by Alzheimer in 1901 [10]. He reported his findings on "the peculiar disease process of the cerebral cortex." The pathology of Alzheimer's disease is still not clear. According to the "amyloid hypothesis," the deposition of beta amyloid plaque in the brain is the main cause of Alzheimer's disease [11]. In the central nervous system, different cells produce amyloid precursor protein (type 1 transmembrane protein). First the amyloid precursor protein (APP) undergoes sequential cleavage into two different pathways, then both the pathways are cleaved by γ-secretase. APP cleaved by alpha secretase produces sAPPα and C83 (membrane bound 83rd amino acid C-terminal), and further cleavage of C83 by γ-secretase results in production of another secreted filament p3 in a non-amyloidogenic pathway. In an amyloidogenic pathway, APP cleavage by beta secretase produces sAPPβ and C99 (membrane bound amino acid 99th C-terminal). C99 cleaved by secretase results in secretion of amyloid beta and AICD [10]. An aggregation of amyloid beta and an imbalance between production and clearance may be involved in the pathophysiology of Alzheimer's disease.

The tubulin associated unit (TAU) protein, first isolated from pig brain extract, is essential for microtubule (MT) assembly. It is a heat stable phosphoprotein, and may be phosphorylated at numerous sites by different protein kinases such as casein kinase type-1 and protein kinases dependent on cyclic AMP. Phosphorylated TAU protein regulates its capacity to stimulate microtubule assembly [12]. In humans, TAU protein is localized on axons and astrocytes or oligodendrocytes in very low amounts [13]. TAU protein helps in aiding assembly and stability of microtubules. Phosphorylation of TAU changes its shape, which regulates biological activity. Hyperphosphorylation lowers the potential of TAU to bind the microtubules and results in formation of neurofibrillary tangles which is another major pathological lesions in the brain [14]. Hyperphosphorylation also leads to loss of binding to microtubules, resulting in synaptic dysfunction and thus impairment in learning and loss of memory.

4.2 Cardiovascular disease

Diseases of the heart and circulatory system are known as cardiovascular diseases such as cerebrovascular disease, coronary artery disease, myocardial infraction, stroke, and cardiac arrhythmias [15]. Cardiovascular disease is one of the most common causes of mortality in developed countries. Globally one out of three deaths may be caused by cardiovascular disease, and it is a chronic disease. Cardiovascular disease evolves during the lives of many people as it is asymptomatic for a long time. In Europe, 49% of death is caused by cardiovascular disease. The European Union spent approximately €190 billion annually on health care costs for cardiovascular disease [16]. Cardiovascular disease may be caused by smoking tobacco, unhealthy eating, high alcohol consumption, low physical activity, oxidative stress, high blood pressure, and obesity. Cardiovascular disease is also caused by other factors such as age, gender, genetic, and environmental factors [15]. Reactive oxygen species and reactive nitrogen species are produced by mitochondrial respiratory chains and other enzymatic processes in the cardiovascular system, and are involved in the pathophysiology of cardiovascular disease. Instead of nitric oxide, superoxide anions are produced by endothelial nitric oxide synthase (eNOS) at higher oxidative stress levels [17]. TNF alpha and interleukin-1 beta promote higher activation and macrophages differentiation to secrete interleukin-6, interleukin-8, and other inflammatory cytokines that show toxic effects on heart muscles and lead to damaged heart muscles [18]. Oxidative stress shows a major impact on cardiovascular disease; therefore, by reducing oxidative stress we can overcome the risk of cardiovascular disease.

4.3 Diabetes

Diabetes mellitus is a group of heterogeneous disorders, with high concentrations of blood glucose [19]. The number of people have diabetes mellitus has increased four-fold in last 30years. One out of 11 adults have diabetes mellitus and it is the 9th major cause of death worldwide [20]. Worldwide, 415 million people had diabetes in 2015 and it is estimated that 193 million people have undiagnosed diabetes [21]. Out of these, type 2 diabetes accounts for 90%. According to data, it is expected that there will be 642 million cases by 2040 [20]. Obesity is one factor in increasing incidence of type 2 diabetes mellitus in children and adolescents. In the pediatric population, type 2 diabetes mellitus accounts for up to 45% of new cases [22]. In Asia, type 2 diabetes is growing at a faster rate, and India and China are two major epicenters. The causes of increasing the prevalence of epidemic of diabetes mellitus are various, and include poor eating habits, urbanization, luxurious life, less physical activity, obesity, oxidative stress, and aging. Pathophysiology of diabetes mellitus includes less secretion of insulin and its receptors not working properly. β-cells of pancreas dysfunction result in low secretion of insulin, not maintaining glucose homeostasis in the blood, whereas insulin resistance leads to increased glucose in the liver and decreased glucose uptake in muscles and adipose tissue. Both β-cell and insulin resistance contribute to the pathophysiology of type 2 diabetes mellitus [20]. As a result of aerobic respiration in cells, reactive oxygen species are formed. Lipid peroxidation is caused by reactive oxygen species and plays a major role in the development of oxidative stress, which is the most important factor in pathogenesis and progress of type 2 diabetes mellitus [23].

4.4 Cancer

Cancer is the uncontrolled growth of cells without cell division, cells contact inhibition is loss and form a mass (tumor) known as benign tumor, sometimes this tumor cell invades different tissues or organs, known as a malignant tumor. The GLOBOCAN 2020 data base, which is part of the IARC Global Cancer Observatory and is available online, contains projection of cancer incidence and mortality in 185 countries for 2020 for 36 types of cancer and all cancer sites combined. In 2020, the worldwide cancer burden is expected to be 19.3 million new cases and 10.0 million deaths. One in every five people develops cancer during their lifetime, and one in every eight men and one in every eleven women dies from it and it is expected to rise to 28.4 million new cases by 2040 (source: https://www.iarc.who.int/news-events/latest-global-cancer-data-cancer-burden-rises-to-19-3-million-new-cases-and-10-0-million-cancer-deaths-in-2020). Breast cancer is the most diagnosed cancer with 2.3 million new cases, followed by lung, colorectal, and prostate cancer. A major cause of cancer death is lung cancer, followed by colorectal, liver, stomach, and breast cancer [24]. Low dose computed tomography methods for lung cancer detection have shown reduction in mortality. The mortality rate of lung cancer is higher than those of other types of cancer such as breast cancer, colorectal cancer, stomach cancer, etc. Men are more affected by lung cancer than women. Asbestos inhalation with tobacco leads to lung cancer [25]. Smoking is one of the foremost causes of lung cancer, and another major cause of lung cancer is air pollution. Based on genetic susceptibility, nearly 15% of smokers have advanced lung cancer. Skin cancer is another type of cancer caused by environmental exposure (ultra-violet radiation); medical treatment and hereditary skin disease make individuals more prone to skin cancer. Gorlin-Goltz syndrome and Rombo syndrome are commonly associated with basal cell carcinoma [26]. Colorectal cancer is the second leading cause of death out of 36 different types of cancer in 185 countries of the world. The etiology of colorectal cancer are environment and genetic factors. Food habits and nutrition play major roles in the development of colorectal cancer [27].

5 Novel therapeutic approaches

5.1 Role of quercetin

Humans include plant-based products in their diet that are rich in flavonoids. Quercetin is a major flavonoid found in our diet. Fruits, vegetables, and beverages such as apple peel, grapes, tomatoes, tea, red onions, red wine, cherries asparagus, red leaf lettuce, and berries are the main sources of quercetin [28,29]. Quercetin is a yellow compound having a lipophilic nature which is insoluble in cold water and poorly soluble in hot water [29]. Generally, quercetin is found in two different forms: aglycone and glycosylated. Aglycone (a form of quercetin) is lipophilic in nature so is readily absorbed through the cell phospholipid bilayer, while quercetin glycosylated (a form that is hydrophilic) in nature is difficult to absorb into the body. The glycosylated form is found in soybean, fruits, and vegetables. Quercetin of the glycosylated form is hydrolyzed by enzymes to form the aglycone form of quercetin. Lactase phlorizin hydrolase enzymes are found in mammalian intestinal mucosal epithelial cells [30]. The aglycone form of quercetin is simply diffused over intestinal mucosal epithelial cells phospholipid bilayer. Monoglucoside is another form of quercetin, absorbed by sodium glucose transporter-1 (SGLT-1), into the cytoplasm of intestinal cells of rats, converted to deglycosylated with β-glucosidase present inside the cytoplasm of intestinal cells [31]. Thus, we can say that quercetin molecules are absorbed through phospholipid bilayer via simple diffusion as well as facilitated diffusion in the intestine by SGLT-1. Many studies shows that quercetin improves learning and memory, suppresses the expression of Aβ secretion, reduces mitochondrial dysfunction, decreases reactive oxygen species production, and also reduces memory loss and improves spatial learning by activating several signaling pathways such as ERK/CREB/BNDF, Nrf2-ARE, Akt, and AMPK or by inhibiting signaling pathways like Ask/JNK/jun, RAGE, etc. [29]. Thus, we can say that quercetin has potential to reduce or delay Alzheimer's disease. Quercetin acts on blood pressure in a dose-dependent manner. Oral intake of quercetin shows antihypertensive actions in spontaneously hypertensive rats. Bioavailability of quercetin is low; therefore, its metabolites may be responsible for this effect [32]. The antiinflammatory activity of quercetin is reported in animal models by inhibiting the activity of cyclooxygenase, and inhibition of NFκβ activation and nitric oxide, which play a role in the pathophysiology of inflammatory disease [33]. Quercetin improves production of sperm and reproductive health by its antioxidant potential; lipid peroxidation caused by oxidative stress damages the sperm membrane and DNA fragmentation, resulting in a decrease in sperm motility and infertility [28].

5.2 Role of curcumin

Plants and their products have been used for medicinal purposes for a very long time. Turmeric is one of the major spices used not only for making curries but also in religious ceremonies. It is also known as "Indian saffron" because of its yellow color. Turmeric is rhizomatous perennial herb, belongs to zingiberacea family [34]. Curcumin is a polyphenolic compound found in turmeric that is commonly used for its health benefits. Turmeric has been consumed as a spice in food from ancient times in India. It is one of the major spices used in India. Curcumin is a bioactive yellow compound present in turmeric. Bioactive compounds of turmeric are used to strengthen the overall body, improve digestion, relieve arthritis and help strengthen bones, dissolve gallstones, get rid of worms, regulate menstruation, treat skin cancer, and assist in wound healing; turmeric has antiinflammatory and antimicrobial properties [34,35]. ROS/cytokine-mediated inflammatory pathways and β-cells (NF-kB)/cyclooxygenase-2 activated by expression of a nuclear factor kappa light chain enhancer, which may be reduced by curcumin, has been reported in several papers and also increased antioxidants by activating nuclear factor erythroid 2 related factor 2 (NrF2) [36]. According to a meta-analysis by Qin Xiang Ng, curcumin has shown an antidepressant effect in depressed patients. According to the monoamine hypothesis, depression is caused by a low level of the monoamines neurotransmitter in the brain. Curcumin may help in increasing the monoamines concentration and also inhibit monoamine oxidase A and B enzymes, regulate the level of neurotransmitter in brain, and increase the antiinflammatory cytokines and the level of brain-derived neurotrophic factor; these mechanisms all help to reduce depression [37]. Curcumin has less solubility, poor absorption from the digestive system, and quick elimination from the body. Absorbed curcumin is broadly transformed to its water-soluble metabolites, glucuronides, and sulfates. Dihydrocurcumin and tetrahydrocurcumin products of NADPH induced reduction of curcumin by microbes, while a large part of ingested curcumin is excreted through feces. For improving absorption of curcumin, different strategies can be used such as a 1.5-fold increase when black pepper extract co-administration with curcumin. The phospholipid curcumin complex increases absorption by 3.4-fold [38]. According to Tabrizi et al., curcumin supplementation decreases TNF-α concentration and caused a significant decrease in IL-6 in patients with metabolic syndrome in six different randomized controlled trials. These results indicate that curcumin shows an antiinflammatory action [39]. Curcumin delays type 2 diabetes mellitus in animals by improving β cell function, averting β cell death. Curcumin also reduced insulin resistance in humans [40]. Exercise-induced oxidative stress and inflammation could be reduced by curcumin supplementation [36]. Liposomal curcumin inhibited tumor angiogenesis and suppressed pancreatic carcinoma in a murine xenograft model [41].

5.3 Role of resveratrol

Resveratrol (RSV) is plant secondary metabolite (a polyphenol) found in fruits such as red grapes, cyanococcus, red wine, peanuts, and soy berries [42,43]. RSV is a phytoalexin, and the level of RSV varies according to food sources, seasons, and batches. RSV acts as an immune modulator; the first result came 18years ago, in an investigation done by Gao et al. on mitogen/antigen-induced proliferation. This study showed that RSV suppress the proliferation of spleen cells induced concanavalin A, interleukin-2 or alloantigens and also inhibit the production of interferon-gamma (IFN-γ) and interleukin-2 by lymphocyte and production of tumor necrosis factor alpha (TNF-α) or interleukin-12 by macrophage [44]. Resveratrol mimics caloric restriction candidate by activating sirtuins. French people have a lower risk of cardiovascular disease with a high fat diet because they consume on average more red wine, a drink that contains resveratrol. RSV was first discovered in white hellebore roots in 1939 [45]. AMPK activated by RSV in muscles increase SIRT1 and PGC-1α protein levels, increases citrate synthase activity, and improves mitochondrial respiration on fatty acid-derived substrate in muscles without alteration in mitochondrial content. RSV decreases circulating glucose, intrahepatic lipid content, inflammation markers, and triglycerides, and increases the intramyocellular lipid level [46]. In various studies, RSV shows antiinflammatory, antioxidant, cardioprotective, immune modulator, and neuroprotective action as well as anti-tumor properties. RSV, like other polyphenols such as quercetin and curcumin, being a lipophilic compound, shows extensive metabolization and rapidly systematic elimination therefore shows low bioavailability [47]. RSV exists in two different geometric forms: the *trans* form has more stability and great biological activity, while the *cis* form is less active. Several reports have observed RSV's useful effects on diabetes mellitus, metabolic syndrome, obesity and age-related diseases [48]. Clinically used drugs in combination with RSV reverse multidrug resistance in cancerous cells [49]. RSV acts as an antiaging agent by maintaining telomerase activity by inducing telomerase maintenance factors. It also activates Werner syndrome helicase without altering cell proliferation [50]. Therefore, endothelial progenitor cells senescence can be delayed by RSV. It also acts as an antioxidant by suppressing ROS production and scavenging free radicals in many clinical in vitro and in vivo studies. The limitations of RSV are that it is a very light-sensitive compound; exposure to light causes about 80%–90% of the *trans* form to the *cis* form of RSV. It can also be degraded by high temperatures and changes in pH [51]. The free form of RSV is not suitable for drug delivery because of its low bioavailability. For improving the bioavailability in recent years, RSV has been combine with β-cyclodextrins and hydroxypropyl-β-cyclodextrins, liposome, and lipid nanoparticles [52].

6 Liposomes

Liposomes are spherical structure composed of lipids, in which a lipophilic bilayer formed a sandwich-like structure between two hydrophilic layers. Phospholipids show a similarity to the cellular membrane lipids. Therefore, they behave in a biocompatible and biodegradable way. Liposomes are widely used for their potential drug delivery associated with their similarity to organize structure [53]. Multiple drug candidates can be incorporated within their hydrophobic and hydrophilic chambers. It is stated that even gases like nitric oxide can be encapsulated into the liposomes to treat pulmonary hypertension [54]. Hydrophobic compounds (drugs and imaging agents) are present in the lipid bilayer while hydrophilic compounds can reside in the aqueous pool [55]. Liposomes fall into the overall category of nanomedicines and play a significant role in health care applications such as cancer, cardiovascular disease, diabetes, inflammation, and neurodegenerative diseases [53]. The encapsulation efficacy of certain drugs depends on the partition coefficient or LogP [56]. Nanomedicines based on lipids are used to protect drugs from in vivo degradation, modify distribution, control the release of drug and site-targeted drug delivery to the disease area, and increase bioavailability and solubility. Drug delivery systems based on lipids are also effective as vaccine adjuvants because they have potential to protect and deliver the antigens (peptides, proteins, and nucleic acid) that stimulate immune responses by the antigen presenting cells [53].

6.1 Charge on liposomes

Liposomes can be positively or negatively charged, or neutral, by adding other materials at different environmental pH. The liposomes having positive charge are composed of cationic lipids, such as *N*-[1(2,3-dioleoyloxy) propyl]-*N,N,N*-trimenthylammonium chloride (DOTAP) and used for delivery of nucleic acid. For the protection from enzymatic degradation, cationic liposomes formed lipoplexes with negatively charged nucleic acids by electrostatic interactions, which also aids internalization of liposomes via electrostatic interaction because the animal cell membrane is negatively charged. Compared with neutral liposomes, charged liposomes are more stable in a dispersed medium due to electrorepulsion between the molecules [55].

6.2 Role of liposomes in various diseases

Humans have used plant-based natural products to combat various diseases since ancient times. Current medicines are primarily derived from plants and herbs on the basis of traditional knowledge and practices. Natural resources provide over a quarter of all key medicinal chemicals and derivates available today. Natural substances with various molecular backgrounds provide a starting point for the development of new medications. Interest in building synthetically accessible lead molecules that mirror their counterpart's chemistry has been a recent trend in natural product-based drug discovery. Natural goods have unique qualities, such as chemical variety, chemical and biological capabilities with macromolecular specificity, and lower toxicity. As a result, they are good candidates for drug discovery. Other computational studies have facilitated an understanding of molecular interactions of drugs and development of next-generation drug inventions such as drug delivery and site-targeted drug discovery systems. Nanomedicine is revealed as the connecting link between the biological and physical sciences by applying nanostructure and nanophases in several areas of science [57]. The potential of therapeutics molecules is limited by their insufficient delivery, accumulation in target tissue, or unwanted side effects causing potentially severe toxicities in a healthy organ [56]. Liposomes formulation can be done with different phospholipids (natural or synthetic) as well as cholesterol. Amphiphilic phospholipids at higher concentrations spontaneously form a lipid bilayer surrounding an aqueous environment. Phase transition temperature of different phospholipids and cholesterol are different. This is an important parameter for altering the drug encapsulation efficiency, storage stability, and in vivo stability. Phospholipids are found in a fluid state, when the environment temperature is higher than the transition temperature; therefore, they are more permeable for water and favorable for drug encapsulation for preparation of liposomes. Liposomes used as drug delivery in vivo are often prepared with phospholipids whose transition temperature is higher than the body temperature; because the transition temperature is lower than the body temperature, the phospholipids in drug loaded liposomes are in a fluid state. This results in premature drug release. For stabilizing the liposomes cholesterol of the lipid bilayer is commonly used. Note that membrane rigidity and size of liposomes may increase by cholesterol in a dose-dependent manner [58]. Generally, liposomes may be classified as unilamellar or multilamellar. Unilamellar liposomes are suitable for loading and delivery of hydrophilic drugs, and are divided into small size (SUV, 50–100nm) or large size (LUV, 100–250nm). Unilamellar liposomes are composed of a single lipid bilayer and a large aqueous core, while multilamellar liposomes (MLVs) have a small space because they are composed of several lipid layers, and are therefore suitable for loading hydrophobic drugs [56]. The easiest and simplest form of liposomes preparation in laboratory is the lipid film hydration method followed by sonication and extrusion. In this method, small homogeneous liposomes are obtained [59]. The liposomal drug delivery system increases the safety and efficacy of chemotherapeutic drugs compared to those without liposomal (free) formulation [60]. Lymphoma and multiple myeloma patients treated with pegylated doxorubicin liposomes showed very good outcomes. These pegylated liposomes improve pharmacokinetics and reduced cardiotoxicity when compared with conventional doxorubicin delivery methods [61]. Novel therapeutic approaches are summarized in Table 1.

TABLE 1 The role of polyphenols like quercetin, curcumin, and resveratrol in treatment of various disease.

Polyphenols	Treatment	Reference
Pegylated doxorubicin liposomes, liposomal curcumin	Myeloid neoplasm Inhibit tumor angiogenesis	[41,61]
Quercetin	Reduces mitochondrial dysfunction, delays Alzheimer's, exhibits antihypersensitive action, decreases ROS, suppresses nitric oxides, improves reproductive health	[28,29]
Curcumin	Depression, skin cancer, wound healing, diabetes, dissolving gall bladder stones, bone strengthening, regulation of menstruation, anti-inflammatory, antimicrobial	[34–37,40]
Resveratrol	Cardio-protective, immune modulator, neuroprotective, diabetes mellitus, obesity, maintaining telomerase activity, antiinflammatory, antioxidant, ROS scavenging	[44,47,48,50]

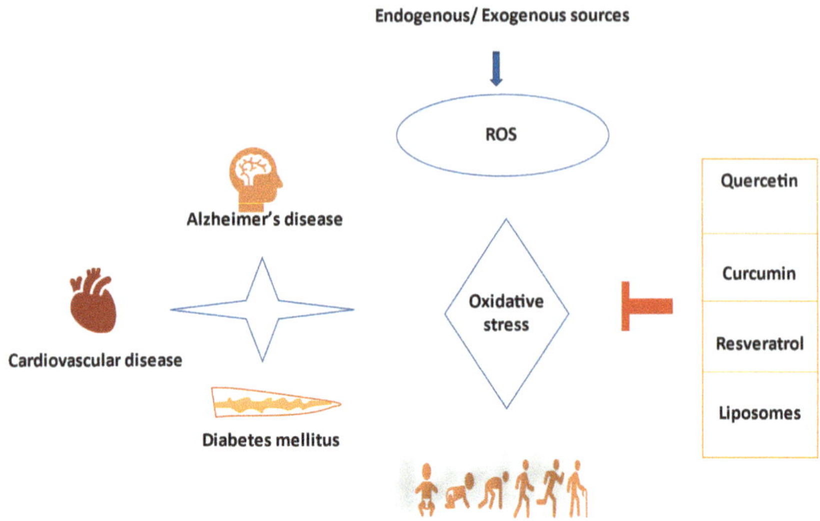

FIG. 1 Endogenous and exogenous sources are responsible for generation of ROS/RNS via the electron transport chain. This leads to oxidative damage to biomolecules, which ultimately results in various diseases such as Alzheimer's disease, cardiovascular disease, and diabetes mellitus. ROS/RNS can be quenched or neutralized with the help of quercetin, curcumin, and resveratrol.

7 Conclusion

Reactive oxygen species and reactive nitrogen species are the utmost cause of aging and age-related diseases. Plant polyphenols always finds its own importance as therapeutic agents in treating many diseases. As shown in this chapter, it is evident that quercetin, curcumin, and resveratrol have therapeutic potential to delay the aging process. Liposomal formulation may increase the bioavailability and absorption of polyphenols in cells, which ultimately aids human health (Fig. 1).

Acknowledgments This study was supported by a Fellowship from the Council of Scientific and Industrial Research (CSIR), the Government of India, to Mr. A.K. (09/1152(0016)/2019-EMR-I). This agency had no role in the interpretation or writing of the manuscript.

References

[1] S.I. Rizvi, P.K. Maurya, Markers of oxidative stress in erythrocytes during aging in humans, Ann. N. Y. Acad. Sci. 1100 (2007) 373–382.
[2] L. Partridge, J. Deelen, P.E. Slagboom, Facing up to the global challenges of ageing, Nature 561 (2018) 45–56.
[3] A. Hernandez-Segura, J. Nehme, M. Demaria, Hallmarks of cellular senescence, Trends Cell Biol. 28 (2018) 436–453.
[4] J. Luo, K. Mills, S. le Cessie, R. Noordam, D. van Heemst, Ageing, age-related diseases and oxidative stress: what to do next? Ageing Res. Rev. 57 (2020), 100982.

[5] P.K. Maurya, S.I. Rizvi, Age-dependent changes in glutathione-s-transferase: correlation with total plasma antioxidant potential and red cell intracellular glutathione, Indian J. Clin. Biochem. 25 (2010) 398–400.
[6] S.I. Rizvi, P.K. Maurya, Alterations in antioxidant enzymes during aging in humans, Mol. Biotechnol. 37 (2007) 58–61.
[7] P.K. Maurya, C. Noto, L.B. Rizzo, A.C. Rios, S.O. Nunes, D.S. Barbosa, S. Sethi, M. Zeni, R.B. Mansur, M. Maes, E. Brietzke, The role of oxidative and nitrosative stress in accelerated aging and major depressive disorder, Prog. Neuro-Psychopharmacol. Biol. Psychiatry 65 (2016) 134–144.
[8] R.A. Zinovkin, V.P. Romaschenko, I.I. Galkin, V.V. Zakharova, O.Y. Pletjushkina, B.V. Chernyak, E.N. Popova, Role of mitochondrial reactive oxygen species in age-related inflammatory activation of endothelium, Aging (Albany NY) 6 (2014) 661–674.
[9] M. Gregori, M. Masserini, S. Mancini, Nanomedicine for the treatment of Alzheimer's disease, Nanomedicine (Lond) 10 (2015) 1203–1218.
[10] J.A. Soria Lopez, H.M. Gonzalez, G.C. Leger, Alzheimer's disease, Handb. Clin. Neurol. 167 (2019) 231–255.
[11] H.W. Querfurth, F.M. LaFerla, Alzheimer's disease, N. Engl. J. Med. 362 (2010) 329–344.
[12] G. Simic, M. Babic Leko, S. Wray, C. Harrington, I. Delalle, N. Jovanov-Milosevic, D. Bazadona, L. Buee, R. de Silva, G. Di Giovanni, C. Wischik, P.R. Hof, Tau protein hyperphosphorylation and aggregation in Alzheimer's disease and other tauopathies, and possible neuroprotective strategies, Biomol. Ther. 6 (2016) 6.
[13] A. Migheli, M. Butler, K. Brown, M.L. Shelanski, Light and electron microscope localization of the microtubule-associated tau protein in rat brain, J. Neurosci. 8 (1988) 1846–1851.
[14] N.N. Naseri, H. Wang, J. Guo, M. Sharma, W. Luo, The complexity of tau in Alzheimer's disease, Neurosci. Lett. 705 (2019) 183–194.
[15] G.D. Flora, M.K. Nayak, A brief review of cardiovascular diseases, associated risk factors and current treatment regimes, Curr. Pharm. Des. 25 (2019) 4063–4084.
[16] S. Francula-Zaninovic, I.A. Nola, Management of measurable variable cardiovascular disease' risk factors, Curr. Cardiol. Rev. 14 (2018) 153–163.
[17] Q. Li, J.Y. Youn, H. Cai, Mechanisms and consequences of endothelial nitric oxide synthase dysfunction in hypertension, J. Hypertens. 33 (2015) 1128–1136.
[18] C.Y. Liu, Y. Zhou, T. Chen, J.C. Lei, X.J. Jiang, AMPK/SIRT1 pathway is involved in arctigenin-mediated protective effects against myocardial ischemia-reperfusion injury, Front. Pharmacol. 11 (2020), 616813.
[19] M. Roden, Diabetes mellitus: definition, classification and diagnosis, Wien. Klin. Wochenschr. 128 (Suppl. 2) (2016) S37–S40.
[20] Y. Zheng, S.H. Ley, F.B. Hu, Global aetiology and epidemiology of type 2 diabetes mellitus and its complications, Nat Rev Endocrinol 14 (2018) 88–98.
[21] S. Chatterjee, K. Khunti, M.J. Davies, Type 2 diabetes, Lancet 389 (2017) 2239–2251.
[22] A. Mohamadi, D.W. Cooke, Type 2 diabetes mellitus in children and adolescents, Adolesc. Med. State Art Rev. 21 (2010) 103–119. x.
[23] K. Rehman, M.S.H. Akash, Mechanism of generation of oxidative stress and pathophysiology of type 2 diabetes mellitus: how are they interlinked? J. Cell. Biochem. 118 (2017) 3577–3585.
[24] H. Sung, J. Ferlay, R.L. Siegel, M. Laversanne, I. Soerjomataram, A. Jemal, F. Bray, Global cancer statistics 2020: GLOBOCAN estimates of incidence and mortality worldwide for 36 cancers in 185 countries, CA Cancer J. Clin. 71 (3) (2021) 209–249, https://doi.org/10.3322/caac.21660.
[25] F. Nasim, B.F. Sabath, G.A. Eapen, Lung cancer, Med. Clin. North Am. 103 (2019) 463–473.
[26] J. Schierbeck, T. Vestergaard, A. Bygum, Skin cancer associated genodermatoses: a literature review, Acta Derm. Venereol. 99 (2019) 360–369.

[27] K. Thanikachalam, G. Khan, Colorectal cancer and nutrition, Nutrients 11 (2019).

[28] E. Moretti, L. Mazzi, C. Bonechi, M.C. Salvatici, F. Iacoponi, C. Rossi, G. Collodel, Effect of Quercetin-loaded liposomes on induced oxidative stress in human spermatozoa, Reprod. Toxicol. 60 (2016) 140–147.

[29] F. Babaei, M. Mirzababaei, M. Nassiri-Asl, Quercetin in food: possible mechanisms of its effect on memory, J. Food Sci. 83 (2018) 2280–2287.

[30] A.J. Day, F.J. Canada, J.C. Diaz, P.A. Kroon, R. McLauchlan, C.B. Faulds, G.W. Plumb, M.R. Morgan, G. Williamson, Dietary flavonoid and isoflavone glycosides are hydrolysed by the lactase site of lactase phlorizin hydrolase, FEBS Lett. 468 (2000) 166–170.

[31] Y. Marunaka, R. Marunaka, H. Sun, T. Yamamoto, N. Kanamura, T. Inui, A. Taruno, Actions of quercetin, a polyphenol, on blood pressure, Molecules 22 (2017).

[32] I. Najmanova, J. Pourova, P. Mladenka, A mixture of phenolic metabolites of quercetin can decrease elevated blood pressure of spontaneously hypertensive rats even in low doses, Nutrients 12 (2020).

[33] M. Guran, G. Sanliturk, N.R. Kerkuklu, E.M. Altundag, A. Suha Yalcin, Combined effects of quercetin and curcumin on antiinflammatory and antimicrobial parameters in vitro, Eur. J. Pharmacol. 859 (2019), 172486.

[34] S. Prasad, B.B. Aggarwal, Turmeric, the golden spice: from traditional medicine to modern medicine, in: I.F.F. Benzie, S. Wachtel-Galor (Eds.), Herbal Medicine: Biomolecular and Clinical Aspects, CRC Press/Taylor & Francis, Boca Raton (FL), 2011.

[35] H. Sasaki, Y. Sunagawa, K. Takahashi, A. Imaizumi, H. Fukuda, T. Hashimoto, H. Wada, Y. Katanasaka, H. Kakeya, M. Fujita, K. Hasegawa, T. Morimoto, Innovative preparation of curcumin for improved oral bioavailability, Biol. Pharm. Bull. 34 (2011) 660–665.

[36] S.A.B. Ms, P.D.H. Waldman, P.D.B. Krings, P.D.J. Lamberth, P.D.J. Smith, P.D.M. McAllister, Effect of curcumin supplementation on exercise-induced oxidative stress, inflammation, muscle damage, and muscle soreness, J. Diet Suppl. 17 (2020) 401–414.

[37] Q.X. Ng, S.S.H. Koh, H.W. Chan, C.Y.X. Ho, Clinical use of curcumin in depression: a meta-analysis, J. Am. Med. Dir. Assoc. 18 (2017) 503–508.

[38] A. Liu, H. Lou, L. Zhao, P. Fan, Validated LC/MS/MS assay for curcumin and tetrahydrocurcumin in rat plasma and application to pharmacokinetic study of phospholipid complex of curcumin, J. Pharm. Biomed. Anal. 40 (2006) 720–727.

[39] R. Tabrizi, S. Vakili, M. Akbari, N. Mirhosseini, K.B. Lankarani, M. Rahimi, M. Mobini, S. Jafarnejad, Z. Vahedpoor, Z. Asemi, The effects of curcumin-containing supplements on biomarkers of inflammation and oxidative stress: a systematic review and meta-analysis of randomized controlled trials, Phytother. Res. 33 (2019) 253–262.

[40] S. Chuengsamarn, S. Rattanamongkolgul, R. Luechapudiporn, C. Phisalaphong, S. Jirawatnotai, Curcumin extract for prevention of type 2 diabetes, Diabetes Care 35 (2012) 2121–2127.

[41] P. Anand, A.B. Kunnumakkara, R.A. Newman, B.B. Aggarwal, Bioavailability of curcumin: problems and promises, Mol. Pharm. 4 (2007) 807–818.

[42] L. Malaguarnera, Influence of resveratrol on the immune response, Nutrients 11 (2019).

[43] J.M. Breuss, A.G. Atanasov, P. Uhrin, Resveratrol and its effects on the vascular system, Int. J. Mol. Sci. 20 (2019).

[44] X. Gao, Y.X. Xu, N. Janakiraman, R.A. Chapman, S.C. Gautam, Immunomodulatory activity of resveratrol: suppression of lymphocyte proliferation, development of cell-mediated cytotoxicity, and cytokine production, Biochem. Pharmacol. 62 (2001) 1299–1308.

[45] M. Springer, S. Moco, Resveratrol and its human metabolites-effects on metabolic health and obesity, Nutrients (2019) 11.

[46] S. Timmers, E. Konings, L. Bilet, R.H. Houtkooper, T. van de Weijer, G.H. Goossens, J. Hoeks, S. van der Krieken, D. Ryu, S. Kersten, E. Moonen-Kornips, M.K.C. Hesselink, I. Kunz, V.B. Schrauwen-Hinderling, E. Blaak, J. Auwerx, P. Schrauwen, Calorie

restriction-like effects of 30 days of resveratrol supplementation on energy metabolism and metabolic profile in obese humans, Cell Metab. 14 (2011) 612–622.

[47] A. Chimento, F. De Amicis, R. Sirianni, M.S. Sinicropi, F. Puoci, I. Casaburi, C. Saturnino, V. Pezzi, Progress to improve oral bioavailability and beneficial effects of resveratrol, Int. J. Mol. Sci. 20 (2019).

[48] M. Koushki, N. Amiri-Dashatan, N. Ahmadi, H.A. Abbaszadeh, M. Rezaei-Tavirani, Resveratrol: a miraculous natural compound for diseases treatment, Food Sci. Nutr. 6 (2018) 2473–2490.

[49] J.H. Ko, G. Sethi, J.Y. Um, M.K. Shanmugam, F. Arfuso, A.P. Kumar, A. Bishayee, K.S. Ahn, The role of resveratrol in cancer therapy, Int. J. Mol. Sci. 18 (2017).

[50] Y.R. Li, S. Li, C.C. Lin, Effect of resveratrol and pterostilbene on aging and longevity, Biofactors 44 (2018) 69–82.

[51] X.D. Zhu, X.P. Lei, W.B. Dong, Resveratrol as a potential therapeutic drug for respiratory system diseases, Drug Des. Dev. Ther. 11 (2017) 3591–3598.

[52] A.R. Neves, M. Lucio, S. Martins, J.L. Lima, S. Reis, Novel resveratrol nanodelivery systems based on lipid nanoparticles to enhance its oral bioavailability, Int. J. Nanomedicine 8 (2013) 177–187.

[53] S. Shah, V. Dhawan, R. Holm, M.S. Nagarsenker, Y. Perrie, Liposomes: advancements and innovation in the manufacturing process, Adv. Drug Deliv. Rev. 154–155 (2020) 102–122.

[54] N. Filipczak, J. Pan, S.S.K. Yalamarty, V.P. Torchilin, Recent advancements in liposome technology, Adv. Drug Deliv. Rev. 156 (2020) 4–22.

[55] H. Zhang, Thin-film hydration followed by extrusion method for liposome preparation, Methods Mol. Biol. 1522 (2017) 17–22.

[56] Y. Fan, Q. Zhang, Development of liposomal formulations: from concept to clinical investigations, Asian J. Pharm. Sci. 8 (2013) 81–87.

[57] J.K. Patra, G. Das, L.F. Fraceto, E.V.R. Campos, M.D.P. Rodriguez-Torres, L.S. Acosta-Torres, L.A. Diaz-Torres, R. Grillo, M.K. Swamy, S. Sharma, S. Habtemariam, H.S. Shin, Nano based drug delivery systems: recent developments and future prospects, J Nanobiotechnol. 16 (2018) 71.

[58] J.M. Lopez-Pinto, M.L. Gonzalez-Rodriguez, A.M. Rabasco, Effect of cholesterol and ethanol on dermal delivery from DPPC liposomes, Int. J. Pharm. 298 (2005) 1–12.

[59] J. Sachaniya, R. Savaliya, R. Goyal, S. Singh, Liposomal formulation of vitamin A for the potential treatment of osteoporosis, Int. J. Nanomedicine 13 (2018) 51–53.

[60] V. Makwana, J. Karanjia, T. Haselhorst, S. Anoopkumar-Dukie, S. Rudrawar, Liposomal doxorubicin as targeted delivery platform: current trends in surface functionalization, Int. J. Pharm. 593 (2021), 120117.

[61] C. Zhang, H. Yao, P.Y. Kong, Y. Liu, L. Gao, L. Gao, Y.Y. Ma, J. Liu, X. Tan, X. Zhang, Pegylated liposomal doxorubicin for myeloid neoplasms, Anti-Cancer Drugs 30 (2019) 948–952.

CHAPTER

5

Novel therapeutic approaches targeting oxidative stress in neurodegenerative diseases

Somu Yadav, Akshay Kumar, Deepika, and Pawan Kumar Maurya

Department of Biochemistry, Central University of Haryana, Mahendergarh, India

1 Introduction

Aging in humans refers to the internal physiological degeneration of an organism due to a persistent reduction in the age-specific fitness components, and it is essentially comprehended as a multidimensional procedure of physical, social, and neuropsychological changes. Aging has been and is being studied at various levels (genetic, molecular), yet the definite underlying biological mechanism has not been found to unravel the secrets of longevity. In this sense, one dimension could decline faster than the others. A complementary concept is the idea of senescence. Leonard Hayflick in 1961 first described the phenomenon of cellular senescence, referring to the limited capacity of isolated cells to proliferate in culture (the Hayflick Limit), but the concept can also be applied to whole organisms. For example, after a near-perfect renewal period (in humans, between the age of 20 and 35 years), there is a physiological decline in the ability to increase homeostatic imbalances, risk of disease, and respond to stress. Contrary to aging, senescence is a purely biological concept. A highly prevalent evidence says that the severe mental disorders could be associated with accelerated aging and earlier onset of age-related medical conditions such as type 2 diabetes mellitus, all forms of atherosclerosis, age-related macular degeneration, hypertension metabolic imbalances, cardiovascular

https://doi.org/10.1016/B978-0-323-90905-1.00005-5

diseases, arterial hypertension, and autoimmunity [1,2]. Furthermore, the association between aging and severe neurodegenerative disorders, especially Parkinson's and Alzheimer's diseases, are well recognized.

Indeed, biological characteristics of aging, such as low-grade chronic inflammatory activation, decrease in neurotrophic support, and amyloid cascades, seem to be more pronounced in severe mental disorders compared to normative aging. Here we examine how the investigation of aging could add to the search for promising and favorable molecular targets in attempts to better comprehend neurodegeneration, and putatively enhance our capacity to diminish clinical advances of neuropathology across clinical staging. Several discoveries on aging in extreme neurodegenerative issues are preparatory, and the readers should consider them.

2 Dimensions of aging in the investigation of neurodegeneration

2.1 Structural changes in Alzheimer's and Parkinson's disease

Alzheimer's disease (AD) is the most predominant dementing disorder which is found in elderly individuals. In spite of the fact that macrostructural atrophy of AD has been completely depicted, however, whether the microstructural changes offer remarkable data autonomous from macrostructural atrophy was unclear. The microstructural differences are unassociated with macrostructural atrophy by using conventional and nonconventional MRI methods (T1-loaded and diffusion-loaded MR scanning), which were used to study retrosplenial and temporal regions, additionally in corticopontine tracts, thalamus, precentral gyrus, and striatum in patients and controls. Volumetric contrasts (free of microstructural adjustments) were studied for the most part in the splenium, posterior cingulum, and entorhinal cortex. A longitudinal, multimodal MRI in AD patients indicated cortical atrophy in the medial temporal and parietal regions. It describes a broad example of white matter harm which includes the cingulum, temporal-occipital, parietal, and corpus callosum, and frontal white matter tracts. Tissue loss was found in parietal, temporal, and frontal regions in AD patients where total gray matter atrophy was observed. In various other studies, it was revealed that atrophy was found in the medial temporal region along with lateral posterior parietal regions, precuneus, retrosplenial cortex, and posterior cingulate [3]. Contrary to this, these regions in elderly individuals where dementia had not occurred no longer display increased atrophy. However, in patients with severe dementia, the rate of accelerated atrophy is heightened and amplified in far-reaching cortical regions, which incorporate the frontal cortex.

In Parkinson's disease (PD), the broad neurodegenerative changes happen all through the brain as the disease advances. Postmortem investigations of PD patients have uncovered the existence of Lewy body dementia causing a pathological reaction and apoptosis. It also includes the reduction of neurotransmitters and interneurons, in the two cortical and subcortical regions. The evaluated pervasiveness of Parkinson's disease with dementia (PDD) is around 30%. PDD is due to neuropathological variation in cortical and subcortical brain regions. PDD patients' brains have been identified with deteriorated subcortical structures. They usually consist of most of the medial piece of the substantia nigra, thalamus, the basal ganglia, and amygdala, nucleus basalis of Meynert, and locus coeruleus. In addition, patients with PDD displayed reduced gray matter volume in various regions of the brain. These regions are those such as the left half of the thalamus, accumbens nuclei, and bilateral putamen. They also include the para-hippocampal region, bilateral hippocampus, and anterior cingulate gyrus. On the right side of the hippocampus, left superior temporal gyrus, and left anterior cingulate gyrus, a loss in gray matter was detected in PD patients. In these patients, it was also reported that hippocampal atrophy is constantly associated with altered memory [4]. Interestingly, hippocampal structural alterations can be of great significance in predicting the forthcoming cerebral reduction in PD. Dementia, cerebral damage, memory loss, and depression have been linked with amygdala atrophy in PD patients [5]. Reduced amygdala volumes and increased atrophy in early-stage PD have been demonstrated. Brain structures, master communications that happen in the thalamus, connect the basal ganglia to cortical regions. They have been reported with significant atrophy. Lower thalamus volumes of patients with PD were compared with controls and investigated [6]. A neuroimaging technique known as voxel-based morphometry has been employed to identify the variations in cortical structure in PD, which can distinguish the whole brain region with gray matter density [7]. Several such investigations have revealed reductions in the frontal lobe, olfactory cortex, anterior cingulate cortex, temporal lobe, limbic and paralimbic areas, intraparietal sulcus, and parietal and occipital cortex [8].

2.2 Amyloid deposition

Deposition of extracellular amyloid peptide and existence of intraneuronal neurofibrillary tangles, threads of neuropil, and anomalistic material on dystrophic nerve cell procedures of neurotic plaques is being considered as a major pathological characteristic of AD. These progressions are found in postmortem investigation of elderly individuals without dementia. One major difference in both demented AD patients and elderly individuals without clinical signs of AD is the distribution pattern of neurofibrillary tangles. AD patients present a pattern of widespread lesions occurring in many areas of the brain, whereas nondemented elderly individuals show Aβ accumulated to definite inclining sites, and neurofibrillary tangles [9]. Changes in amyloid peptide plasma levels in AD patients were recently described, and these changes were also reported in major depressive disorders (MDDs). In patients with mood disorders, when compared with the healthy population, the risk of dementia and cerebral loss has been increased [10].

PD, characterized by continuous motor and cerebral damage, is further classified into PDD and dementia with Lewy bodies (DLB). DLB has been reported as the second major neurodegenerative disorder and adds up to 20% of dementia cases [11]. The "one-year rule" differentiates the two types of dementias. Parkinsonism is described as DLB possessing dementia before or after a year of progression into the disorder. By utilizing molecular imaging technology, assimilation of Aβ-amyloid plaques has been detected in both PDD and DLB. Major revelations in the setup of PD, PDD, and DLB have been studied with the help of amyloid PET imaging, representing an increased quantity of neocortical amyloid deposition in DLB than PDD more frequently [12]. Interestingly, in elderly individuals, increased amounts of amyloid deposition occur rarely in PD with mild cognitive impairment (MCI) in comparison to the levels reported in cognitively normal individuals.

2.3 Oxidative stress

Imbalances in levels of oxidants and antioxidants result in a condition known as oxidative stress (OS). This plays an important role in the pathophysiology of neurodegenerative diseases which include AD and PD. Raised oxidative stress is a hallmark of the aging process. Naturally, in healthy young individuals, immune cells use free radicals to mount an immune defense and clear pathogens and tumor cells with the typical "respiratory burst." In addition, reactive oxygen species (ROS) are caused by neuro-inflammation occurring when the brain is dealing with neuronal damage. However, prolonged higher levels of oxidant species than antioxidant defenses can be deleterious and add to the pathogenesis of numerous diseases. The accumulation of free radicals induces damage in lipids, proteins, and DNA leading to the malfunction of cell systems, but specially neuroimmune-endocrine systems. The etiology of AD and PD is dependent on multiple factors. It includes factors such as genetic and environmental factors that are regarded as risk factors for both diseases. In all aerobic organisms, ROS like nitric oxide, hydroxyl radical ($^{\cdot}OH$), superoxide radical ($O_2^{\cdot-}$), peroxynitrite ($ONOO^-$), and hydrogen peroxide (H_2O_2) are normally produced and exist inside the cell by maintaining the balance between antioxidant enzymes and antioxidant molecules [13]. An increase in the endogenous ROS levels or antioxidant defense induction can be clearly and rapidly indicated from the OS. Production and collection of ROS are a shared factor in numerous diseases and environmental insults, which can lead to severe damage to cells, causing cell death and physiological dysfunction in nearly all aerobes [14]. Environmental conditions such as xenobiotics, extreme temperatures, heavy metals, herbicides, wounding, toxins, and air pollutants can cause increased ROS formation. Free radicals can also be generated and stimulated by injured cells and tissues. A schematic representation of role of oxidative stress, amyloid deposition has been shown in Fig. 1.

Oxidative imbalances have been found in several illnesses and neurodegenerative diseases. Extensive studies pointed out the role of $O_2^{\bullet-}$, H_2O_2, $^{\bullet}OH$, and NO in oxidative stress-mediated neurodegeneration in AD [15]. Microglia activation generates excessive superoxide radicals. Higher

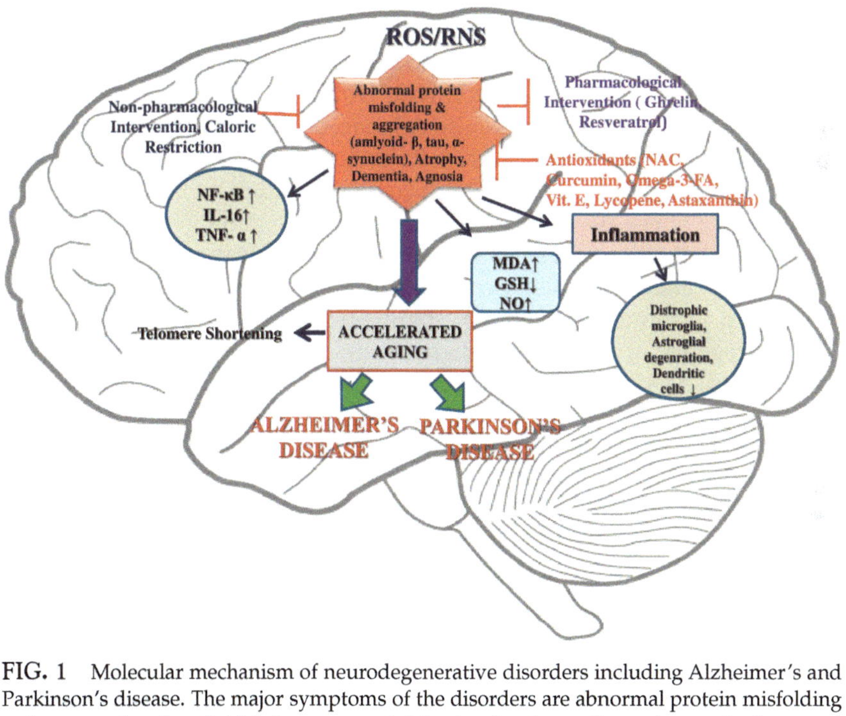

FIG. 1 Molecular mechanism of neurodegenerative disorders including Alzheimer's and Parkinson's disease. The major symptoms of the disorders are abnormal protein misfolding and aggregation (amyloid-β, tau, α-synuclein), atrophy, dementia, and agnosia. Increased oxidative stress leads to accelerated aging due to increased levels of reactive oxygen species (ROS) and reactive nitrogen species (RNS). ROS and RNS are determined by the oxidative stress biomarkers such as elevated malondialdehyde (MDA), reduced glutathione (GSH), and nitrous oxide (NO) levels. Signaling pathways such as nuclear factor-κB (NF-κB), tumor necrosis factor-α (TNF-α), and interleukin-16 are up-regulated during neurodegenerative diseases. Telomere shortening in turn is the cause of aging. In such diseases, inflammation in immune cells leads to dystrophic microglia, astroglial degeneration, and reduction in the number of dendritic cells. Therapies of Alzheimer's and Parkinson's disease include antioxidant therapy (*N*-acetylcysteine (NAC), curcumin, omega-3 fatty acids, vitamin E, lycopene, astaxanthin), use of ghrelin, resveratrol, and calorie restriction regimes.

metabolic demand and the postmitotic nature of glial cells and neurons make them more susceptible to OS. The low rate of brain regeneration and insufficient antioxidant potential further favors OS. A balance between pro-oxidant-antioxidant, progressive protein products, and malondialdehyde and 4-hydroxynonenal (lipid peroxide products) along with the oxidative stress biomarkers are relevant. A reduction in GSH is seen in the aging brain, which is interconnected with cognitive dysfunction and is accelerated in many central nervous system (CNS) diseases including AD. Acrolein, a by-product of lipid peroxidation has been previously presented as a factor involved in the AD pathological process. In some recent reports, it has been reported that the levels of acrolein-conjugated protein (Acr-PC) of AD patients in the cerebrospinal fluid (CSF) and plasma samples are found to be notably increased compared to control, but 3-hydroxypropyl mercapturic acid (3-HPMA) and urinary acrolein metabolite levels are remarkably reduced in AD patients. The crucial antioxidant enzyme and the GPx, which is in direct correlation with circulatory selenium, is reported in meta-analysis and is found to be decreased in AD patients when compared to control. In the AD patients, compared to control, the GSH/GSSG ratio is reduced, whereas DNA damage is reported to be increased.

Oxidative stress play an important role in the pathogenesis of PD. In recent concepts, the course of PD includes somatization, peripheral inflammation, and depression occurring from nitrostative stress, oxidative stress, and immune-inflammatory pathways interspersed from PD [16]. Circadian genes, aryl hydrocarbon receptor, lipid peroxide, cyclic adenosine monophosphate, and sirtuins further afflicted individuals with PD. Higher susceptibility for peroxide formation in a specific region manner is due to the change in the overall content of fatty acids in the anterior cingulate cortex (ACC) of the PD. In PD patients, a recent report has revealed that an increased level of oxidative stress is possibly related to the various markers with gender and disease stage. Neural degeneration in substantia nigra additionally happens amid sporadic PD [17]. PD patients demonstrate increased amounts of lipid peroxidation and decreased levels of GSH. In the cortical areas and the hippocampus, the GSH level is reduced while in the temporal and cerebral cortex, lipid peroxide levels are increased. OS and mitochondrial dysfunction can cause reduced levels of GSH. A harmful form of alpha-synuclein (SNCA) accumulates due to increased OS [18]. iPLA2γ, calcium-independent phospholipase A2γ hydrolyzes glycerol-phospholipids to liberate free fatty acids. iPLA2γ-deficiency has been associated with abnormal mitochondrial function. More importantly, the iPLA2 family consists of causative proteins in mitochondrial neurodegenerative disorders such as PD. Significantly increased activities of SOD, Cu/ZnSOD, and GST, reduced GR activity, and an increase of MDA concentration are observed in the striatum of PD rats, compared to the control group. Various biomarkers have been studied and summarized in Table 1 for both AD and PD.

TABLE 1 Oxidative stress biomarkers in Alzheimer's and Parkinson's disease.

Mechanism	Biomarkers	Alzheimer's or Parkinson's disease	References
Lipid peroxidation	Malondialdehyde (MDA) Thiobarbituric acid-reactive substances (TBARS) 4-Hydroxynonenal (HNE) F2-Isoprostanes	Elevated both AD and PD	[19]
Protein oxidation	Protein carbonyl content	Elevated in AD	[20]
Nucleic acid (DNA and RNA) damage	The comet assay or single cell gel electrophoresis (SCGE) 8-Hydroxyguanosine	Elevated in AD	[21]
Enzymatic antioxidants	Superoxide dismutase Glutathione peroxidase Glutathione reductase Glutathione-S-transferase	Elevated in both AD and PD	[22]
Nonenzymatic antioxidants	Albumin Vitamin D Vitamin A Vitamin C Vitamin E	Low levels in both AD and PD	[23]

2.4 Immunosenescence

Immunosenescence refers to the functional changes in the immune system that occur with aging, especially the sustained low-grade inflammatory activity, sometimes referred to as "inflamm-aging." In addition, elderly people also show a difficulty with or slow response to acute immune challenges, which explains the vulnerability of elderly people to common infections, such as influenza. From an immune point of view, successful aging is the capacity to respond quickly and strongly to the immunological challenges without the loss of efficiency of antiinflammatory networks. The concept of inflamm-aging has been largely accepted as a possible explanation of the effect of age on risk for several inflammation-associated diseases including cardiovascular diseases, cancer, and type 2 diabetes. In addition, inflamm-aging also has been postulated as a co-cause of Alzheimer's disease, based on the fact that people with this kind of dementia also have increased inflammatory activity in comparison to controls [24]. Inflamm-aging manifests changes in both adaptive and innate immunity. These changes may cause alterations in the thymus and bone marrow, resulting in a change in the composition of the immune cell range.

Pathogenesis and progression of AD are not solely due to neuronal dysfunction, but can also be due to neuronal inflammation. Inflammation occurs in pathological susceptible regions inside the brain cells in AD. Both innate immunity cells (primarily microglia and astrocytes) and adaptive immunity cells (mediated by B and T cells) produce responses resulting in inflamm-aging in AD. Microglia are immune cells in the CNS that interact with neurons, astrocytes, and oligodendrocytes [25]. Microglia cells cause acute insult to the CNS, result in neuroinflammation, and act as neuroprotective cells. The microglial response may be crucial in tripling the balance causing tau-related pathology and progress dementia by triggering interactions between tau and Aß. Normal microglial cells are mononuclear phagocyte systems; they display dendritic morphology and possess a neuroprotective function, i.e., the potential to transform into a macrophage, by production and secretion of various neurotrophic factors such as NGF, TGF-β, GDNF, etc., by diverting noxious particles like free iron and harmful substances away from the neuron, and finally by regulating neuronal connectivity and synaptic surveillance [26]. In an aged brain, microglial cells are transformed into dysfunctional dystrophic microglia displaying abnormal morphological features. Hence, microglia dystrophy or dystrophic microglia protect neurons during synapse loss and dementia in AD brain. Inflammatory response by innate immune cells is represented in Fig. 2.

Astrocytes are also associated with CNS possessing various functions, i.e., brain homeostasis, assistance in neurogenesis, neuroprotection, and shaping the gray matter by the process of tiling. In the AD brain, both reactive and

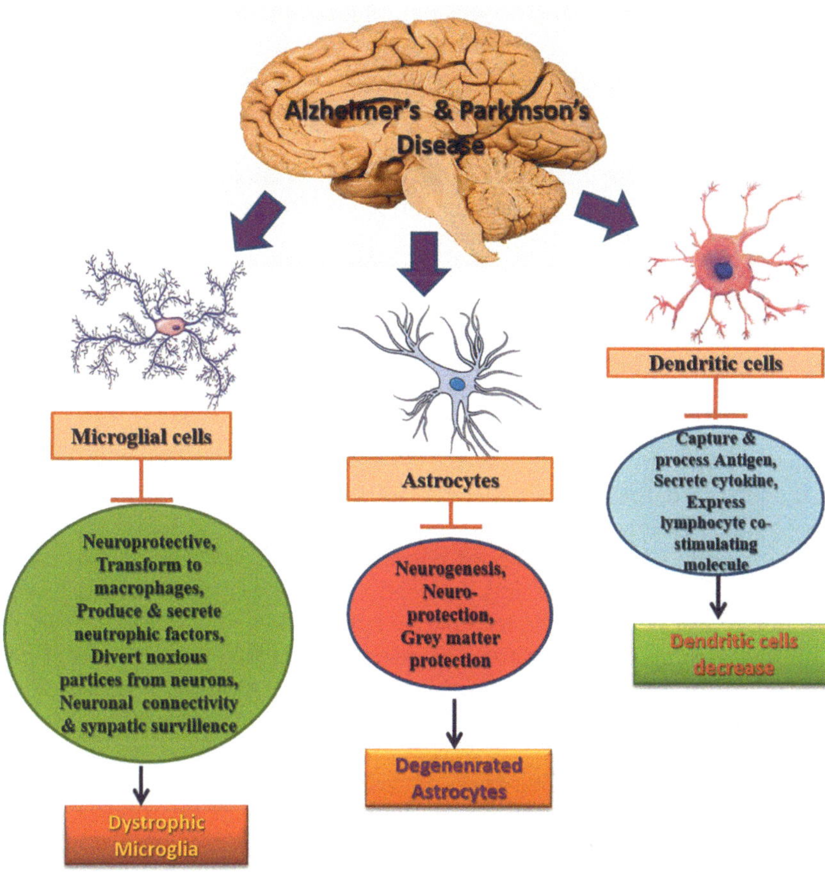

FIG. 2 Inflammation in innate immune cells is caused by neurodegenerative disorders including Alzheimer's and Parkinson's disease. Microglial cells are responsible for neuroprotection, transform to macrophages, production and secretion of neutrophic factors, diversion of noxious particles from neurons, neuronal connectivity, and synaptic surveillance. Astrocytes maintain neurogenesis, neuro-protection, and gray matter, while dendritic cells capture and process antigen, secrete cytokine, and expresses lymphocyte co-stimulating molecules. In neurodegeneration, microglial dystrophy occurs, along with degeneration of astrocytes, and the number of dendritic cells declines.

atrophic changes occur in astrocytes, leading to the formation of plaques and tangles [27]. The reactive astrocytes are involved in cytokine release, production of proinflammatory factors, and up-regulation of inducible nitric oxide synthase (iNOS) resulting in the release of ROS/NO [28]. Due to morphological changes in astrocytes (astroglial degeneration) marked by neurodegeneration, there is deposition of Aβ in brain parenchyma as well in blood vessel walls and formation of tangles from abnormal tau filaments [29]. In 1910, A. Alzheimer himself suggested the importance of glial cells.

Although the hallmark of AD pathogenesis is innate immune cells, adaptive immune cells also contribute to the disease development and progression. The loss of B and T cells contributes in the neuroinflammatory outcome of microglia while reducing the phagocytic activity, and also accelerates Aβ pathogenesis. In AD patients the number of T-lymphocytes increases, activating microglia contributing to increased expression of MHC I and II, allowing migration of T-lymphocytes. The number of B cells decreases in AD patients when compared with healthy individuals. Both B and T cells cause an immune response, as represented in Table 2.

In the PD brain, the neurodegenerative process leads to autoimmune reactions due to increased immune response. Innate immune cells involved in PD are dendritic cells (DC), microglia, and astrocytes. DC plays a key role in immunosenescence in PD; the amount of DC decreases the aged innate immune system. Altered DC function causes effects such as reduced T cell proliferation rate and increased proinflammatory cytokine. In the PD brain, the number of activated microglia increases, causing neurodegenerative effects and resulting in inflammation-induced oxidative stress. Astrocytes causing neural damage such as senescence-associated β-galactosidase (SA β-gal) and reduced survival of neurons lead to the progression of PD [38]. The toll-like receptor (a key constituent of the innate immune system) activates the adaptive immune system of PD patients, leading to the progression of the disease. Aging causes changes in the hematopoietic stem cells, leading to a decrease in the number of naive T and B cells in bone marrow, which subsequently reduces the ability to recognize new antigens. The immune cells involved in the PD progression are listed in Table 3.

2.5 Telomere shortening

The interrelation between mental health, physical health, and genetic factors can be recognized by using a novel approach by measuring telomere length (TL). Telomeres possess DNA tandem repeats of $(TTAGGG/CCCTAA)_{n,}$ which are located at the chromosome ends and also include proteins that maintain that region. The telomeres appear to have developed with a specific end goal to beat the purported "end replication issue."

This issue prevails because of three prerequisites: (1) DNA polymerase includes nucleotides just toward the direction 5′- 3′; (2) it requires RNA primers to begin the replication; and (3) the RNA primer must be expelled eventually from the concluding product. Considering these premises, there will be a reduction of DNA similar to the size of the primer at the 5′- end of the recently incorporated strand. Neurological disorders are generally related to telomere lengths of leukocytes, which results in cellular aging. Since telomeres become shorter in length after each cell division, cellular aging is directly proportional to the length of the telomere. However,

TABLE 2 Modification in T cells and B cells in Alzheimer's disease.

T cell or B cell and their products or complexes	Lymphocyte cell type	Change	Significance	Reference
CD8 + CD71 +	Total T cell	Increase	Activation in MHC class-I CD8 restricted cells, and indication of iron homeostasis	[30]
CD8 + CD28 –	Total T cell	Decrease	Decrease in the activity of suppressor	[31]
CD8 + CD28 +	Total T cell	Increase	Activation of CD8 compartment while a reducing number of cytotoxic CD8 + CD28 – cells	[32]
CD8 +	Total T cell	Significant increase	$CD8^+$ expanded clonally in CSF of AD patients	[33]
CD4 +	Total T cell	Slight increase	Elevated rate of apoptosis, immunoreactive for Aβ	[34]
CD3 +	Total T cell	increase	$CD3^+$ T-cell number increased in hippocampus, correlation with tau pathology	[35]
$CD4 + CD25^{high}$	Activated CD4 + T cells	Decrease	Due to shift toward effector memory T cells	[36]
CD19 +	Total B cell	Decrease	Total number of B cell significantly decrease	[37]

TABLE 3 Modification in T cells and B cells in Parkinson's disease.

T cell or B cell and their products or complexes	Lymphocyte cell type	Change	Significance	Reference
CD4 +	T helper cell	Decrease	These molecules are reactive against the neuronal antigens which presented on MHC I molecules	[39]
CD19 +	B cell	Decrease	Correlated with clinical severity, reduction in cell population due to apoptosis	[40]
CD4 + CD29 +	Memory T cell	Decrease	Proportion of remaining CD4 + cells expressing the CD29 + activation marker, the overall decrease in population compared with controls	[41]
CD4 + CD45RO +	Memory T cell	Decrease	More selective decrease	[42]
CD8 +	T cell	Decrease	CD8 T cells that are reactive against neuronal antigens presented on MHC-I molecules	[43]
Cytokines				
TNFα	Involved in apoptotic and inflammation	Increase	By activating brain microglia, it mediates cell death in a responsive population of DA neurons	[44]
IFN γ	Activates macrophages	Increase	By up-regulating microglial activity involved in the death of DA neurons	[45]
IL-1β	Proinflammatory	Increase	Sustained proinflammatory levels, it has a toxic effect in the substantia nigra	[46]
TGF-β1	Antiinflammatory	Increase	Generation of high amounts of Th17 cause additional inflammation due to conjunction with IL-6	[47]
IL-6	Proinflammatory	Increase	It accelerates muscle catabolism leading to poor physical performance	[48]
IL-10	Antiinflammatory	Decrease	LPS-induced loss of dopaminergic neurons is aided in protection against	[49]
IL-4	Stimulate naive T cell and active B cell	Increase	Increased concentration was found in PD pfatients	[50]

some factors influence telomere shortenings such as OS and inflammation. Inflammation increases the levels of ROS which then affects telomere length through various pathways. For example, PGE2 increases telomere shortening and decreases telomerase activity, whereas ROS damages the DNA, specifically the telomeric DNA, therefore the GGG triplet is at high risk of OS [51]. The result is cellular senescence and decreased proliferative capacity.

Variation in the TL is being studied more and more due to potential risk factors for AD, whereas studies stated previously on TL in AD represent inconsistent results. The disintegration of telomeres is an imperative mechanism for the regulation of the aging process by constraining cell proliferation. Studies evaluating TL in AD observed shorter TL frequently in association with AD development [52]. An evaluation uncovered a quickened rate of telomere shortening in AD, and also a solid impact of gender and ApoE status on TL with regards to this disease. A meta-analysis demonstrated reliable proof of shorter telomeres in AD patients and epigenomic markers related to neurodegeneration.

Shortening of TL has been involved in PD; however, with the available information, the relationship between TL and PD cannot be concluded [53]. As per the meta-analysis report on TL in PD, consistent results have not been obtained, suggesting the importance of further study on TL in AD to analyze and assess TL in a different brain region. A study suggests that neurodegeneration related to PD and telomeric and subtelomeric structural alterations in structure occur. Such structural telomere adjustments in all probability happen auxiliary to the increasing speed of aging-related telomeric changes and the quickened loss of cells bearing short telomeres. Another investigation demonstrated that in PD patients, telomere shortenings are for the most part dependent on the plasmatic concentration of carbonyl proteins, and a negative affiliation is seen between plasma carbonyl protein levels and activity of SOD. TL in PD patients and control did not differ significantly. In addition, very short TL is not observed in PD patients than controls.

3 Clinical implications

3.1 Antioxidants and antiaging agents' development as drugs

As OS is associated with neurodegenerative disease pathogenesis, antioxidant compounds are therefore being examined as therapeutic tools. In the treatment of neurodegenerative diseases, the advancement of novel medications and the synergistic impact of antioxidants should be created. A few antioxidants, which include dietary flavonoids, are utilized as antiaging agents. These substances utilize antioxidant and dietary compounds.

N-Acetylcysteine (NAC) is a precursor of glutathione, which inflects glutamate, reduces levels of inflammatory cytokines, and is engaged in oxidative homeostasis. NAC is a precursor of reduced GSH and L-cysteine; it can act as a mucolytic agent and is a tool compound. Tool compounds are biocompatible, selective small molecules which can be applied in biological system to study a process. Inside cells, NAC acts as a sulfhydryl group source and free radical scavenger by contacting ROS molecules, e.g., OH and H_2O_2. However, NAC has been used as an antidote in cases of an overdose of acetaminophen, but several proofs are associated with the antioxidant system of the human body and NO system framework in the state of toxicity, infection, inflammation, and stress [54]. Enhanced cell protection can be achieved against OS, with the help of NAC. NAC acts as a useful agent to scavenge free radicals; likewise it can manage the GSH levels inside the cell, reducing the OS effect of ROS.

The compounds that can cross the blood–brain barrier (BBB) are considered critical for the treatment of dysfunction of brain parenchyma. The capacity of NAC to cross the BBB is dependent on dose and mode of administration.

Studies have announced intravenous (jugular) and intrablood vessel (intracarotid) organization of 14C-NAC brought about the great BBB porousness. Rather than NAC, NAC-amide, a dynamic inference of NAC has been estimated in the brain after oral or intraperitoneal organization of NAC. At the point when NAC was supplanted with NAC ethyl ester, a fast increase in the brain levels of NAC and cysteine was recognized because of hydrolysis of NAC ethyl ester.

Various neural frameworks are associated with sporadic PD. Cardinal histopathology in PD includes alpha-synuclein in immunopositive Lewy bodies and Lewy neurites. Injuries at first happen in the vagal nerves and the anterior olfactory nucleus and the dorsal motor nucleus of the glossopharyngeal nerve [55]. Mitochondria-targeted nanocarriers of NAC alone and a combination of NAC and iron chelator deferoxamine (DFO) possess a precautionary impact in a cellular rotenone model of PD [56]. One continuous randomized clinical trial to assess the part of NAC as a neuroprotective agent in PD is at present enrolling members (NCT01470027).

In AD patients, clinical trials involving administering NAC blocked OS. Novel treatments with natural antioxidants and plant produce/compounds with neuroprotective properties are being used for treatment combinations. Curcumin is a polyphenolic metabolite and a dynamic part of turmeric (Curcuma longa), a dietary zest utilized as a part of Indian food and as a pharmaceutical. Curcumin has antioxidant, antiinflammatory, and antimalignancy properties. It crosses the BBB and is neuroprotective in neurological disease. Antiinflammatory medications reduce the progression and occurrence of AD. The deficient phagocytosis of Aβ prompts the downregulation of β-1,4-mannosyl-glycoprotein and 4-β-*N*-acetylglucosaminyltransferase

(MGAT3), and MGAT3 prevented transcription. Toll-like receptor (TLR)-3, TLR-5, TLR-7, TLR-8, and TLR-9, and TLR-10 transcription, in the mononuclear cell are downregulated in AD patients when compared with control.

Bisdemethoxycurcumin, a curcuminoid compound, can successfully improve the deficient phagocytosis of Aβ and also the interpretation of MGAT3 and TLRs, and the interpretation of TLR2–4. This suggests that bisdemethoxycurcumin can remedy the immune deformities in AD patients. Decreasing the Aβ accumulation in the brains represents AD pathogenesis regression, which can be achieved by administering free curcumin or its derivatives [57].

It was reported that curcumin can cross the BBB and labels senile plaques and brain amyloid angiopathy in a mouse model of AD with the help of multiphoton microscopy [58]. Curcumins are exceptionally successful to reduce OS and inflammation in the brain and a low dosage (approx. 160 ppm of curcumin administered orally for half a year) diminished the level of insoluble and dissolvable Aβ and plaque load in numerous influenced brain parts; however, the high measurement of 5000 ppm was not able to change the Aβ level. Curcumin additionally reduces phosphorylated c-jun N-terminal kinase (JNK) and insulin receptor substrate-1 (IRS-1) in the AD model animal [59]. Curcumin may potentially be a helpful therapeutic for the treatment of AD pathogenesis.

In PD, the effect happens in the substantia nigra of the ventral midbrain due to unique degenerating dopaminergic neurons, therefore depleting the dopamine levels in the striatum. The target for the treatment of PD is the improvement of dopaminergic neurons of methods for the precursor of dopamine, levodopa, or dopamine D2 receptor agonist. The neuroprotection action of curcumin with its antioxidant properties, increases striatal dopamine level, and chelates $Fe2^{+}$ in the 6-hydroxydopamine in a rat model of PD.

Thus, the protection of brain damage from mitochondria and peroxynitrite-mediated nitrosative stress is due to antioxidant activity. By activating the Wnt/β-catenin signaling pathway, curcumin can increase viability, adhesive ability, and survival, and reduce brain cell apoptosis. Increased state of Wnt3a and β-catenin mRNA and protein expressions and c-myc and cyclinD1 mRNA expressions improved SOD and GSH-Px levels, diminished MDA levels and increased the mitochondrial membrane potential. This exhibited that curcumin protected mice against OS related to the PD rodent model. It has likewise been considered that curcumin supplementation kept the LPS-induced up-regulation in the protein action of transcription factor NF-κB, proinflammatory cytokines (TNF-α, IL-1β, and IL-1α), inducible nitric oxide synthase (iNOS), and in addition the directing particles of the characteristic apoptotic pathway (Bax, Bcl-2, Caspase 3, and Caspase 9).

Curcumin attaches to the oligomeric form of α-synuclein and prohibits further α-synuclein fibrillation; in turn, it mediates the advancement of disease by agent modification in the PD therapeutics and prognosis.

Therefore, curcumin may affect mitochondrial dysfunction, oxidative stress, and mTOR and MAPK expression [60].

Several studies suggest that including dietary supplements such as omega-3 plays an essential role in controlling a chronic neurodegenerative pathology [61]. Omega-3 at low measurements shows antiinflammatory and neuroprotective properties. The neuroprotective omega-3 shields neurons from OS and inflammation, and may ensure the shield by expanding glutathione reductase action and diminishing lipid peroxide and ROS [62]. Omega-3 possesses an antiinflammatory effect and activity; it is obtained from sources containing alpha-tocopherol in very small amounts, but the efficacy of alpha-tocopherol is only viewed at higher doses. The antiinflammatory impact of omega-3 can be visualized due to resolvin D2 (obtained from omega-3), in consideration of the compound withholding inflammatory intermediatory expression in the LPS-induced PD rat model. Past investigations have indicated that resolvin D1 constrains the formation of LPS-actuated microglia inflammatory response in the form of IL-1beta, iNOS, and TNF-alpha expression. Therefore, the procedure for the impact incorporates resolvin D1, resulting in downregulation of NF-kB and succeeded by proinflammatory cytokines. Endogenous omega-3 fatty acid-derived lipids mediate to promote the inflammation resolution for the resolvin D series [63].

Vitamin E is a powerful antioxidant that may have advantageous impacts in neurodegenerative diseases in managing OS and Aβ-related free radicals [64]. Vitamin E is a gathering of eight unique compounds that comprise four types of tocopherols and four types of tocotrienols appended with alpha (α), beta (β), gamma (γ), and delta (δ) vitamins which bolster antioxidants in the body. It was identified that α-tocopherol is the most bioavailable antioxidant type of vitamin E and is useful in AD and PD. Alpha-tocopherol was found to be defensive against lipid peroxidation. Experts found that all types of tocopherols improved the Aβ generation and diminished the Aβ degradation. Tocopherols increased the levels of Aβ protein. Various clinical examinations referred to the advantages of vitamin E treatment in postponing the movement of neurodegenerative diseases. Vitamin E has likewise been discovered to be compelling in overseeing acetylcholinesterase action. It directly removes O^{2-}, quenches singlet oxygen and superoxide dismutase (SOD), and establishes an antioxidant system in the human body together with glutathione peroxidase (GSH-Px). Through the reaction with lipid oxygen radicals and lipid peroxyl free radicals, vitamin E exhibits antioxidant activity by providing the protons to interrupt the lipid peroxidation chain reaction.

Studies show that lycopene can effectively protect the damage of neurons; administration of lycopene for a rotenone-induced mouse model with Parkinson's disease could result in a significant increase in the number of dopaminergic neurons and reduced activity of oxidative stress indicators such as MDA, SOD, GSH-PX, and CAT, indicating that the damage level of oxidative stress is mitigated due to the application of lycopene [65]. Research based on the nutrition and health survey indicated a relationship between the levels of lycopene and lutein in serum and the risk of Alzheimer's disease, which indicates that the increased dietary intake of foods rich in lycopene and lutein can reduce the risk of Alzheimer's disease.

Astaxanthin is reported to have anti-aging and antiinflammatory activities by researchers. In the aging model mice induced by D-galactose, astaxanthin treatment can recover the activities of GSH-Px and SOD, enhance GSH content and reduce oxidative stress, improve pathological injury of the hippocampus, and increase the expression level of BNDF, thus achieving the anti-aging role [66]. In vitro studies have found that astaxanthin treatment can also reduce the activity of ROS and the expression of SP1 in PC12 cells induced by NR1MPP+toxi. Similarly, the expression of SP1 and NR1 under oxidative stress conditions can lead to the death of neuronal cells. This suggests that astaxanthin may have protective roles in patients with Parkinson's disease and delay the progress of this disease. Previous studies have also demonstrated that astaxanthin not only has the function of oxidation resistance but also plays an important role in antiinflammation. Rat macrophages were treated with synthetic astaxanthin and shrimp extract and it was found that astaxanthin can significantly inhibit the generation of superoxide anion free radicals such as O^{2-} and NO, and can also inhibit the secretion of proinflammatory cytokines such as TNF-α. Polyphenol quercetin present in red wine and red onion helps protect against oxidative damage by inducing SIRT1 in telomere repair modulating NADPH [67].

3.2 Other pharmacological interventions

Ghrelin, synthesized in the stomach as an inactive form, is acylated in the endoplasmic reticulum by the ghrelin *O*-acyltransferase (GOAT) enzyme to acyl ghrelin. The acyl ghrelin is delivered in the bloodstream, which can cross the blood–brain barrier (BBB) and bind to receptors of ghrelin, also called growth hormone secretagogue receptor1a (GHSR), in the brain. The metabolic effect of ghrelin is neuroprotective in PD. The neuroprotective mechanisms include reducing apoptosis and suppressing microglial activation and local inflammatory responses in the substantia nigra [68]. In addition, postprandial plasma ghrelin concentrations are lower in human PD patients.

Resveratrol (3,5,4′-trihydroxy trans-stilbene) is a natural phenol, found in red wine and the skin of red grapes. It has potential benefits in age-related diseases and aging as the molecule has been screened *in vitro* to activate SIRT1 and other related pathways. It can enhance the pathology of age-related AD. Administration of resveratrol for a long time decreases the number and level of amyloid plaques in the hippocampus and APP/PS1 mice in the medial cortex; on the other side, memory loss is prevented. Resveratrol by implication actuates SIRT1 because of its impact on the signaling of cyclic adenosine 3′,5′- monophosphate (cAMP). Resveratrol increases the level of cAMP due to competitive inhibition between cAMP-degrading phosphodiesterases, which leads to the AMPK pathway activation, finally resulting in the enhanced activity of SIRT1. Researchers have reported that amyloid burden can be decreased with resveratrol. BBB can also be protected by resveratrol by enhancing the Claudin-5 (a tight junction protein) expression and reducing matrix metalloprotein-9 (MMP-9) expression [69]. Therefore, resveratrol can be used for treating age-related diseases, specifically AD, as the molecule is potent to prevent cognitive decline and against the formation of amyloid plaque. It positively affects telomere length by regulating the telomere maintaining factor, ATP-dependent helicase, thereby maintaining the efficiency of mitochondria and OS defense against the telomere. Resveratrol regulates telomere length by activating p53, which represses the transcription of peroxisome proliferator-activated receptor-gamma coactivator-1 alpha (PGC-1α), which further modulates sirtuin (SIRT) activation [70].

3.3 Nonpharmacological interventions

Caloric restriction (CR) is being reported repeatedly as the only intercession to increase life expectancy; it is defined as reduced food intake but without causing malnutrition. The effects of CR have been compiled and documented in various studies involving a broad range of species, consisting of primates and rodents [71]. Despite the life span extending effect in humans, there is ambiguity as evidences suggest that CR enhances metabolic parameters, including (but not constrained to) fat mass, insulin sensitivity, body weight, and lipid profile [72]. In addition, age-related pathological processes, including chronic diseases such as cardiovascular disease, cancer, and diabetes and their risk factors caused by inflammation, hormone imbalance, or metabolic disorder, can likely be reversed or avoided with CR, as proved previously [73]. Among these impacts, other researchers reported that CR had a causative effect on the monoamine level in the part of the brain which is involved in the regulation of mood and reward; it simultaneously influenced the ideal model of behavior for depression anxiety [74].

CR also affects the molecular mechanism, which includes important signaling pathways significant to nutrient sensing relating to nitrosative/oxidative balance, and mitochondrial function [75]. It has been reported in various studies that not only does CR reduce prooxidant and gain antioxidant markers, but also it standardizes age-related changes in OS gene expression, markers, and related [76]. CR impacts likewise extended to phenotypical aging process expression, e.g., the thickness of carotid wall and endothelial function.

The decrease in the intake of energy, but not causing low nutrition, which is combined with the regular intake of minerals, vitamins, and important biomolecules, can result in stopping the decrease in life span due to age-related disorders and increase the life expectancy in various organisms. CR in mammals is related to different medical advantages: decrease in the levels of ROS, proinflammatory cytokines, and blood atherosclerotic lipids; fat mass decrease; improved insulin sensitivity; and reduced metabolic rate and body temperature [77].

Decreased assimilation of amyloid β (αβ) plaques and astrocytes activation can be achieved in the young AD transgenic mouse model, which acts as a subject for short-term CR that was previously studied. Long-standing CR was observed to be related to downregulating genes of the immune system and up-regulating genes related to neuroplasticity and neurogenesis. Long-standing CR intake causes a reducing effect on the assimilation of αβ peptide and accumulation of neuritic plaque in the brain of a mouse model (Tg2576 and APP/PS1) in the neuronal pathophysiology of AD post-CR dietary regimen. The Tg2576 mouse conveys the human amyloid precursor protein (APP) twofold KM670/671NL mutation and the APP/PS1 display contains the human APP KM670/671NL mutations and presenilin 1 (PS1) L166P mutation. The advantageous impacts in Tg2576 mice could be because of advancement of hostile amyloidogenic α-secretase action or by the diminished articulation of PS1 and presenilin enhancer 2, segments of the c-secretase complex. PS1 and PS2 (double knockout) mouse models were likewise tried. A study including 4-month CR administration reported enhanced memory, caspase-3 activation, lessened ventricle enlargement, decreased tau hyperphosphorylation, and astrogliosis [78]. Dietary administrations have turned out to be useful in other neurodegenerative diseases including Parkinson's disease. Dietary limitation improved the loss of dopaminergic neurons and shortages in the function of the motor in a 1-methyl-4-phenyl-1,2,3,6-tetrahydropyridine (MPTP)-prompted rat model of PD. The valuable impact of CR was discovered when 2-deoxy-D-glucose (2-DG) (nonmetabolizable glucose analog) was administered. This enhanced behavior and reduced the damage in the substantia nigra of dopaminergic neurons.

CR is the most robust mechanism to date to boost life span and promote healthy aging. The exact mechanism is still unclear, but various theories include altered stress response pathways, along with altered signaling pathways involving SIRT1, FOXO, UCP2, and AMPK along with altered homeostasis of metabolic hormones such as ghrelin and insulin. It is considered that CR induces a low level of stress to compensate for metabolism to improve intracellular mitochondrial health. It enhances metabolism and reduces neurodegeneration.

4 Implications for research

The conceptualization of severe neurodegenerative disorders has challenged the traditional view that they are brain illnesses. Gathering evidence focuses on the contribution of a dynamic myriad of pathophysiological pathways, interacting mutually, and with the environment. This implies the need for new theoretical models. The accelerated aging paradigm expands the previous hypothesis to explain the role of systemic changes in neurons and offers new insights when including neurodegeneration in a set of multisystemic changes as a relevant piece in the puzzle of the neurobiology of severe mental disorders. Finally, the recognition of aging processes as part of a neurodegenerative disorder leads to the possibility of treatments targeting specific pathways—the so-called anti-aging strategies, including nutritional approaches and pharmacological and nonpharmacological interventions.

Acknowledgment

This study was supported by a Fellowship from the Council of Scientific and Industrial Research (CSIR), the Government of India to S.Y. (Reference No: 09/1152(0013)/2019-EMR-I), A.K. (Reference No: 09/1152(0016)/2019-EMR-I) and from the University Grant Commission to Deepika (Reference No: 201610000784).

Conflicts of interest/competing interests

The authors report no conflicts of interest.

References

[1] A.J.P.O. de Almeida, M.S. de Almeida Rezende, S.H. Dantas, S. de Lima Silva, J.C.P.L. de Oliveira, F. de Lourdes Assunção Araújo de Azevedo, R.M.F.R. Alves, G.M.S. de Menezes, P.F. dos Santos, T.A.F. Gonçalves, V.B. Schini-Kerth, I.A. de Medeiros, Unveiling the role of inflammation and oxidative stress on age-related cardiovascular diseases, Oxidative Medicine and Cellular Longevity 2020 (2020) 1954398.

[2] M. Ganguli, J.C. Beer, J.M. Zmuda, C.M. Ryan, K.J. Sullivan, C.-C.H. Chang, R.H. Rao, Aging, diabetes, obesity, and cognitive decline: a population-based study, J. Am. Geriatr. Soc. 68 (5) (2020) 991–998.

[3] M. Edde, B. Dilharreguy, G. Theaud, S. Chanraud, C. Helmer, J.-F. Dartigues, H. Amieva, M. Allard, M. Descoteaux, G. Catheline, Age-related change in episodic memory: role of functional and structural connectivity between the ventral posterior cingulate and the parietal cortex, Brain Struct. Funct. 225 (7) (2020) 2203–2218.

[4] N. Betrouni, R. Lopes, L. Defebvre, A.F.G. Leentjens, K. Dujardin, Texture features of magnetic resonance images: a marker of slight cognitive deficits in Parkinson's disease, Mov. Disord. 35 (3) (2020) 486–494.

[5] X. Jia, Z. Wang, T. Yang, Y. Li, S. Gao, G. Wu, T. Jiang, P. Liang, Entorhinal cortex atrophy in early, drug-naive Parkinson's disease with mild cognitive impairment, Aging Dis. 10 (6) (2019) 1221–1232.

[6] Y. Chen, G. Zhu, D. Liu, Y. Liu, T. Yuan, X. Zhang, Y. Jiang, T. Du, J. Zhang, The morphology of thalamic subnuclei in Parkinson's disease and the effects of machine learning on disease diagnosis and clinical evaluation, J. Neurol. Sci. 411 (2020), 116721.

[7] G. Donzuso, R. Monastero, C.E. Cicero, A. Luca, G. Mostile, L. Giuliano, R. Baschi, M. Caccamo, C. Gagliardo, S. Palmucci, M. Zappia, A. Nicoletti, Neuroanatomical changes in early Parkinson's disease with mild cognitive impairment: a VBM study; the Parkinson's disease cognitive impairment study (PaCoS), Neurol. Sci. (2021).

[8] X. Xu, Q. Han, J. Lin, L. Wang, F. Wu, H. Shang, Grey matter abnormalities in Parkinson's disease: a voxel-wise meta-analysis, Eur. J. Neurol. 27 (4) (2020) 653–659.

[9] M. Malek-Ahmadi, K. Chen, S.E. Perez, E.J. Mufson, Cerebral amyloid Angiopathy and Neuritic plaque pathology correlate with cognitive decline in elderly nondemented individuals, J. Alzheimers Dis. 67 (2019) 411–422.

[10] K. Ritchie, I. Carrière, S. Gregory, T. Watermeyer, S. Danso, L. Su, C.W. Ritchie, J.T. Brien, Trauma and depressive symptomatology in middle-aged persons at high risk of dementia: the PREVENT dementia study, Journal of neurology Neurosurgery & Psychiatry 92 (1) (2021) 16.

[11] I.G. McKeith, T.J. Ferman, A.J. Thomas, F. Blanc, B.F. Boeve, H. Fujishiro, K. Kantarci, C. Muscio, J.T. O'Brien, R.B. Postuma, D. Aarsland, C. Ballard, L. Bonanni, P. Donaghy, M. Emre, J.E. Galvin, D. Galasko, J.G. Goldman, S.N. Gomperts, L.S. Honig, M. Ikeda, J.B. Leverenz, S.J.G. Lewis, K.S. Marder, M. Masellis, D.P. Salmon, J.P. Taylor, D.W. Tsuang, Z. Walker, P. Tiraboschi, Research criteria for the diagnosis of prodromal dementia with Lewy bodies, Neurology 94 (17) (2020) 743–755.

[12] P. Kanel, M.L.T.M. Müller, N.I. Bohnen, Parkinsonian dementias: PET findings, in: R.A.J.O. Dierckx, A. Otte, E.F.J. de Vries, A. van Waarde, K.L. Leenders (Eds.), PET and SPECT in Neurology, Springer International Publishing, Cham, 2021, pp. 491–513.

[13] R. Wadhwa, R. Gupta, P.K. Maurya, Oxidative stress and accelerated aging in neurodegenerative and neuropsychiatric disorder, Curr. Pharm. Des. 24 (40) (2018) 4711–4725.

[14] T. Aggarwal, R. Wadhwa, N. Thapliyal, K. Sharma, V. Rani, P.K. Maurya, Oxidative, inflammatory, genetic, and epigenetic biomarkers associated with chronic obstructive pulmonary disorder, J. Cell. Physiol. 234 (3) (2019) 2067–2082.

[15] A. Singh, R. Kukreti, L. Saso, S. Kukreti, Oxidative stress: a key modulator in neurodegenerative diseases, Molecules 24 (8) (2019) 1583.

[16] A. Rizor, E. Pajarillo, J. Johnson, M. Aschner, E. Lee, Astrocytic oxidative/nitrosative stress contributes to Parkinson's disease pathogenesis: the dual role of reactive astrocytes, Antioxidants 8 (8) (2019) 265.

[17] G. Bjørklund, M. Dadar, G. Anderson, S. Chirumbolo, M. Maes, Preventive treatments to slow substantia nigra damage and Parkinson's disease progression: a critical perspective review, Pharmacol. Res. 161 (2020), 105065.

[18] V.S. Van Laar, J. Chen, A.D. Zharikov, Q. Bai, R. Di Maio, A.A. Dukes, T.G. Hastings, S.C. Watkins, J.T. Greenamyre, C.M. St Croix, E.A. Burton, α-Synuclein amplifies cytoplasmic peroxide flux and oxidative stress provoked by mitochondrial inhibitors in CNS dopaminergic neurons in vivo, Redox Biol. 37 (2020), 101695.

[19] C. Peña-Bautista, C. Vigor, J.-M. Galano, C. Oger, T. Durand, I. Ferrer, A. Cuevas, R. López-Cuevas, M. Baquero, M. López-Nogueroles, M. Vento, D. Hervás, A. García-Blanco, C. Cháfer-Pericás, Plasma lipid peroxidation biomarkers for early and non-invasive Alzheimer disease detection, Free Radic. Biol. Med. 124 (2018) 388–394.

[20] A. Tramutola, C. Lanzillotta, M. Perluigi, D.A. Butterfield, Oxidative stress, protein modification and Alzheimer disease, Brain Res. Bull. 133 (2017) 88–96.

[21] C. Guo, P. Ding, C. Xie, C. Ye, M. Ye, C. Pan, X. Cao, S. Zhang, S. Zheng, Potential application of the oxidative nucleic acid damage biomarkers in detection of diseases, Oncotarget 8 (43) (2017) 75767–75777.

[22] L. Miguélez, I. Martinez de Toda, M. De la Fuente, P-178—oxidative stress in blood cells of men and women with Alzheimer's and Parkinson's diseases, Free Radic. Biol. Med. 120 (2018) S98.
[23] G. Bivona, B. Lo Sasso, G. Iacolino, C.M. Gambino, C. Scazzone, L. Agnello, M. Ciaccio, Standardized measurement of circulating vitamin D [25(OH)D] and its putative role as a serum biomarker in Alzheimer's disease and Parkinson's disease, Clin. Chim. Acta 497 (2019) 82–87.
[24] P.K. Maurya, L.B. Rizzo, Neuroprogression and accelerated ageing in severe psychiatric disorders, in: Neuroprogression in Psychiatry, Oxford Medicine Online, 2019, p. 21.
[25] A.M. Ibrahim, F.H. Pottoo, E.S. Dahiya, F.A. Khan, J.B.S. Kumar, Neuron-glia interactions: molecular basis of Alzheimer's disease and applications of neuroproteomics, Eur. J. Neurosci. 52 (2) (2020) 2931–2943.
[26] C.S. Subhramanyam, C. Wang, Q. Hu, S.T. Dheen, Microglia-mediated neuroinflammation in neurodegenerative diseases, Semin. Cell Dev. Biol. 94 (2019) 112–120.
[27] A. Verkhratsky, J.J. Rodrigues, A. Pivoriunas, R. Zorec, A. Semyanov, Astroglial atrophy in Alzheimer's disease, Pflügers Arch. 471 (10) (2019) 1247–1261.
[28] H. Chun, H. Im, Y.J. Kang, Y. Kim, J.H. Shin, W. Won, J. Lim, Y. Ju, Y.M. Park, S. Kim, S.E. Lee, J. Lee, J. Woo, Y. Hwang, H. Cho, S. Jo, J.-H. Park, D. Kim, D.Y. Kim, J.-S. Seo, B.J. Gwag, Y.S. Kim, K.D. Park, B.-K. Kaang, H. Cho, H. Ryu, C.J. Lee, Severe reactive astrocytes precipitate pathological hallmarks of Alzheimer's disease via H2O2− production, Nat. Neurosci. 23 (12) (2020) 1555–1566.
[29] G.G. Kovacs, Astroglia and tau: new perspectives, Front. Aging Neurosci. 12 (2020) 96.
[30] K.J. Bryson, M.A. Lynch, Linking T cells to Alzheimer's disease: from neurodegeneration to neurorepair, Curr. Opin. Pharmacol. 26 (2016) 67–73.
[31] U. Kawai, N. Uchida, M. Oono, M. Fujita-Nakata, M. Nakanishi, M. Sanada, S. Nagayama, M. Matsui, Relationship of systemic inflammation and cellular immunity with advancement of cognitive decline in patients with Alzheimer's disease, Clin. Exp. Neuroimmunol. 9 (3) (2018) 182–188.
[32] L. Speciale, E. Calabrese, M. Saresella, C. Tinelli, C. Mariani, L. Sanvito, R. Longhi, P. Ferrante, Lymphocyte subset patterns and cytokine production in Alzheimer's disease patients, Neurobiol. Aging 28 (8) (2007) 1163–1169.
[33] D. Gate, N. Saligrama, O. Leventhal, A.C. Yang, M.S. Unger, J. Middeldorp, K. Chen, B. Lehallier, D. Channappa, M.B. De Los Santos, A. McBride, J. Pluvinage, F. Elahi, G.K.-Y. Tam, Y. Kim, M. Greicius, A.D. Wagner, L. Aigner, D.R. Galasko, M.M. Davis, T. Wyss-Coray, Clonally expanded CD8 T cells patrol the cerebrospinal fluid in Alzheimer's disease, Nature 577 (7790) (2020) 399–404.
[34] H. Solleiro-Villavicencio, S. Rivas-Arancibia, Effect of chronic oxidative stress on neuroinflammatory response mediated by CD4+T cells in neurodegenerative diseases, Front. Cell. Neurosci. 12 (114) (2018).
[35] M. Merlini, T. Kirabali, L. Kulic, R.M. Nitsch, M.T. Ferretti, Extravascular CD3+ T cells in brains of Alzheimer disease patients correlate with tau but not with amyloid pathology: an immunohistochemical study, Neurodegener. Dis. 18 (1) (2018) 49–56.
[36] C. D'Angelo, D. Goldeck, G. Pawelec, L. Gaspari, A. Di Iorio, R. Paganelli, Exploratory study on immune phenotypes in Alzheimer's disease and vascular dementia, Eur. J. Neurol. 27 (10) (2020) 1887–1894.
[37] J. Amin, D. Boche, Z. Clough, J. Teeling, A. Williams, Y. Gao, L. Chudley, L. Lau, F. Smith, S. Harris, C. Holmes, Peripheral immunophenotype in dementia with Lewy bodies and Alzheimer's disease: an observational clinical study, J. Neurol. Neurosurg. Psychiatry 91 (11) (2020) 1219–1226.
[38] M.-L. Xia, X.-H. Xie, J.-H. Ding, R.-H. Du, G. Hu, Astragaloside IV inhibits astrocyte senescence: implication in Parkinson's disease, J. Neuroinflammation 17 (1) (2020) 105.

[39] H. González, F. Contreras, R. Pacheco, Regulation of the neurodegenerative process associated to Parkinson's disease by CD4+ T-cells, J. NeuroImmune Pharmacol. 10 (4) (2015) 561–575.

[40] H. Kobo, A. Bar-Shira, D. Dahary, Z. Gan-Or, A. Mirelman, O. Goldstein, N. Giladi, A. Orr-Urtreger, Down-regulation of B cell-related genes in peripheral blood leukocytes of Parkinson's disease patients with and without GBA mutations, Mol. Genet. Metab. 117 (2) (2016) 179–185.

[41] C.H. Stevens, D. Rowe, M.-C. Morel-Kopp, C. Orr, T. Russell, M. Ranola, C. Ward, G.M. Halliday, Reduced T helper and B lymphocytes in Parkinson's disease, J. Neuroimmunol. 252 (1) (2012) 95–99.

[42] J.A.H. Saunders, K.A. Estes, L.M. Kosloski, H.E. Allen, K.M. Dempsey, D.R. Torres-Russotto, J.L. Meza, P.M. Santamaria, J.M. Bertoni, D.L. Murman, H.H. Ali, D.G. Standaert, R.L. Mosley, H.E. Gendelman, CD4+ regulatory and effector/memory T cell subsets profile motor dysfunction in Parkinson's disease, J. NeuroImmune Pharmacol. 7 (4) (2012) 927–938.

[43] G.T. Kannarkat, J.M. Boss, M.G. Tansey, The role of innate and adaptive immunity in Parkinson's disease, J. Parkinsons Dis. 3 (4) (2013) 493–514.

[44] H. Sasidharakurup, L. Nair, K. Bhaskar, S. Diwakar, in: Computational modelling of TNFα pathway in Parkinson's disease—a systemic perspective, International Conference on Complex Networks and Their Applications, Springer International Publishing, Cham, 2020, pp. 762–773.

[45] D. Rathnayake, T. Chang, P. Udagama, Selected serum cytokines and nitric oxide as potential multi-marker biosignature panels for Parkinson disease of varying durations: a case-control study, BMC Neurol. 19 (1) (2019) 56.

[46] M.C. Leal, J.C. Casabona, M. Puntel, F.J. Pitossi, Interleukin-1β and tumor necrosis factor-α: reliable targets for protective therapies in Parkinson's disease? Front. Cell. Neurosci. 7 (2013).

[47] L.P. Diniz, I. Matias, A.P.B. Araujo, M.N. Garcia, F.G.Q. Barros-Aragão, S.V. Alves-Leon, J.M. de Souza, D. Foguel, C.P. Figueiredo, C. Braga, L. Romão, F.C.A. Gomes, α-Synuclein oligomers enhance astrocyte-induced synapse formation through TGF-β1 signaling in a Parkinson's disease model, J. Neurochem. 150 (2) (2019) 138–157.

[48] B. Veselý, M. Dufek, V. Thon, M. Brozman, S. Királová, T. Halászová, E. Koriťáková, I. Rektor, Interleukin 6 and complement serum level study in Parkinson's disease, J. Neural Transm. 125 (5) (2018) 875–881.

[49] R. Kim, H.-J. Kim, A. Kim, M. Jang, A. Kim, Y. Kim, D. Yoo, J.H. Im, J.-H. Choi, B. Jeon, Peripheral blood inflammatory markers in early Parkinson's disease, J. Clin. Neurosci. 58 (2018) 30–33.

[50] B. Brodacki, J. Staszewski, B. Toczyłowska, E. Kozłowska, N. Drela, M. Chalimoniuk, A. Stępien, Serum interleukin (IL-2, IL-10, IL-6, IL-4), TNFα, and INFγ concentrations are elevated in patients with atypical and idiopathic parkinsonism, Neurosci. Lett. 441 (2) (2008) 158–162.

[51] S. Bielskutė, J. Plavec, P. Podbevšek, Impact of oxidative lesions on the human telomeric G-quadruplex, J. Am. Chem. Soc. 141 (6) (2019) 2594–2603.

[52] S.-H. Koh, S.H. Choi, J.H. Jeong, J.-W. Jang, K.W. Park, E.-J. Kim, H.J. Kim, J.Y. Hong, S.J. Yoon, B. Yoon, J.-H. Kang, J.-M. Lee, H.-H. Park, J. Ha, Y.J. Suh, S. Kang, Telomere shortening reflecting physical aging is associated with cognitive decline and dementia conversion in mild cognitive impairment due to Alzheimer's disease, Aging 12 (5) (2020) 4407–4423.

[53] T. Levstek, S. Redenšek, M. Trošt, V. Dolžan, K.T. Podkrajšek, Assessment of the telomere length and its effect on the symptomatology of Parkinson's disease, Antioxidants 10 (1) (2021) 137.

[54] R. Eshrati, M. Jafari, S. Gudarzi, A. Nazari, E. Samizadeh, M.G. Hesami, Comparison of ameliorative effects of Taraxacum syriacum and N-acetylcysteine against acetaminophen-induced oxidative stress in rat liver and kidney, J. Biochem. (2020).

[55] S. Sasaki, Y. Horie, Association between olfactory impairment and disease severity and duration in Parkinson's disease, Mov. Disord. Clin. Pract. 7 (7) (2020) 820–826.
[56] L. Mursaleen, B. Noble, S.H.Y. Chan, S. Somavarapu, M.G. Zariwala, N-acetylcysteine Nanocarriers protect against oxidative stress in a cellular model of Parkinson's disease, Antioxidants 9 (7) (2020) 600.
[57] E. Chainoglou, D. Hadjipavlou-Litina, Curcumin in health and diseases: Alzheimer's disease and curcumin analogues, derivatives, and hybrids, Int. J. Mol. Sci. 21 (6) (2020) 1975.
[58] B. Salehi, D. Calina, A.O. Docea, N. Koirala, S. Aryal, D. Lombardo, L. Pasqua, Y. Taheri, C.M.S. Castillo, M. Martorell, N. Martins, M. Iriti, H.A.R. Suleria, J. Sharifi-Rad, Curcumin's nanomedicine formulations for therapeutic application in neurological diseases, J. Clin. Med. 9 (2) (2020) 430.
[59] T.K. Das, S.K. Chakrabarti, I.N. Zulkipli, M.R.W. Abdul Hamid, Curcumin ameliorates the impaired insulin signaling involved in the pathogenesis of Alzheimer's disease in rats, J. Alzheimer's Dis. 3 (2019) 59–70.
[60] Z. Wang, C. Yang, J. Liu, B. Chun-Kit Tong, Z. Zhu, S. Malampati, S. Gopalkrishnashetty Sreenivasmurthy, K.-H. Cheung, A. Iyaswamy, C. Su, J. Lu, J. Song, M. Li, A curcumin derivative activates TFEB and protects against parkinsonian neurotoxicity in vitro, Int. J. Mol. Sci. 21 (4) (2020) 1515.
[61] J. Giacobbe, B. Benoiton, P. Zunszain, C.M. Pariante, A. Borsini, The Antiinflammatory Role of Omega-3 Polyunsaturated Fatty Acids Metabolites in Pre-Clinical Models of Psychiatric, Neurodegenerative, and Neurological Disorders, Front. Psychiatry 11 (122) (2020).
[62] J. Shi, W. Wang, G. Sang, H. Xi, Y. Sun, C. Lu, H. Ye, L. Huang, Short term usage of omega-3 polyunsaturated fatty acids ameliorate lipopolysaccharide-induced inflammatory response and oxidative stress in the neonatal rat hippocampal tissue, Front. Nutr. 7 (238) (2020).
[63] T. Ishihara, M. Yoshida, M. Arita, Omega-3 fatty acid-derived mediators that control inflammation and tissue homeostasis, Int. Immunol. 31 (9) (2019) 559–567.
[64] E. Singh, G. Devasahayam, Neurodegeneration by oxidative stress: a review on prospective use of small molecules for neuroprotection, Mol. Biol. Rep. 47 (4) (2020) 3133–3140.
[65] H.-B. Man, W.-P. Bi, Protective effect of lycopene in a mouse model of Parkinson's disease via reducing oxidative stress and apoptosis, analytical quantitative cytopathology, Histopathology 40 (2018) 253–258.
[66] V. Sorrenti, S. Davinelli, G. Scapagnini, B.J. Willcox, R.C. Allsopp, D.C. Willcox, Astaxanthin as a putative gerhoprotector: molecular basis and focus on brain aging, Mar. Drugs 18 (7) (2020) 351.
[67] C. Iside, M. Scafuro, A. Nebbioso, L. Altucci, SIRT1 activation by natural phytochemicals: an overview, Front. Pharmacol. 11 (1225) (2020).
[68] H. Wang, S. Dou, J. Zhu, Z. Shao, C. Wang, B. Cheng, Ghrelin protects dopaminergic neurons against MPTP neurotoxicity through promoting autophagy and inhibiting endoplasmic reticulum mediated apoptosis, Brain Res. 1746 (2020), 147023.
[69] B. Vairappan, M. Sundhar, B.H. Srinivas, Resveratrol restores neuronal tight junction proteins through correction of Ammonia and inflammation in CCl4-induced cirrhotic mice, Mol. Neurobiol. 56 (7) (2019) 4718–4729.
[70] S.D. Gutlapalli, V. Kondapaneni, I.A. Toulassi, S. Poudel, M. Zeb, J. Choudhari, I. Cancarevic, The effects of resveratrol on telomeres and post myocardial infarction remodeling, Cureus 12 (11) (2020) 11482.
[71] F. Pifferi, F. Aujard, Caloric restriction, longevity and aging: recent contributions from human and non-human primate studies, Prog. Neuro-Psychopharmacol. Biol. Psychiatry 95 (2019), 109702.

[72] S. Moon, J. Kang, S.H. Kim, H.S. Chung, Y.J. Kim, J.M. Yu, S.T. Cho, C.-M. Oh, T. Kim, Beneficial effects of time-restricted eating on metabolic diseases: a systemic review and meta-analysis, Nutrients 12 (5) (2020) 1267.
[73] F. Madeo, D. Carmona-Gutierrez, S.J. Hofer, G. Kroemer, Caloric restriction mimetics against age-associated disease: targets, mechanisms, and therapeutic potential, Cell Metab. 29 (3) (2019) 592–610.
[74] M. Portero-Tresserra, D. Rojic-Becker, C. Vega-Carbajal, G. Guillazo-Blanch, A. Vale-Martínez, M. Martí-Nicolovius, Caloric restriction modulates the monoaminergic system and metabolic hormones in aged rats, Sci. Rep. 10 (1) (2020) 19299.
[75] C. Carvalho, P.I. Moreira, Cognitive impairment in obesity and diabetes, in: J. Faintuch, S. Faintuch (Eds.), Obesity and Diabetes: Scientific Advances and Best Practice, Springer International Publishing, Cham, 2020, pp. 399–414.
[76] C. Tonini, M. Segatto, F. Martino, L. Cigliano, M. Nazzaro, L. Barberio, M. Mandalà, V. Pallottini, Effects of late-life caloric restriction on age-related alterations in the rat cortex and Hippocampus, Nutrients 13 (1) (2021) 232.
[77] K.A. Escobar, N.H. Cole, C.M. Mermier, T.A. VanDusseldorp, Autophagy and aging: maintaining the proteome through exercise and caloric restriction, Aging Cell 18 (1) (2019), e12876.
[78] J. Most, L.M. Redman, Impact of calorie restriction on energy metabolism in humans, Exp. Gerontol. 133 (2020), 110875.

CHAPTER

6

The role of oxidative stress in kidney diseases

Vikas Saini

Biomedical Sciences Program, Department of Vocational Studies & Skill Development, Central University of Haryana, Mahendergarh, India

1 Introduction

Kidney diseases are becoming a major global health burden with more than 850 million people currently estimated to be suffering from some form of kidney disease [1]. This number is double that of diabetic patients (422 million, [2]), 20 times that of cancer patients (42 million [3]), and 23 times that of people suffering with AIDS (36.7 million [4]). Hence, kidney diseases have become one of the commonly occurring diseases, globally.

The term "kidney diseases" characterizes the condition in which an individual kidney, due to injury or a pathological condition, is unable to filter the blood properly. This state causes buildup of waste in the blood that further leads to other health problems, such as heart disease. There are more than 200 conditions identified for kidney diseases, which can broadly be grouped into acute kidney diseases and disorders (AKD) and chronic kidney disease (CKD) [5].

Acute kidney diseases and disorders are described as the conditions in which slow deterioration of kidney functions is observed due to irreversible loss of nephron and kidney cells. This kidney dysfunction, when it persists for more than 3 months, leads to chronic kidney disease (CKD) [6].

Involvement of reactive oxygen species (ROS) has been found in almost all the lifestyle diseases, and kidney diseases are no different. Several research studies have revealed that elevated intracellular levels of ROS lead to the pathogenesis of kidney disease [7,8]. In this chapter, we will focus on the studies indicating the involvement of reactive oxygen species in the progression of kidney diseases.

https://doi.org/10.1016/B978-0-323-90905-1.00006-7

2 Kidney architecture at the cellular level

The kidney is mainly responsible for functions:

(i) maintenance of electrolyte homeostasis by filtering the blood plasma;
(ii) elimination of metabolic products; and.
(iii) production and secretion of rennin and erythropoietin hormone.

The filtration of blood plasma to form the concentrate of waste product is done by dedicated structures called nephrons [9]. The nephron is composed of renal corpuscle, juxtaglomerular apparatus, and tubular segment. Blood is received via the afferent arteriole in the renal corpuscle, where the blood is filtered. The tubular segment is dedicated to the reabsorption and secretion of electrolytes and other biological molecules to form the final concentrate to be released into the collecting tubule to form urine. Each kidney has around 200,000–1.8 million nephrons and the formation of nephrons is completed by 36 weeks of the development of an infant. The number of nephrons in each individual depends on factors like birth weight, adult height, body surface area, glomerular volume, and ethnic groups [10].

Based on its structure and function, the tubular segment of nephron is further divided into the proximal convoluted tubule (PCT), loop of Henle, distal convoluted tubule (DCT), and collecting duct [11]. The primary function of PCT is to reabsorb sodium ion (Na^+) and H_2O along with important biomolecules like glucose, phosphate, lactate, amino acids, and chloride and bicarbonate ions [12]. The loop of Henle follows the PCT in the tubule and plays a role in reabsorption of calcium and magnesium ion and maintaining significant osmolality in the renal medullary interstitium [13]. The high gradient of osmolality in the region provides the downstream collecting duct to form the concentrated urine by the process of osmosis. The DCT maintains the acid–base balance and brings small changes in the concentration of phosphate, sodium, potassium, and calcium ions [14]. The fluid from DCT goes to the collecting duct where sodium ions are reabsorbed by the action of aldosterone followed by water reabsorption to form hypertonic urine [11].

The renal corpuscle is composed of the glomerulus, Bowman's capsule, glomerular basement membrane, podocytes, intraglomerular mesangial cells, and filtration slits. Associated with the renal corpuscle is the juxtaglomerular apparatus, which is composed of the macula densa, juxtaglomerular cells, mesangium, and extraglomerular mesangial cells [15]. Juxtaglomerular (JG) cells are specialized smooth muscle cells present in the walls of afferent arterioles and are responsible for the synthesis of enzyme renin. JG cells in response to two conditions secrete renin [16]: when stretch receptors or baroreceptors in the vascular walls sense a drop in afferent arteriole perfusion pressure; and when macula densa cells detect a decrease in sodium and chloride concentration in tubular fluid and stimulate juxtaglomerular cells to release renin. Extraglomerular mesangial cells also play an important role in the regulation of systemic blood pressure through the renin-angiotensin system (RAAS) [16,17].

The glomerulus consists of a network of tiny capillaries called a tuft. It is located at the starting point of the nephron (Bowman's capsule) and forms a junction for filtration of blood plasma. The space between these blood capillaries is called the mesangium and is filled with cells known as mesangial cells [17]. The mesangial cells present inside and outside of the Bowman's capsule are called intraglomerular and extraglomerular mesangial cells, respectively. The extraglomerular mesangial cells are one of the components of the juxtaglomerular complex. Intraglomerular mesangial cells are macrophages derived from smooth muscle cells rather than from monocytes. Their function includes phagocytosis of immunoglobulins, structural support to the tuft, regulation of glomerular filtration rate, formation of mesangial matrix, synthesis of the extracellular matrix (ECM), and monitoring capillary lumen glucose concentration [17,18]. Intraglomerular mesangial cells regulate the glomerular filtration rate by their ability to contract. Underlying these cells is a basement membrane made up of endothelium cells present in glomerular capillaries. Contraction of mesangial cells is coupled with contraction of the basement membrane of the endothelium, causing shrinkage in the surface area of the basement membrane, leading to decreased glomerular filtration rate [18]. Specialized cells called podocytes and their foot extensions surround the blood capillaries of the glomerulus. Foot extensions of the podocytes are called pedicels. These pedicels have gaps or slits between them [19]. The blood is filtered through these slits, which are known as filtration slits. The size of these slits forms the basis of filtration; slits being too small for the protein molecules, red blood cells (RBCs), etc. do not allow them to move into PCT whereas small molecules, amino acids, electrolytes, and water etc. pass through the slits toward the proximal convoluted tubule [18]. In total, the following components of the renal corpuscles act as a filter to carry out filtration of the blood: (i) glomerular basement membrane; (ii) endothelial layer of glomerulus; (iii) mesangial cells; and (iv) podocytes or filtration slit. Injury to or death of these cells results in loss of kidney function and can lead to kidney failure. Over the last 30 years, major advances have been made to understand the causes and mechanism of injury in these layers and nephrons.

3 Nephrotoxicity and renal pathology

The nephron is at the risk of injury due to continuous exposure to the toxic metabolites, therapeutic drugs, environmental pollutants, and changes in hemodynamics, inflammation, or obstruction of renal excretion. When the chemical in the form of therapeutic drugs, toxic metabolites, or environmental pollutants adversely affects the kidney function, it is known as nephrotoxicity [20], whereas when kidney damage is diagnosed to be induced by pathological events such as hypertension, diabetes, obesity, sepsis, or liver failure, it is referred to as renal pathology [21]. However, fundamental to both these conditions are structural changes in the nephron and kidney cell death. In addition, the mechanism of kidney cell death in both conditions is similar. For instance, acute kidney injury induced by both ischemia and cisplatin (cancer chemotherapeutic drug) involves oxidative stress, PCT cell death, loss of brush border membrane, and polarity [22,23]. Nephrotoxicity is caused by accumulation of xenobiotics or toxic metabolites bio-transformed in the liver and intrarenal bioactivation of xenobiotics to reactive metabolites.

3.1 Source of nephrotoxicity

The number of serious adverse effects on kidney is growing faster; as a result, interest in nephrotoxicity and its sources has grown rapidly in the last decade. Mentioned below are the sources of therapeutic agents, diagnostic agents, environmental toxicants, or other health products that have been found to cause nephrotoxicity [24–26].

3.1.1 Therapeutic agents

Therapeutic agents causing nephrotoxicity can be classified into chemotherapeutics, antimicrobial, analgesics, immunosuppressive, and miscellaneous.

Cancer therapeutics: platins, mitomycin, gemcitabine, methotrexate, pentostatin, interleukin 2, ifosfamide, crizotinib, clofarabine, pemetrexed, carfilozomib, ipilimumab, and antiangiogenesis agents (anti-VEGF Ab, sVEGFR, tyrosine kinase inhibitors).

Antimicrobial agents: aminoglycosides, antiviral agents, amphotericin B, colistin, polymyxin B, sulfadiazide, quinolones, vancomycin.

Analgesics: NSAIDs, phenacetin, selective COX-2 inhibitors.

Immunosuppressive: calcineurin inhibitors, sirolimus, everolimus.

Miscellaneous: ACE inhibitors, mesalamine, topiramate, methoxyflurane, sucrose, pamidronate, statins, SGLT-2 inhibitors.

3.1.2 Environmental pollutants/toxins

Heavy metals: uranium, copper, bismuth mercury, cadmium, arsenic, lead.

Solvents: hydrocarbons.

Others: silicon, germanium, trichloroethylene, bromate, brominated-flame retardants, ethylene glycol, diglycolic acid.

3.1.3 *Diagnostic agents*

Radiocontrast: high and low osmolar, isoosmolar.
Other agents: gadolinium (high concentration), oral NaP solution.

3.1.4 *Alternative/health products*

Herbal remedies: mycotoxins (ochratoxin, fumonisin B1, and citrinin), aristolochic acids, glycyrrhiza, *Ephedra* sp., *Taxus celebica*, uno degatta, *Datura* sp.
Adulterants: dichromate, phenybutazone, melamine, mefenamic acid.

4 Classification of kidney diseases

Kidney diseases can be classified based on pathological symptoms or clinical manifestations caused by different agents. Broadly, we can divide them into acute kidney injury and chronic kidney disease [20].

4.1 Acute renal injury (ARI)

Acute renal injury can be observed in different regions of the kidney. Based on the injury pattern and location, there can be multiple type of acute renal injury.

(i) **Acute tubular necrosis (ATN):** The most common cause of acute renal injury is acute tubule necrosis. It occurs due to ischemia, toxin exposure, or sepsis. Here, the term tubular necrosis is a misnomer as the injury is not limited only to tubules.
(ii) **Acute interstitial nephritis (AIN):** This is characterized by inflammation in the kidney interstitium. It produces lesion in the kidney and causes a decline in renal function. It is an allergic response toward inflammatory infiltrate.
(iii) **Glomerulonephritis:** Injury in the glomeruli is called glomerulonephritis. It can be acute or chronic.
(iv) **Tubular injury:** This is the condition in which injury occurs directly to epithelial cells of tubules. It is restricted only to the tubules and no inflammation is involved in it.

4.2 Chronic kidney disease (CKD)

Any individual can develop CKD. However, people with diabetes, high blood pressure (hypertension), or heart disease, those who are

African-American, Hispanic, Native American, or Asian, the elderly (older than 60 years), and those who have kidney disease in the family are more prone to having CKD. Vice versa, individuals with CKD are also prone to developing heart disease, anemia, bone disease, and high levels of potassium and calcium [10].

Based on the functioning of kidneys to filter the blood, clinically, CKD has been divided into five stages from mild condition in stage 1 to complete renal failure in stage 5. The characterization of renal damage and its staging remains the same regardless of whether the injury is caused by nephrotoxicity or renal pathology [6].

The rate at which blood is filtered through the glomeruli is used to measure the overall kidney function. This is called the glomerular filtration rate. The estimated glomerular filtration rate (eGFR) in stage 1 chronic kidney disease is 90 or greater. This stage represents mild damage, but the kidneys may be working well. Biomarkers like protein in the urine reflect physical damage to the kidneys. eGFR in stage 2 chronic kidney disease is 60–89. An eGFR of 30–59 represents stage 3 and reflects some damage and dysfunctionality in the kidneys. Stage 4 CKD with eGFR of 15–29 represents moderately or severely damaged kidneys. This stage is the last stage before renal failure. An eGFR of less than 15 means the kidneys are close to failure or have failed completely. As a result of kidney failure, waste builds up in the blood and leads to other health problems [27].

5 The chemical species of oxidative stress

Oxygen-based reactive oxygen species (ROS) and nitrogen-based reactive nitrogen species (RNS) are responsible for oxidative stress at the cellular level. These chemical molecules are formed by the incomplete reduction of molecular oxygen and nitrogen molecules. When the concentration of ROS is increased above the threshold level, it leads to oxidative stress [27]. Reactive oxygen species includes superoxide anion (O_2^-), nitric oxide (NO), hydrogen peroxide (H_2O_2), hydrogen radical (OH), singlet oxygen ($^1O_2$), peroxynitrite (ONOO-), and nitrogen dioxide (NO_2). Sources of ROS primarily include mitochondria along with peroxisomes, cytoplasm, and nucleus [28]. In addition, enzymes—namely, cytochrome 450 oxygenase, xanthine oxygenase, lipoxygenase, cyclooxygenase, and nitric oxide synthase—are all involved in the synthesis of ROS [29–32]. A minimal concentration of ROS is healthy for a cell and acts as a signal for constitutive functions such as cell proliferation and mitogenesis. These molecules can react with macromolecules or cellular structures like lipids and DNA, and lead to cellular damage or injury.

5.1 Cellular mechanism for the formation of reactive oxygen species

In normal cellular conditions, one molecule of oxygen reacts with four protons generating two molecules of water ($O_2+4H^++4e^-=2H_2O$). However, if reduction is not complete, it can generate several reactive oxygen species [31]. In a cell, ROS are generated by spontaneous reaction or by enzymes present in cytoplasm, plasma membrane, or organelles like mitochondria and peroxisomes. ROS are also the catalytic by-products of metabolic reaction carried out by enzymes; for example, superoxide ion is produced by the action of xanthine oxidase on O_2 to form water coupled with the formation of uric acid [31]. Cell membrane-bound NADPH oxidase (NOX) is another key enzyme involved in ROS production in phagocytes. Multiple inflammatory molecules activate the NADPH oxidase and lead to oxidative bursts in these cells. Let us look at the chemical reaction of the formation of ROS species in a cell.

(i) Superoxide anion ($O_2^{\cdot-}$)

Reduction of molecular oxygen by a single election forms a superoxide anion $\left(O_2+1e^-=O_2^{\ -}\right)$. This superoxide anion has both a free radical and an electron in its outer orbital. It can act both as a reducing agent and an oxidizing agent. For instance, as a reducing agent, it can reduce ferricytochrome *c*, and as an oxidizing agent, it gains an extra electron to form hydrogen superoxide ($O_2^{\ -}+2H^++1e^-=H_2O_2$) [30]. Under normal cellular conditions, O^{2-} is produced in small quantities as a product of cellular metabolism. It is produced by enzymatic reaction or autoxidation of reduced electron transferring compounds like catecholamines, tetrahydropterins, leucoflavins, and hydroquinone. In the cytosol, it is formed by the action of aldehyde oxidase and xanthine oxidase, whereas in the mitochondria, enzyme NADH dehydrogenase, ubiquinone-cytochrome *b* complex, and electron transport chain (ETC) located on the inner mitochondrial membrane are involved in the formation of superoxide anion. This superoxide anion is also the primary product of cytochrome P450 and cytochrome *b*5 present in the membrane of endoplasmic reticulum in the presence of NADPH and NADH, respectively [31].

(i) Hydrogen peroxide (H_2O_2)

Hydrogen peroxide is produced in a cell in two ways.

(a) spontaneous reduction of superoxide anion [32]; or.

(b) catalysis of superoxide anion by superoxide dismutase enzyme.

In general, hydrogen peroxide is generated along with superoxide anion, which serve as a substrate for the formation of H_2O_2. Therefore, H_2O_2 presence is observed in mitochondria, cytosol, and peroxisomes [33]. However, the maximum concentration of H_2O_2 is observed in mitochondria and peroxisomes, where highly reduced respiratory chain carriers and oxidases, respectively, are present in high concentrations. Hydrogen peroxides serve as a cytotoxic agent in immune cells, involved in the signaling pathway and pathology of several diseases [33].

(ii) Hydroxyl radical (OH)

$(H_2O_2 + O_2{}^- = {}^1O_2 + OH^- + OH)$.

This reaction is catalyzed Fe^{3+} or other trace elements that can act as oxidation–reduction catalysts. Cyclooxygenase and lipoxygenase enzymes, which oxidize arachidonic acid, also led to the synthesis of hydroxyl radicals [34].

6 Cellular antioxidant defense system

The cell's infrastructure is not damaged by low concentrations of ROS. Cells have an antioxidant system in the form of antioxidant enzysmes and antioxidant molecules, which neutralizes these anionic or radical reactive molecules to nonreactive molecules and protects the cell from damage [35]. The key enzymes are superoxide dismutase and glutathione peroxidase.

6.1 Superoxide dismutase (SOD)

In the kidneys, enzymes superoxide dismutase (SOD) is present in the cytoplasm and mitochondria. These enzymes are of two types based on the presence of a cofactor (Cu and Zn or Mn) in the active site. Principally, SOD is not a detoxifying enzyme as it converts superoxide ion into hydrogen peroxide, which is also a ROS. However, H_2O_2 is acted upon by catalase and generates H_2O and O_2. Catalase is also present in kidney cells, particularly in peroxisomes and cytosol [32].

6.2 Glutathione peroxidase (GSHPX)

Another enzyme present in the kidneys and that protects cellular macromolecules from oxidation is glutathione peroxidase (GSHPX). GSHPX protects cell from excess H_2O_2 by oxidation of reduced glutathione (GSH) into its oxidized state (GSSG) [32,35]:

$$(2GSH + H_2O_2 = GSSG + 2H_2O)$$

The GSSG formed is further reduced into two molecules of GSH by glutathione reductase in the presence of NADPH:

$(GSSG + NADPH + H^+ = 2GSH + NADP^+)$

This enzyme is present in both the cortex and medulla regions of the kidneys. Selenium is the cofactor of glutathione whereas mercury inhibits its activity. Several other minerals like Zn, Cu and vitamins, tocopherol, and riboflavin are also important for the strong cellular antioxidant system as they function as cofactors of antioxidant enzymes [33].

7 Sources of ROS in the kidneys

Several studies have recognized the role of ROS and RNS for the initiation and progression of toxic-induced nephropathy, ischemia–reperfusion injury (IRI), kidney fibrosis, and renal pathological conditions [36–38]. In the context of pathogenesis of diabetic kidney disease, the roles of xanthine oxidase, NADPH oxidase, glycolysis, polyol pathway, and nitric oxide synthase have been identified for ROS production [39]. The principal source of oxidative species in the kidney cells are NADPH oxidase and mitochondria. Both ROS and RNS are generated in both renal tubular and vascular cells during cellular stress [31,36]. At low concentrations, both ROS and RNS are important for constitutive physiological constitutive functions, whereas, under stress conditions, production of ROS and RNS is up-regulated, and reduced antioxidant activity is observed. This results in the accumulation of these molecules, and renal cellular structure and function is compromised. ROS is generated in renal cells by the following two mechanisms.

1. A superoxide ion is generated by the action of xanthine oxidase on molecular oxygen to water coupled with the formation of uric acid.
2. Cyclooxygenase and lipoxygenase enzyme, which oxidizes arachidonic acid, leads to the formation of hydroxyl radical (OH). Cyclooxygenase is present in all cells of the kidneys, predominantly in the glomeruli, renal vessels, collecting duct, and medullary interstitium where lipoxygenase is restricted to the glomeruli [38].

7.1 NADPH oxidase as a source of ROS

NADPH oxidase (NOX) is located in plasma membrane and generates the formation of ROS by transferring electrons from NADPH to O_2 to form $O_2^{\cdot-}$ that is rapidly transformed into H_2O_2. Kidney cells, particularly PCT, DCT, macula densa, mesangium, and endothelium, express the genes of NOX family that are required for the functioning of NADPH oxidase [39]. NADPH oxidase activity can be activated by interferon gamma (IFNc) [40], and angiotensin-II [41]. Glomerular mesangial cells generate superoxide ion by membrane-bound NADPH oxidase. This formation of superoxide ion is in response to cytokines and is not dependent on phagocytosis [42]. NOX4 is the principal isoform, which is involved in constitutive production of hydrogen peroxide [43]. The NOX4 isoform has been found to be localized in the mitochondria of rat kidney cortex, which allows NADPH oxidase-generated superoxide to directly prompt mitochondrial generation of ROS [43].

7.2 Mitochondrial ROS

About 0.2%–2% of oxygen consumed by mitochondria produces reactive oxygen species [44,45]. In kidney cells, mitochondria constitutively maintain low levels of ROS to sense low levels of oxygen and carry out signaling for constitutive functions. However, the ROS production in the mitochondria is significantly augmented by a range of renal pathological conditions such as chronic hyperglycemia and signals from TNFα, angiotensin II, oxidized low-density lipoprotein (LDL), and superoxide generated by NADPH oxidase [41,42,46]. The increase in mitochondrial superoxide production is stimulated by a rise in intracellular calcium, opening of mitochondrial potassium-adenosine triphosphate (ATP) channels, and a change in the potential gradient across the membrane. Moreover, disruption of iron–sulfur clusters, inhibitory interactions with cytochrome *c* oxidase, and relative alteration in expression of ETC components have been proved to be associated to the increase in mitochondrial ROS generation [44].

In normal conditions, Mn-SOD (SOD2) converts NADPH oxidase-generated superoxide to hydrogen peroxide and secretes it in the cytoplasm. In this way, mitochondria are protected from oxidative damage [44]. However, if superoxide ions are not detoxified, this results in disruption of iron–sulfur centers, releasing iron, and activating signaling pathways that worsen mitochondrial function. In mice, ablation of SOD2 resulted in glomerular sclerosis, enhanced tubular damage and renal interstitial inflammation, and accelerated renal cellular senescence. Furthermore, excess ROS in mitochondria opens the mitochondrial membrane channels [47], resulting in breakdown of the mitochondria membrane potential [48], and accumulation of uncoupling protein-2 (UCP2). The increase in UCP-2 results in proton leakage that restricts ATP synthase functioning and finally lowers the synthesis of ATP [49]. Dysfunction of ATP synthase affects the ATP-supported reabsorption process in tubular epithelial cells. If the impairment goes beyond the threshold level, it will trigger cell death [50]. Hence, increased mitochondrial production of ROS in response may be the key initiator of mitochondrial dysfunction and further progression into nephropathy [39].

8 Reactive oxygen species and kidney diseases

Substantial evidence suggests that free radical reactions make a significant contribution to the kidney disease pathology. In renal pathophysiology, ROS formation is a secondary event to the primary mechanism of the disease process. For instance, activated neutrophils produce H_2O_2, O_2^-, and HOCI to kill the pathogens in kidney [51]. However, if many phagocytes become stimulated in a confined area, they can produce tissue damage by excess of ROS. Several studies have demonstrated the role of ROS in kidney disease, particularly in the pathogenesis mediated by the immune molecules, toxic chemicals, and ischemia kidney [51]. The sources of ROS for renal injury can be both kidney cells and immune cells. An imbalance between production of ROS and expression of antioxidant enzymes by kidney cells or immune cells creates toxic conditions for damage to the kidneys. Excess of ROS can stimulate direct injury to kidney cells (nephrotoxicity) or indirect damage by activation and amplification of several inflammatory mediators. Various studies have explored the role of ROS in kidney diseases that include toxic renal disease, ischemic acute renal failure, acute glomerulonephritis, and renal graft rejection [51,52]. Moreover, renal injury/disease can occur in any compartment of the kidney by nephrotoxins or pathological factors like obesity, diabetes, cardiovascular diseases, hypertension, and liver failure, which potentiate the renal kidney damage. Furthermore, almost any pathological condition can be expected to have involvement of enhanced ROS [52]. In the next section, we will focus on the direct effect ROS along with associated renal pathophysiology on different compartments of the kidneys, i.e., renal tubules, renal vasculature, and glomerulus.

9 Oxidative stress in the renal tubules

9.1 Proximal tubule

The proximal tubule (PT) is the key site of the production of ROS. The reason is huge production of ATP in PT cells, which is required to maintain the Na^+/K^+ ATPase pump. Na^+/K^+ ATPase sustains the electrochemical gradient across the membrane that governs the active transport of electrolytes and molecules across the epithelial cells [52]. As is well known, all ATP is generated by highly oxidative glucose metabolism that exposes proximal tubule cells to oxidative stress which is neutralized by an antioxidant system. However, in comparison to other tubular cells, proximal tubule cells do not synthesize glutathione; instead, they are dependent on freely circulating glutathione to neutralize mitochondrial-generated ROS [53,54]. The vulnerability of proximal tubule cells to oxidative injury is seen in renal pathology. During diabetes and hyperglycemia, PT cells filters and reabsorb excessive glucose via sodium-dependent glucose transporters [52]. The high availability of glucose in PT cells increases glycolysis, which results in the generation of large amounts of superoxide and ROS in mitochondria [55]. NADPH oxidase, particularly NOX4, is highly expressed in PT cells. Increased uptake of glucose in PT cells also activates protein kinase C, which further triggers NADPH oxidase to produce superoxide [41]. In the context of hypertension, high levels of ANG II upregulate NADPH oxidase-derived superoxide production in PT cells, which further inhibits the proximal tubule fluid reabsorption by increasing the expression of the Na^+/H^+ exchanger regulatory factor (NHERF2) in hypertensive rats [56]. Panico et al. suggested reabsorption of Na^+ and fluid reabsorption changes along the nephron during hypertension. Reabsorption in the PT is lower than in downstream tubules where reabsorption increases, which impacts oxygen usage for Na^+ uptake and secretion.

In addition to hypertension and diabetic nephropathy, the PT cells can be injured by various nephrotoxins. PT cells extensively uptake nephrotoxins via both apical and basolateral transport systems by endocytic mechanisms [57]. These nephrotoxins include antibacterial drugs (e.g., cephaloridine) and antivirals (e.g., tenofovir), which affects the S1–S2 segments of PT, whereas anticancer drugs and heavy metals (e.g., hexavalent chromium, mercury, HCBD, palladium) directly injure the S3 segment proximal tubule [57]. These heavy metals exhibit nephrotoxicity by reacting with the antioxidant enzymes/system and cofactors, hindering their activity. Heavy metals are also proven to diminish antioxidant systems or prompt the formation of hydroxyl-like radicals by the Fenton reaction. Halogenated alkene (HCBD) conjugates with GSH to form *S*-(pentachlorobutadienyl)-L-cysteine, which on cleavage by β-lyase forms nephrotoxic thioketene [58] Chromium can accumulate in the renal cell by an anion transport system. Inside the cell, it is subjected to intracellular reduction or reaction with hydrogen peroxide to form hydroxyl radicals, resulting in DNA damage [59]. Like HCBD, cisplatin can be metabolized by GSH-conjugate intermediate into a nephrotoxin. This nephrotoxin is excreted into the bile [60]. Furthermore, the role of nitrosative stress particularly by peroxynitrite anion (ONOO −) has been demonstrated in vitro and in vivo in cisplatin-induced renal damage [61]. ONOO − is produced by reaction of nitric oxide and superoxide anion in a cell. The enhanced oxidative stress, independent of the mechanism of nephrotoxicity, results in protein oxidation, DNA damage, and lipid peroxidation; the end point is always cell death, by either apoptosis or necrosis [62].

Other nephrotoxic substances include polycationic aminoglycosides, various complex sugars, and starches. Aminoglycosides such as gentamicin and amikacin are found to be accumulated (around 10%) in the kidneys. These drugs bind to the acidic phospholipids on the brush border membrane of the epithelial cells of the proximal tubule. Recently, a giant endocytic receptor protein named megalin present on the apical membrane was identified for binding and endocytosis of aminoglycosides in PT cells. Once these drugs enter the cytoplasm, they further translocate into the lysosomal compartment where they inhibit the lysosomal enzymes and damage the organelle, and subsequently form myeloid bodies with the membrane fragments and damaged organelle [63].

Aminoglycosides are taken in by PT cells via apical and basolateral pathways. Uptake via the apical pathway leads to accumulation of aminoglycoside in the cells, which initiates an injury pathway leading to tissue injury or death. The basolateral pathway is predominantly used for uptake of endogenous and exogenous organic ions by peritubular capillaries [64,65]. Transport protein, human organic anion, and cation transporters present on the PT cells are involved in the transportation of toxins, acyclic nucleotide phosphonates, and cisplatin, respectively. Transport of toxicants within the intracellular space and their subsequent exit is regulated by various carrier proteins and apical transport proteins, respectively [65]. If apical secretory transporters suffer loss of function mutations, this promotes accumulation of toxic substances within PT cells, which results in poor toxin efflux from the cells into urine, further causing cellular injury via necrosis or apoptosis. This extensive movement of substances increases renal tubular exposure, and risk for kidney injury occurs when other risk factors are introduced. Several complex sugars such as sucrose, dextrans, hydroxyethyl starch, and contrast media have also been demonstrated to cause osmotic nephrosis [66].

Proteinuria is an independent risk factor for progression of kidney disease. Proteinuria has been found to induce glia maturation factor-β (GMF-B) in PT cells of normal and proteinuria disease models [67,68]. In vitro studies done on the PT cell line and NIH383 revealed that angiotensin-II, tumor necrosis factor, and other oxidative agents like buthionine sulfoximine and H_2O_2 induce the overexpression of GMF-B, which can result in oxidative injury, F-actin reorganization, and apoptosis. Furthermore, prolonged increase in H_2O_2 leads to oxidative damage with the involvement of p38 protein [69]. The role of ROS in PT cells has been established via activation of the MAP, PKB/Akt, and JK phosphorylation pathway, which upregulates expression of TGF-b by activation of transcription factors: NF-kB, and STAT [70]. TGF-β induces mesenchymal transition into myofibroblasts via the SMAD pathway. Myofibroblasts, once formed, produce excessive ECM proteins, including collagens leading to tubulointerstitial fibrosis and impairment of renal function [71].

9.2 Medulla

A study was performed on Sprague Dawley rats to determine the tissue levels of H_2O_2 in the cortex and medulla of the kidneys. It was found that the basal level of H_2O_2 in the renal medulla is twice that of the renal cortex, suggesting the higher levels of oxidative stress in renal medulla under normal physiological conditions. The same study also showed that H_2O_2

in the renal medulla exerts a modest action on renal basal vascular tone and excretory function. A decrease of basal renal medullary H_2O_2 concentration by administration of catalase increased medullary blood flow (MBF), water excretion, and Na^+ excretion by 10%, 12%, and 16%, respectively [72]. Further, direct delivery of H_2O_2 into the medulla decreased MBF and water and Na^+ excretion, suggesting that excessive H_2O_2 in the renal medulla produces vasoconstriction and antinatriuresis, contributing to the development of hypertension [73].

Albumin, as a plasma protein, contributes to the maintenance of plasma osmotic pressure, buffering of acid–base, and transport of biomolecules. Various congenital and acquired renal disorders induce ROS in PT cells. Albumin has been observed to contribute and prevent oxidative stress in the proximal tubules. It works as a survival factor at physiological concentrations and shows antioxidant capabilities, and during pathological conditions such as proteinuria, exposure of the PT cells to an excess of albumin induces oxidative stress and causes renal damage. In proximal tubular cells, higher concentration of albumin upregulates the intracellular ROS by IFN-γ–independent activation of STATs and membrane-bound NADPH oxidase system. In addition, the low activity of antioxidant enzymes, glutathione peroxidase and catalase, may be responsible for the accumulation of intracellular ROS. Hence, along with ROS generation, low expression of ROS neutralizing mechanisms contributes to the induction of ROS by albumin [74].

9.3 Distal and collecting duct

The cortical collecting duct represents the last part of the nephron and plays an important role in determining the hormone-regulated final concentration of sodium ion in the urine. In these tubules, Na^+ is primarily reabsorbed via the epithelial Na^+ channel (ENaC) present on the apical membrane. Studies done in rat cortical ducts have reported that ANG II activates ENaC through a Ca^{2+}-independent PKC pathway by up-regulating NOX expression and thereby enhancing O_2^- generation. Furthermore, it was reported that the inhibition of NOX with diphenyleneiodonium chloride stopped the activation of ENaC by ANG II, signifying the NADPH oxidase-dependent generation of ROS [75]. An increase in either apical ENaC number boosts $Na+$absorption and can lead to Liddle's syndrome, an inherited form of hypertension [76,77]. Aldosterone is the principal regulator of ENaC and has been demonstrated to increase O_2^- production substantially in A6 distal nephron cells. The role of interaction between superoxide ion and nitric oxide (NO) is important for the regulation of renal ENaC in distal kidney cells, and ENaC could be inhibited by either an increase in NO production or a decrease in O_2^- generation [78].

10 Oxidative stress in the glomerulus

Effects of ANG II on kidney cells have been demonstrated in the context of hypertension [78,79]. In the glomerulus, mesangial cells and podocytes have been observed to generate ROS in response to angiotensin II and aldosterone, resulting in ROS-dependent podocyte injury and mesangial cell proliferation [80]. Podocyte injury is linked as an initial event to the hardening of the glomerulus (glomerulosclerosis) and increased levels of protein in urine (proteinuria) [81]. Nephrin is a cell adhesion protein synthesized by glomerular epithelial cells. It forms a slit diaphragm which anchors it to podocytes. ROS generation in podocytes has been found to reduce the expression of nephrin, and induce autophagy and apoptosis through mitochondrial generation of ROS [82,83]. Another study demonstrated induction of adenosine 1 receptors A1AR-dependent hypertension, proteinuria, and podocyte injury and reduced caveolin-1 phosphorylation and nephrin expression on prolonged administration of angiotensin II in rats [84]. In podocytes, mice given aldosterone and a high-sodium diet showed increased blood pressure independent of oxidative stress, podocyte injury, and albuminuria [90]. In mesangial cells, aldosterone upregulated NOX2, p47phox, p67phox, Rac-1, and Sgk1, and elevated BP and ROS production. ROS mediated the proliferation, migration, and production of extracellular matrix (ECM) in mesangial cells induced by angiotensin II or aldosterone. ANG II increases NADPH oxidase-dependent superoxide ion generation and increases production of ECM proteins such as fibronectin that are implicated in the growth of glomerulosclerosis. The perpetuating generation of ROS altered the gene expression in mesangial cells by the involvement of PKC [91], PKB/Akt [85], PDK-1 [86], and JNK signaling pathways [87]. The increased levels of ROS induced mesangial cells and podocytes to generate angiotensin II and reinforce a response that promotes more ROS generation. Thus, intrarenal angiotensin II and aldosterone signaling via the mesangial cells elicit oxidative stress and are widely associated in the pathogenesis of renal diseases. ROS can also directly alter the glomerular filtration barrier by changing the permeability of the glomerular basement membrane, leading to proteinuria [88]. Furthermore, an increase in ROS generation leads to glomerulonephritis [89], glomerulosclerosis, and activates the cell death pathways via autophagy and apoptosis in podocytes [79,82,83] and mesangial cells [90,91].

11 Oxidative stress in the renal vasculature

NOX, xanthine oxidase, eNOS uncoupling, and mitochondrial ROS are the primary sources of oxidative stress in the renal vasculature [92] and contribute to vascular disease by promoting division of vascular smooth muscle cells (VSMCs), macrophage recruitment, altering vascular tone and activation of matrix metalloproteinases (MMPs) [93]. As a result, vascular remodeling and increases in preglomerular resistance are observed, and form the central mechanisms in the pathogenesis of renal and cardiovascular disease [94,95]. Renal vasculature is formed by various cells such as endothelial cells, adventitial fibroblast, VSMCs, and inflammatory cells, and produces ROS and RNS. Both have direct effects on vasoreactivity in renal vessels [96]. A few studies have shown that ROS, especially O_2^- and high levels of H_2O_2, are responsible for vasoconstriction in afferent arterioles in the kidneys. Furthermore, the balance between O_2^- and NO generate vasoconstrictor and vasodilator responses, respectively, in the blood vessels. In renal pathological conditions such as hypertension, NOX increases ROS in the kidneys [97], which weakens the formation of functional endothelium-derived nitric oxide by uncoupled NOS enzymes, hampering the ability of NO to balance vasodilators and vasoconstrictor inducers such as endothein-1 and angiotensin, respectively. This set the stage for the progression of renal hemodynamic pathogenesis. Furthermore, if the bioavailability of NO continues to decrease, attenuation of natriuresis [98,99] and endothelial dysfunction start, leading to renal diseases [100–103].

The endothelium is sensitive to the changes in the level of ROS. At high concentrations, endothelium promotes inflammation, and increases cytokine production and expression of surface adhesion molecules [104]. Endothelial nitric oxide synthase (eNOS) remains bound to the plasma membrane of endothelial cells and cytosolic face of the outer mitochondrial membrane [105] and produces nitric oxide (NO). At normal physiological conditions, NO induces expression of antioxidative genes, protects the renal endothelium [106,107], and promotes normal renal hemodynamics; however, at higher concentrations, it inhibits cytochrome *c* oxidase [108], potentially altering mitochondrial ROS generation and resulting in renal disease. A moderate increase in mitochondrial ROS results in stabilization of hypoxia-inducible factor (HIF) in endothelial cells [109], and stimulates nuclear factor E2-related factor (NRF-2) [110]. Normally, NRF-2 remains in bound form with Kelch-like ECH-associated protein-1 (Keap-1); however, during oxidative stress, Keap-1 is oxidized, followed by release of NRF-2, which acts as a transcription factor and promotes expression of antioxidative enzymes such as glutathione and thioredoxin systems, SOD, and catalase. Reduced activity of NRF-2 and its role in the progression of renal failure have been demonstrated in a mouse model of chronic kidney disease [111]. In mice VSMCs, superoxide ion causes vasoconstriction and enhances basal tone in the medullary circulation [112].

12 Role of ROS in ischemia and reperfusion (I/R)-induced acute kidney injury

Ischemia is a vascular condition in which blood flow is obstructed or reduced to a part of body or tissue, resulting in decreased oxygen and nutrient supply to the tissues. Reperfusion is the restoration of blood flow to an ischemic tissue. The role of ROS in ischemia and reperfusion (I/R)-induced acute kidney injury, and development of kidney fibrosis, has been demonstrated following kidney transplantation and pathological conditions such as hypertension and diabetes. Development of kidney fibrosis following transplantation is now widely recognized as a contributor to chronic allograft loss [113,114]. Apart from oxidative stress, the involvement of angiotensin, changes in hemodynamics, and inflammatory pathway have been observed in the progression of kidney fibrosis [114,115]. During the process of I/R induced renal fibrosis, several phenotypic changes at molecular and cellular level are seen. These include: the loss of epithelial cell membrane transporters and their functions; H_2O_2 production and nitrotyrosine production; decreased expression of SODs and glucose 6-phosphate dehydrogenase; production of extracellular matrix elements, such as collagen and fibronectin; apoptosis of tubular epithelial cells; and expression of SMA commonly found in myofibroblasts [37,116–118]. In conclusion, this study demonstrates that ROS/oxidative stress contributed to the progression of kidney fibrosis following I/R.

References

[1] V.A. Luyckx, Z. Al-Aly, A.K. Bello, E. Bellorin-Font, R.G. Carlini, J. Fabian, G. Garcia-Garcia, A. Iyengar, M. Sekkarie, W. Van Biesen, I. Ulasi, Sustainable development goals relevant to kidney health: an update on progress, Nat. Rev. Nephrol. 17 (1) (2021) 15–32.
[2] A. Kaplan, E. Abidi, A. El-Yazbi, A. Eid, G.W. Booz, F.A. Zouein, Direct cardiovascular impact of SGLT2 inhibitors: mechanisms and effects, Heart Fail. Rev. 23 (3) (2018) 419–437.
[3] https://ourworldindata.org/cancer.
[4] http://www.who.int/gho/hiv/en/.
[5] https://www.kidney.nyc/types-of-kidney-disease.
[6] P. Romagnani, G. Remuzzi, R. Glassock, et al., Chronic kidney disease, Nat. Rev. Dis. Primers. 3 (1) (2017), 17088.
[7] M.V. Irazabal, V.E. Torres, Reactive oxygen species and redox signaling in chronic kidney disease, Cell 9 (6) (2020) 1342.
[8] X.C. Ling, K.-L. Kuo, Oxidative stress in chronic kidney disease, Ren. Replace. Ther. 4 (2018) 53.
[9] P.L. Jørgensen, Structure, function and regulation of Na, K-ATPase in the kidney, Kidney Int. 29 (1) (1986) 10–20.
[10] W.E. Hoy, R.N. Douglas-Denton, M.D. Hughson, A. Cass, K. Johnson, J.F. Bertram, A stereological study of glomerular number and volume: preliminary findings in a multiracial study of kidneys at autopsy, Kidney Int. 63 (2003) S31–S37.

[11] W.G. Guder, B.D. Ross, Enzyme distribution along the nephron, Kidney Int. 26 (2) (1984) 101–111.
[12] M. Imai, D.W. Seldin, J.P. Kokko, Effect of perfusion rate on the fluxes of water, sodium, chloride and urea across the proximal convoluted tubule, Kidney Int. 11 (1) (1977) 18–27.
[13] T.L. Pannabecker, Structure and function of the thin limbs of the loop of Henle, Compr. Physiol. 2 (3) (2012) 2063–2086.
[14] B. Glaudemans, N.V. Knoers, J.G. Hoenderop, R.J. Bindels, New molecular players facilitating Mg2+ reabsorption in the distal convoluted tubule, Kidney Int. 77 (1) (2010) 17–22.
[15] R.J. Young, W.E. Hoy, P. Kincaid-Smith, A.E. Seymour, J.F. Bertram, Glomerular size and glomerulosclerosis in Australian aborigines, Am. J. Kidney Dis. 36 (3) (2000) 481–489.
[16] W.H. Park, Y.H. Han, S.H. Kim, S.Z. Kim, Pyrogallol, ROS generator inhibits As4. 1 juxtaglomerular cells via cell cycle arrest of G2 phase and apoptosis, Toxicology 235 (1–2) (2007) 130–139.
[17] H. Ha, H.B. Lee, Reactive oxygen species as glucose signaling molecules in mesangial cells cultured under high glucose, Kidney Int. 58 (2000) S19–S25.
[18] S.V. Mclennan, E. Fisher, S.Y. Martell, A.K. Death, P.F. Williams, J.G. Lyons, D.K. Yue, Effects of glucose on matrix metalloproteinase and plasmin activities in mesangial cells: possible role in diabetic nephropathy, Kidney Int. 58 (2000) S81–S87.
[19] M.G. Farquhar, The primary glomerular filtration barrier—basement membrane or epithelial slits? Kidney Int. 8 (4) (1975) 197–211.
[20] M.A. Perazella, Renal vulnerability to drug toxicity, Clin. J. Am. Soc. Nephrol. 4 (7) (2009) 1275–1283.
[21] L.M. Barnett, B.S. Cummings, Nephrotoxicity and renal pathophysiology: a contemporary perspective, Toxicol. Sci. 164 (2) (2018) 379–390.
[22] P. Devarajan, Update on mechanisms of ischemic acute kidney injury, J. Am. Soc. Nephrol. 17 (6) (2006) 1503–1520.
[23] T. Ludwig, C. Riethmüller, M. Gekle, G. Schwerdt, H. Oberleithner, Nephrotoxicity of platinum complexes is related to basolateral organic cation transport, Kidney Int. 66 (1) (2004) 196–202.
[24] M.A. Perazella, Pharmacology behind common drug nephrotoxicities, Clin. J. Am. Soc. Nephrol. 13 (12) (2018) 1897–1908.
[25] M.A. Perazella, H. Izzedine, New drug toxicities in the onco-nephrology world, Kidney Int. 87 (5) (2015) 909–917.
[26] M.A. Perazella, Onco-nephrology: renal toxicities of chemotherapeutic agents, Clin. J. Am. Soc. Nephrol. 7 (10) (2012) 1713–1721.
[27] P.K. Maurya, P. Kumar, P. Chandra, Biomarkers of oxidative stress in erythrocytes as a function of human age, World J. Methodol. 5 (4) (2015) 216.
[28] M.L. Genova, M.M. Pich, A. Bernacchia, et al., The mitochondrial production of reactive oxygen species in relation to aging and pathology, Ann. N. Y. Acad. Sci. 1011 (2004) 86–100.
[29] B. Halliwell, M.V. Clement, L.H. Long, Hydrogen peroxide in the human body, FEBS Lett. 486 (1) (2000) 10–13.
[30] E.G. Hrycay, S.M. Bandiera, The monooxygenase, peroxidase, and peroxygenase properties of cytochrome P450, Arch. Biochem. Biophys. 522 (2) (2012) 71–89.
[31] N.S. Zinkevich, D.D. Gutterman, ROS-induced ROS release in vascular biology: redox-redox signaling, Am. J. Physiol. Heart Circ. Physiol. 301 (3) (2011) H647–H653.
[32] S. Dikalov, K.K. Griendling, D.G. Harrison, Measurement of reactive oxygen species in cardiovascular studies, Hypertension 49 (4) (2007) 717–727.
[33] M. Sundaresan, Z.X. Yu, V.J. Ferrans, K. Irani, T. Finkel, Requirement for generation of H2O2 for platelet-derived growth factor signal transduction, Science 270 (5234) (1995) 296–299.

[34] J.I. Abe, B.C. Berk, Fyn and JAK2 mediate Ras activation by reactive oxygen species, J. Biol. Chem. 274 (30) (1999) 21003–21010.
[35] S.B. Maggirwar, D.N. Dhanraj, S.M. Somani, V. Ramkumar, Adenosine acts as an endogenous activator of the cellular antioxidant defense system, Biochem. Biophys. Res. Commun. 201 (2) (1994) 508–515.
[36] V. Cachofeiro, M. Goicochea, S.G. de Vinuesa, P. Oubina, V. Lahera, J. Luno, Oxidative stress and inflammation, a link between chronic kidney disease and cardiovascular disease, Kidney Int. Suppl. (2008) S4–S9.
[37] J. Kim, Y.M. Seok, K.J. Jung, K.M. Park, Reactive oxygen species/oxidative stress contributes to progression of kidney fibrosis following transient ischemic injury in mice, Am. J. Physiol. Renal Physiol. 297 (2009) F461–F470.
[38] L. Shalamanova, F. McArdle, A.B. Amara, M.J. Jackson, R. Rustom, Albumin overload induces adaptive responses in human proximal tubular cells through oxidative stress but not via angiotensin II type 1 receptor, Am. J. Physiol. Renal Physiol. 292 (2007) F1846–F1857.
[39] J.M. Forbes, M.T. Coughlan, M.E. Cooper, Oxidative stress as a major culprit in kidney disease in diabetes, Diabetes 57 (2008) 1446–1454.
[40] L. Malaguarnera, Influence of resveratrol on the immune response, Nutrients 11 (5) (2019) 946.
[41] A. Sachse, G. Wolf, Angiotensin II-induced reactive oxygen species and the kidney, J. Am. Soc. Nephrol. 18 (2007) 2439–2446.
[42] H.H. Radeke, B. Meier, N. Topley, J. Flöge, G.G. Habermehl, K. Resch, Interleukin 1-α and tumor necrosis factor-α induce oxygen radical production in mesangial cells, Kidney Int. 37 (2) (1990) 767–775.
[43] V. Thallas-Bonke, K.A. Jandeleit-Dahm, M.E. Cooper, Nox-4 and progressive kidney disease, Curr. Opin. Nephrol. Hypertens. 24 (2015) 74–80.
[44] E. Cadenas, K.J. Davies, Mitochondrial free radical generation, oxidative stress, and aging, Free Radic. Biol. Med. 29 (3–4) (2000) 222–230.
[45] S. Raha, B.H. Robinson, Mitochondria, oxygen free radicals, disease and ageing, Trends Biochem. Sci. 25 (10) (2000) 502–508.
[46] S.R. Thomas, P.K. Witting, G.R. Drummond, Redox control of endothelial function and dysfunction: molecular mechanisms and therapeutic opportunities, Antioxid. Redox Signal. 10 (2008) 1713–1765.
[47] J.J. Lemasters, A.L. Nieminen, T. Qian, L.C. Trost, S.P. Elmore, Y. Nishimura, R.A. Crowe, W.E. Cascio, C.A. Bradham, D.A. Brenner, B. Herman, The mitochondrial permeability transition in cell death: a common mechanism in necrosis, apoptosis and autophagy, Biochim. Biophys. Acta 1366 (1998) 177–196.
[48] W. Wang, H. Fang, L. Groom, A. Cheng, W. Zhang, J. Liu, X. Wang, K. Li, P. Han, M. Zheng, J. Yin, M.P. Mattson, J.P. Kao, E.G. Lakatta, S.S. Sheu, K. Ouyang, J. Chen, R.T. Dirksen, H. Cheng, Superoxide flashes in single mitochondria, Cell 134 (2008) 279–290.
[49] E. Noiri, F. Addabbo, M. Goligorsky, Reactive oxygen and nitrogen species, oxidative and nitrosative stress, and their role in the pathogenesis of acute kidney injury, in: T. Miyata, K. Eckardt, M. Nangaku (Eds.), Studies on Renal Disorders, Humana Press, New York, 2011, pp. 161–177.
[50] T.R. Pushpa-Rekha, A.L. Burdsall, L.M. Oleksa, G.M. Chisolm, D.M. Driscoll, Rat phospholipid-hydroperoxide glutathione peroxidase. cDNA cloning and identification of multiple transcription and translation start sites, J. Biol. Chem. 270 (1995) 26993–26999.
[51] B. Halliwell, Free radicals, reactive oxygen species and human disease: a critical evaluation with special reference to atherosclerosis, Br. J. Exp. Pathol. 70 (6) (1989) 737.
[52] L.J. Mandel, Metabolic substrates, cellular energy production, and the regulation of proximal tubular transport, Annu. Rev. Physiol. 47 (1985) 85–101.
[53] K.N.M. Khan, C.L. Alden, Kidney, in: W.M. Haschek, C.G. Rousseaux, M.A. Wallig (Eds.), Handbook of Toxicological Pathology, second ed., Academic Press, San Diego, CA, 2002, pp. 255–330.

[54] T.M. Visarius, D.A. Putt, J.M. Schare, D.M. Pegouske, L.H. Lash, Pathways of glutathione metabolism and transport in isolated proximal tubular cells from rat kidney, Biochem. Pharmacol. 52 (1996) 259–272.

[55] T. Nishikawa, D. Edelstein, X.L. Du, S. Yamagishi, T. Matsumura, Y. Kaneda, M.A. Yorek, D. Beebe, P.J. Oates, H.P. Hammes, I. Giardino, M. Brownlee, Normalizing mitochondrial superoxide production blocks three pathways of hyperglycaemic damage, Nature 404 (2000) 787–790.

[56] C. Panico, Z. Luo, S. Damiano, F. Artigiano, P. Gill, W.J. Welch, Renal proximal tubular reabsorption is reduced in adult spontaneously hypertensive rats: roles of superoxide and Na+/H+ exchanger 3, Hypertension 54 (2009) 1291–1297.

[57] C. Marzano, A. Trevisan, L. Giovagnini, D. Fregona, Synthesis of a new platinum (II) complex: anticancer activity and nephrotoxicity in vitro, Toxicol. In Vitro 16 (4) (2002) 413–419.

[58] M. Koob, W. Dekant, Bioactivation of xenobiotics by formation of toxic glutathione conjugates, Chem. Biol. Interact. 77 (2) (1991) 107–136.

[59] J. Aiyar, H.J. Berkovits, R.A. Floyd, K.E. Wetterhahn, Reaction of chromium (VI) with glutathione or with hydrogen peroxide: identification of reactive intermediates and their role in chromium (VI)-induced DNA damage, Environ. Health Perspect. 92 (1991) 53–62.

[60] M.H. Hanigan, E.D. Lykissa, D.M. Townsend, C.N. Ou, R. Barrios, M.W. Lieberman, γ-Glutamyl transpeptidase-deficient mice are resistant to the nephrotoxic effects of cisplatin, Am. J. Pathol. 159 (5) (2001) 1889–1894.

[61] Y.I. Chirino, J. Pedraza-Chaverri, Role of oxidative and nitrosative stress in cisplatin-induced nephrotoxicity, Exp. Toxicol. Pathol. 61 (2009) 223–242.

[62] P. Cristofori, E. Zanetti, D. Fregona, A. Piaia, A. Trevisan, Renal proximal tubule segment-specific nephrotoxicity: an overview on biomarkers and histopathology, Toxicol. Pathol. 35 (2) (2007) 270–275.

[63] J. Nagai, M. Takano, Molecular aspects of renal handling of aminoglycosides and strategies for preventing the nephrotoxicity, Drug Metab. Pharmacokinet. 19 (2004) 59–70.

[64] A. Enomoto, H. Endou, Roles of organic anion transporters (OATs) and a urate transporter (URAT1) in the pathophysiology of human disease, Clin. Exp. Nephrol. 9 (3) (2005) 195–205.

[65] G. Ciarimboli, T. Ludwig, D. Lang, H. Pavenstädt, H. Koepsell, H.J. Piechota, J. Haier, U. Jaehde, J. Zisowsky, E. Schlatter, Cisplatin nephrotoxicity is critically mediated via the human organic cation transporter 2, Am. J. Pathol. 167 (6) (2005) 1477–1484.

[66] M. Dickenmann, T. Oettl, M.J. Mihatsch, Osmotic nephrosis: acute kidney injury with accumulation of proximal tubular lysosomes due to administration of exogenous solutes, Am. J. Kidney Dis. 51 (3) (2008) 491–503.

[67] M. Takenaka, E. Imai, T. Kaneko, T. Ito, T. Moriyama, A. Yamauchi, M. Hori, S. Kawamoto, K. Okubo, Kidney Int. 53 (1998) 562–572.

[68] H. Nakajima, M. Takenaka, J.-Y. Kaimori, Y. Nagasawa, A. Kosugi, S. Kawamoto, E. Imai, M. Hori, K. Okubo, Gene expression profile of renal proximal tubules regulated by proteinuria, Kidney Int. 61 (5) (2002) 1577–1587.

[69] J.Y. Kaimori, M. Takenaka, H. Nakajima, T. Hamano, M. Horio, T. Sugaya, T. Ito, M. Hori, K. Okubo, E. Imai, Induction of glia maturation factor-β in proximal tubular cells leads to vulnerability to oxidative injury through the p38 pathway and changes in antioxidant enzyme activities, J. Biol. Chem. 278 (35) (2003) 33519–33527.

[70] T. Hu, S.P. Ramachandrarao, S. Siva, C. Valancius, Y. Zhu, K. Mahadev, I. Toh, B.J. Goldstein, M. Woolkalis, K. Sharma, Reactive oxygen species production via NADPH oxidase mediates TGF-beta-induced cytoskeletal alterations in endothelial cells, Am. J. Physiol. Renal Physiol. 289 (2005) F816–F825.

[71] A. Zhang, Z. Jia, X. Guo, T. Yang, Aldosterone induces epithelial-mesenchymal transition via ROS of mitochondrial origin, Am. J. Physiol. Renal Physiol. 293 (2007) F723–F731.

[72] Y.F. Chen, A.W. Cowley Jr., A.P. Zou, Increased H2O2 counteracts the vasodilator and natriuretic effects of superoxide dismutation by tempol in renal medulla, Am. J. Physiol. Regul. Integr. Comp. Physiol. 285 (4) (2003) R827–R833.
[73] A. Makino, M.M. Skelton, A.P. Zou, A.W. Cowley Jr., Increased renal medullary H2O2 leads to hypertension, Hypertension 42 (2003) 25–30.
[74] J. Iglesias, V.E. Abernethy, Z. Wang, W. Lieberthal, J.S. Koh, J.S. Levine, Albumin is a major serum survival factor for renal tubular cells and macrophages through scavenging of ROS, Am. J. Physiol. Renal Physiol. 277 (5) (1999) F711–F722.
[75] P. Sun, P. Yue, W.H. Wang, Angiotensin II stimulates epithelial sodium channels in the cortical collecting duct of the rat kidney, Am. J. Physiol. Renal Physiol. 302 (6) (2012) F679–F687.
[76] P.M. Snyder, M.P. Price, F.J. McDonald, C.M. Adams, K.A. Volk, B.G. Zeiher, J.B. Stokes, M.J. Welsh, Mechanism by which Liddle's syndrome mutations increase activity of a human epithelial Na+ channel, Cell 83 (6) (1995) 969–978.
[77] R.A. Shimkets, D.G. Warnock, C.M. Bositis, C. Nelson-Williams, J.H. Hansson, M. Schambelan, J.R. Gill Jr., S. Ulick, R.V. Milora, J.W. Findling, C.M. Canessa, Liddle's syndrome: heritable human hypertension caused by mutations in the β subunit of the epithelial sodium channel, Cell 79 (3) (1994) 407–414.
[78] L. Yu, H.F. Bao, J.L. Self, D.C. Eaton, M.N. Helms, Aldosterone-induced increases in superoxide production counters nitric oxide inhibition of epithelial Na channel activity in A6 distal nephron cells, Am. J. Physiol. Renal Physiol. 293 (2007) F1666–F1677.
[79] Z. Ren, W. Liang, C. Chen, H. Yang, P.C. Singhal, G. Ding, Angiotensin II induces nephrin dephosphorylation and podocyte injury: role of caveolin-1, Cell. Signal. 24 (2) (2012) 443–450.
[80] K. Block, A. Eid, K.K. Griendling, D.Y. Lee, Y. Wittrant, Y. Gorin, Nox4 NAD(P)H oxidase mediates Src-dependent tyrosine phosphorylation of PDK-1 in response to angiotensin II: role in mesangial cell hypertrophy and fibronectin expression, J. Biol. Chem. 283 (2008) 24061–24076.
[81] B.L. Wharram, M. Goyal, J.E. Wiggins, S.K. Sanden, S. Hussain, W.E. Filipiak, T.L. Saunders, R.C. Dysko, K. Kohno, L.B. Holzman, R.C. Wiggins, Podocyte depletion causes glomerulosclerosis: diphtheria toxin–induced podocyte depletion in rats expressing human diphtheria toxin receptor transgene, J. Am. Soc. Nephrol. 16 (10) (2005) 2941–2952.
[82] J. Jia, G. Ding, J. Zhu, C. Chen, W. Liang, N. Franki, P.C. Singhal, Angiotensin II infusion induces nephrin expression changes and podocyte apoptosis, Am. J. Nephrol. 28 (3) (2008) 500–507.
[83] G. Ding, K. Reddy, A.A. Kapasi, N. Franki, N. Gibbons, B.S. Kasinath, P.C. Singhal, Angiotensin II induces apoptosis in rat glomerular epithelial cells, Am. J. Physiol. Renal Physiol. 283 (1) (2002) F173–F180.
[84] S. Terryn, O. Devuyst, Oxidative stress in the kidney: Proximal tubule disorders, in: T. Miyata, K. Eckardt, M. Nangaku (Eds.), Studies on Renal Disorders, Humana Press, New York, 2011, pp. 179–203.
[85] Y. Gorin, J.M. Ricono, N.H. Kim, B. Bhandari, G.G. Choudhury, H.E. Abboud, Nox4 mediates angiotensin II induced activation of Akt/protein kinase B in mesangial cells, Am. J. Physiol. Renal Physiol. 285 (2003) F219–F229.
[86] M. Briet, E.L. Schiffrin, Aldosterone: effects on the kidney and cardiovascular system, Nat. Rev. Nephrol. 6 (5) (2010) 261–273.
[87] G. Ding, A. Zhang, S. Huang, X. Pan, G. Zhen, R. Chen, T. Yang, ANG II induces c-Jun NH2-terminal kinase activation and proliferation of human mesangial cells via redox-sensitive transactivation of the EGFR, Am. J. Physiol. Renal Physiol. 293 (2007) F1889–F1897.

[88] K. Sverrisson, J. Axelsson, A. Rippe, D. Asgeirsson, B. Rippe, Acute reactive oxygen species (ROS)-dependent effects of IL-1beta, TNF-alpha, and IL-6 on the glomerular filtration barrier (GFB) in vivo, Am. J. Physiol. Renal Physiol. 309 (2015) F800–F806.
[89] A.K. Rehan, K.J. Johnson, R.C. Wiggins, R.G. Kunkel, P.A. Ward, Evidence for the role of oxygen radicals in acute nephrotoxic nephritis, Lab. Invest. 51 (4) (1984) 396–403.
[90] T. Bohler, J. Waiser, H. Hepburn, J. Gaedeke, C. Lehmann, P. Hambach, K. Budde, H.H. Neumayer, TNF-alpha and IL-1alpha induce apoptosis in subconfluent rat mesangial cells. Evidence for the involvement of hydrogen peroxide and lipid peroxidation as second messengers, Cytokine 12 (2000) 986–991.
[91] V. Moreno-Manzano, Y. Ishikawa, J. Lucio-Cazana, M. Kitamura, Selective involvement of superoxide anion, but not downstream compounds hydrogen peroxide and peroxynitrite, in tumor necrosis factor-alpha-induced apoptosis of rat mesangial cells, J. Biol. Chem. 275 (2000) 12684–12691.
[92] C.F. Mueller, K. Laude, J.S. McNally, D.G. Harrison, ATVB in focus: redox mechanisms in blood vessels, Arterioscler. Thromb. Vasc. Biol. 25 (2) (2005) 274–278.
[93] K.K. Griendling, M. Ushio-Fukai, NADH/NADPH oxidase and vascular function, Trends Cardiovasc. Med. 7 (8) (1997) 301–307.
[94] K. Bedard, K.H. Krause, The NOX family of ROS-generating NADPH oxidases: physiology and pathophysiology, Physiol. Rev. 87 (1) (2007) 245–313.
[95] B. Lassegue, R.E. Clempus, Vascular NAD(P)H oxidases: specific features, expression, and regulation, Am. J. Physiol. Regul. Integr. Comp. Physiol. 285 (2) (2003) R277–R297.
[96] W.J. Welch, H. Baumgartl, D. Lubbers, C.S. Wilcox, Nephron pO2 and renal oxygen usage in the hypertensive rat kidney, Kidney Int. 59 (2001) 230–237.
[97] M. Araujo, C.S. Wilcox, Oxidative stress in hypertension: role of the kidney, Antioxid. Redox Signal. 20 (2014) 74–101.
[98] C.S. Wilcox, Oxidative stress and nitric oxide deficiency in the kidney: a critical link to hypertension? Am. J. Physiol. Regul. Integr. Comp. Physiol. 289 (2005) R913–R935.
[99] N.D. Vaziri, B. Rodriguez-Iturbe, Mechanisms of disease: oxidative stress and inflammation in the pathogenesis of hypertension, Nat. Clin. Pract. Nephrol. 2 (2006) 582–593.
[100] S. Fichtlscherer, S. Breuer, A.M. Zeiher, Prognostic value of systemic endothelial dysfunction in patients with acute coronary syndromes: further evidence for the existence of the "vulnerable" patient, Circulation 110 (2004) 1926–1932.
[101] J.P. Halcox, W.H. Schenke, G. Zalos, R. Mincemoyer, A. Prasad, M.A. Waclawiw, K.R. Nour, A.A. Quyyumi, Prognostic value of coronary vascular endothelial dysfunction, Circulation 106 (2002) 653–658.
[102] C.A. Hamilton, M.J. Brosnan, M. McIntyre, D. Graham, A.F. Dominiczak, Superoxide excess in hypertension and aging: a common cause of endothelial dysfunction, Hypertension 37 (2001) 529–534.
[103] M. Jameson, F.X. Dai, T. Luscher, J. Skopec, A. Diederich, D. Diederich, Endothelium-derived contracting factors in resistance arteries of young spontaneously hypertensive rats before development of overt hypertension, Hypertension 21 (1993) 280–288.
[104] F. Liano, E. Junco, J. Pascual, R. Madero, E. Verde, The spectrum of acute renal failure in the intensive care unit compared with that seen in other settings. The Madrid Acute Renal Failure Study Group, Kidney Int. Suppl. 66 (1998) S16–S24.
[105] S. Gao, J. Chen, S.V. Brodsky, H. Huang, S. Adler, J.H. Lee, N. Dhadwal, L. Cohen-Gould, S.S. Gross, M.S. Goligorsky, Docking of endothelial nitric oxide synthase (eNOS) to the mitochondrial outer membrane: a pentabasic amino acid sequence in the autoinhibitory domain of eNOS targets a proteinase K-cleavable peptide on the cytoplasmic face of mitochondria, J. Biol. Chem. 279 (2004) 15968–15974.
[106] K.F. Beck, G. Guder, L. Schaefer, M. Pleskova, A. Babelova, M.H. Behrens, D. Mihalik, M. Beck, R.M. Schaefer, J. Pfeilschifter, Nitric oxide upregulates induction of PDGF receptor-alpha expression in rat renal mesangial cells and in anti-Thy-1 glomerulonephritis, J. Am. Soc. Nephrol. 16 (2005) 1948–1957.

[107] E. Dreieicher, K.F. Beck, S. Lazaroski, M. Boosen, W. Tsalastra- Greul, M. Beck, I. Fleming, L. Schaefer, J. Pfeilschifter, Nitric oxide inhibits glomerular TGF-beta signaling via SMOC-1, J. Am. Soc. Nephrol. 20 (2009) 1963–1974.
[108] G.C. Brown, Nitric oxide regulates mitochondrial respiration and cell functions by inhibiting cytochrome oxidase, FEBS Lett. 369 (1995) 136–139.
[109] R.D. Guzy, B. Hoyos, E. Robin, H. Chen, L. Liu, K.D. Mansfield, M.C. Simon, U. Hammerling, P.T. Schumacker, Mitochondrial complex III is required for hypoxiainduced ROS production and cellular oxygen sensing, Cell Metab. 1 (2005) 401–408.
[110] C.A. Piantadosi, H.B. Suliman, Mitochondrial transcription factor A induction by redox activation of nuclear respiratory factor 1, J. Biol. Chem. 281 (2006) 324–333.
[111] H.J. Kim, T. Sato, B. Rodriguez-Iturbe, N.D. Vaziri, Role of intrarenal angiotensin system activation, oxidative stress, inflammation, and impaired nuclear factor-erythroid-2-related factor 2 activity in the progression of focal glomerulosclerosis, J. Pharmacol. Exp. Ther. 337 (2011) 583–590.
[112] A.P. Zou, N. Li, A.W. Cowley Jr., Production and actions of superoxide in the renal medulla, Hypertension 37 (2001) 547–553.
[113] R.B. Colvin, Chronic allograft nephropathy, N. Engl. J. Med. 349 (2003) 2288–2290.
[114] A. Djamali, Oxidative stress as a common pathway to chronic tubulointerstitial injury in kidney allografts, Am. J. Physiol. Renal Physiol. 293 (2007) F445–F455.
[115] I. Lautenschlager, A. Soots, L. Krogerus, H. Kauppinen, O. Saarinen, C. Bruggeman, J. Ahonen, Effect of cytomegalovirus on an experimental model of chronic renal allograft rejection under triple-drug treatment in the rat, Transplantation 64 (1997) 391–398.
[116] Y.Y. Ng, T.P. Huang, W.C. Yang, Z.P. Chen, A.H. Yang, W. Mu, D.J. NikolicPaterson, R.C. Atkins, H.Y. Lan, Tubular epithelial-myofibroblast transdifferentiation in progressive tubulointerstitial fibrosis in 5/6 nephrectomized rats, Kidney Int. 54 (1998) 864–876.
[117] K.M. Park, J.Y. Byun, C. Kramers, J.I. Kim, P.L. Huang, J.V. Bonventre, Inducible nitric-oxide synthase is an important contributor to prolonged protective effects of ischemic preconditioning in the mouse kidney, J. Biol. Chem. 278 (2003) 27256–27266.
[118] O. Skalli, P. Ropraz, A. Trzeciak, G. Benzonana, D. Gillessen, G. Gabbiani, A monoclonal antibody against alpha-smooth muscle actin: a new probe for smooth muscle differentiation, J. Cell Biol. 103 (1986) 2787–2796.

CHAPTER

7

The role of oxidative stress and antioxidants across the spectrum of acute coronary syndrome

Wahid Ali, Neha Srivastava, and Priyanka Sharma
Department of Pathology, King George's Medical University, Chowk, Lucknow, Uttar Pradesh, India

1 Introduction

Oxidative stress is caused by the reactive oxygen species (ROS) which are produced when the cell is exposed to such an environment that is not familiar to the cell. Thus, oxidative stress affects several physiological and pathological conditions. Oxidative stress alters the lipid profiling, protein content, and the DNA of the cell [1]. Several studies have reported that generation of ROS causes many cardiovascular infections such as ischemia–reperfusion heart injury, arrhythmia, etc. [2]. To cope with this oxidative stress, several remediation strategies have been adopted. Cancer prevention agents, messenger RNA (mRNA) utilizing methodologies, and nanomedicine have been widely used for the treatment of several cardiovascular diseases. This chapter gives a point-by-point depiction of oxidative pressure, and attempts to connect oxidative stress with cardiovascular diseases. Finally, the chapter focuses on the clinical remediation strategies of oxidative pressure in cardiovascular infections.

Mounting evidence shows that oxidative pressure has an indispensable part in the turn of events and pathology of different infections [3]. Overproduction of reactive oxygen species, such as superoxide anion ($O_2^{\cdot-}$), singlet oxygen, ozone (O_3) hydrogen peroxide (H_2O_2), and alkoxyl ($RO^{\cdot}$) hydroxyl ion (OH) peroxyl revolutionaries (ROO), causes oxidative stress, which interferes with the homeostasis of the cell and its guard framework [3]. Overproduction of ROS has an inconvenient job in the natural

https://doi.org/10.1016/B978-0-323-90905-1.00007-9

framework by not just focusing on organic particles, like lipid, protein, and DNA, but also acting as a second courier in cell flagging [4]. Through focusing on administrative pathways, ROS brings about cell provocative signs enactment or modified cell demise cardiovascular sicknesses are the main source of bleakness and mortality around the world. Research shows that oxidative pressure plays a significant part in the movement of different cardiovascular sicknesses, like atherosclerosis and cardiovascular breakdown (HF) [5].

2 Oxidative stress

Endogenously produced ROS mainly target mitochondria being part of the electron transport chain (ETC) and oxidative phosphorylation. Mitochondria, including both external and internal mitochondrial layers, is the destination of adenosine triphosphate (ATP) production. During the Krebs cycle, produced acetyl-CoA reaches the mitochondria where it eventually binds with ATP. This results in neurotic conditions and leads to disturbed mitochondrial respiratory chain, and the electron reacts to oxygen to produce superoxide.

Several catalysts are involved in the production of ROS. A mitochondrial porousness progress pore (mPTP) opening occurs due to ROS and affects the solidity of mitochondrial film, which works with the arrival of ROS from the mitochondrial grid into the cytosol.

Several chemical agents, such as myeloperoxidase (MPO), xanthine oxidase (Xo), lipoxygenase, uncoupled nitric oxide synthase (NOS), and nicotinamide adenine dinucleotide phosphate (NADPH) oxidase (NOX), are placed outside of the mitochondrial membrane and are involved in the generation of ROS. Nicotinamide adenine dinucleotide phosphate oxidase has been reported as a main player in the generation of ROS. There are seven members of the NOX family, such as NOX1, NOX2, NOX4, and NOX5, reported in the cardiovascular framework. NOX2, also called gp-91phox, is the most extensively examined one. The enactment of NOX2 relies upon different subunits of the NOX family including p22phox, p67phox, p40phox, and p47phox. Phosphorylation at Ser303, 304, 328, and p47phox occurs due to NOX2. Phosphorylation at p47phox facilitates the binding of p47phox with p22phox, resulting in p40phox and p67phox being open for the binding of Nox2. After activation of Nox2, NADPH binds with the intracellular C-end, which results in the production of ROS by transferring the electrons from NADPH to oxygen. The produced reactive oxygen species induce mitochondrial dysfunction, Xo activation, and NOS uncoupling, which are part of secondary oxidase.

Exogenous production of ROS has been induced by the ROS, which is produced by endogenously and environmental factors. Environmental pollutants, xenobiotics, alcohol, ultraviolet radiation, and smoking are the main sources of exogenously produced ROS. The external environment disturbs the homeostasis of the cell that enhances the endogenous ROS. ROS production has been increased by cigarette smoke stimulating the NOX, whereas UV radiation produces ROS by interacting with water. Antioxidants and cytoprotective enzymes have been inhibited by alcohol, which also leads to production of ROS.

3 Endothelial dysfunction and oxidative stress

Several cardiovascular disorders such as dyslipidemia, hypertension, chronic kidney disease, diabetes, atherosclerosis, ischemia/reperfusion injury, and heart failure are caused by endothelial dysfunction due to oxidative stress [6]. Impaired antiinflammatory and antithrombotic endothelial characteristics, altered modulation of vasomotor tone, vascular growth, and disturbed vascular modeling resulting in endothelial dysfunction are caused by oxidative stress which is produced by an imbalance between NO bioavailability and ROS [7].

A simple squamous layer of cells, the endothelium, discriminate between the vascular wall and blood circulation. When a cell faces vascular stress, endothelium-dependent vasorelaxation has been released by healthy endothelium provides which regulate the aggregation of platelet and permeability of vascular system [8]. It is very reactive to mechanical stimuli, chemical factors, and humoral agents by producing several mediators, such as NO, to maintain vasomotor tone and structural integrity. NO is a potent antiinflammatory, vasodilatory, and antithrombotic, and thus plays an important role in endogenous antioxidant defense. Several enzymes are responsible for the production of NO; cytochrome *p*450 reductase-like enzyme and endothelial nitric oxide synthase (eNOS) produced NO from L-arginine by using tetrahydrobiopterin [9]. The bioavailability of NO has been decreased by degradation of NO and NO degradation caused by down-regulation of eNOS, excessive ROS, and deficiency of substrates or cofactors for eNOS, and inappropriate activation of eNOS due to impaired cellular signaling [10].

4 Molecular effects of oxidative stress

Adhesion of molecules on the vascular endothelium results in the secretion of several proinflammatory mediators which causes a multisystemic, chronic, and progressive inflammatory disease named atherosclerosis. This disease is characterized by the interaction of endothelial and immune cells [11]. The imbalance between the antioxidant mechanism and the generation of excess reactive oxygen species (ROS) leads to enhanced oxidative stress, resulting in the formation of atherogenic oxidized low-density lipoprotein (Ox-LDL), which is a major determinant of atherogenesis [12]. Several sources including lipoxygenase, xanthine oxidase, nNOs, iNOS, eNOS, and nicotinamide adenine dinucleotide phosphate oxidase play an important role in ROS production, and synthesized ROS causes destruction of mitochondrial capacity which results in dysfunction of mitochondria [8]. This in turn enhances ROS synthesis and thus causes oxidative stress. This results in the initiation, formation, and progression of an atherosclerotic lesion [13]. Excess synthesis of NADPH results in endothelial cells when free fatty acids enter the tricarboxylic acid cycle, and excess NADPH is a vital component of ROS formation during oxidative phosphorylation. Several studies have reported that obesity, age, hypertension, diabetes, dyslipidemia, and smoking enhance the production of ROS in mitochondria.

5 Oxidative stress in coronary artery disease

In atherosclerosis, dyslipidemia and endothelial dysfunction have also been reported due to an imbalance between enzymatic and nonenzymatic antioxidant protection systems and ROS production [2]. Numerous studies have been reported in patients with coronary artery disease (CAD) due to impaired balance of prooxidants and antioxidants. In the present scenario, produced ROS and thus oxidative stress is a new risk factor responsible for the development of CAD that affects the onset, prognosis, quality of life, and survival of patients [4].

In atherosclerosis, oxidative stress causes modification in the oxidative system and causes lipid peroxidation at deoxyribonucleic acids (DNA) and protein level; this results in serious negative effects on the function and structure of the vascular system. Including classical free oxygen radicals hydrogen peroxide (H_2O_2), superoxide radical (O_2^-), peroxyl (RO), hydroxyl (OH), and hydroperoxyl (HRO^-), ROS has also been involved in the oxidative stress process of nitrogen species (RNS), especially peroxynitrite ($ONOO^-$) [5]. Signaling molecules are damaged by ROS distorting the signaling pathway, which is involved in vascular inflammation, and it destroys essential biomolecules in cells and participates in oxidative modification of lipids that make them an atherogenic environment. Phagocytic isoforms of NADPH oxidase (Nox2 and Nox1) are the most important sources of oxidative stress with their regulatory subunit xanthine oxidase (XO), p47phox, and dysregulated eNOS [1].

Different signaling pathways induce activation of NADPH oxidase, which results in common risk factors including hyperglycemia, hyperlipidemia, hypoxia, smoking, etc. Mitochondrial enzymes and NADPH oxidases increase ROS production and cause fibrosis, and increase metalloproteinase and cardiomyocyte hypertrophy [7]. The PKC$\alpha/\beta 2$ signaling pathway is involved in the activation of NADPH, in which protein kinase C plays a key role. For activation of Nox2, phosphorylation of the p47phox subunit is required, which is the major Nox2 (gp91phox) [11]. Patients with CAD have high expression of p47phox and are overweight by about 60%, and obese patients with CAD are overweight by about 80%. So far, XO is known to be significantly elevated in CAD patients, and overweight is thought to be a potent driver of the enhanced XO expression [12].

6 Coronary microvascular dysfunction

Coronary microvascular brokenness (CMD) is an issue that prompts the improvement of myocardial ischemia, in spite of the fact that there is no demonstrated deterrent in coronary veins on coronary angiography [14]. Hazard factors that can trigger oxidative pressure in coronary microvascular brokenness are overweight, dyslipidemia, diabetes, and the metabolic condition. A few issues—for example, hypertrophic cardiomyopathy, hypertensive coronary illness, myocarditis, and vasculitis—are models where myocardial ischemia can create without the presence of coronary supply route block [15]. Notwithstanding these conditions, primary and utilitarian adjustments in the coronary microcirculation might be liable for the event of myocardial ischemia, in up to 20% of patients with intense coronary disorder (ACS) and up to half of patients with persistent coronary conditions (CCS) [16].

7 Oxidative stress and lipids, protein, and DNA damage

Intracellular ROS are profoundly dynamic and temperamental. They could cause changes of protein, lipid, and core, and adjust the guideline of protein capacity and sign pathways. Lipids, especially polyunsaturated unsaturated fats (PUFAs) and cholesterol, are viewed as significant objective substrates of oxidative pressure since cell films are comprised of lipids [17]. In addition, the lipids are significant metabolites in the cell. Lipid oxidation can be isolated into three stages: commencement, engendering, and end [18]. The inception begins with the development of free radicals by the unsaturated lipid particle losing a hydrogen iota. During the spread cycle, peroxyl radicals are shaped by the response between lipid revolutionaries and oxygen. The created peroxyl revolutionaries at that point assault different lipids to frame more lipid peroxyl extremists [4]. At the point when the hydrogen source is totally spent, a great deal of nonradical products are delivered. Dynamic aldehydes, for example, 4-hydroxynonenal (4-HNE) and malondialdehyde (MDA), are significant results of lipid peroxidation that have a significant job in the pathogenesis of numerous infections. The degrees of dynamic aldehydes in the blood are prescient of the infection movement [12].

8 Therapies for oxidative stress-associated cardiovascular diseases

8.1 Antioxidant molecules

8.1.1 Nutritional supplements

Gathering confirmations show that numerous wholesome enhancements have cell reinforcement properties. Nutrient A will be a progression of unsaturated healthful natural mixtures and can respond with oxidative species. Nutrient A has appeared to alter the impact of apolipoproteins on the danger of myocardial localized necrosis. Note that carotenoids are the forerunners of nutrient A, and astaxanthin, quite possibly the most prominent carotenoid, is viewed as a ROS scrounger [19]. Preclinical investigations of astaxanthin exhibited the defensive impact in I/R injury and thrombotic sicknesses in creature models. Additionally, astaxanthin has been shown to diminish pulse in suddenly hypertensive rodents. In a randomized twofold dazed clinical preliminary, the admission of astaxanthin decreased the serum lipid peroxidation biomarker levels while expanding the SOD level [7].

Vitamin C can detoxify exogenous and endogenous ROS as well as the modified proteins and lipids that were modified by the ROS. Studies have shown that vitamin C could control endothelial cell proliferation and apoptosis and smooth muscle-mediated vasodilation, which all three are equally important in the pathogenesis of cardiovascular diseases. Vitamin E inhibits superoxide production by impairing assembly of NOX enzymes. Experimental studies have shown that vitamin E can reduce the risk of coronary heart disease and decrease cardiovascular complications. A meta-analysis of 400,000 patients indicated that the rate of coronary heart disease was decreased by vitamin E and vitamin C intake.

Omega-3 addresses an appealing technique to decrease the weakness to oxidative pressure injury in myocardial cells by tweaking redox pathways. In a rodent model of myocardial localized necrosis, rodents enhanced with omega-3 showed lower infarct size. Flavanols, such as quercetin, are used to diminish oxidative pressure markers and improve heart work in both creature models and patients with cardiovascular illnesses. A clinical preliminary with 805 elderly individuals revealed that mortality from coronary illness was diminished by high flavonoid consumption.

Then again, a meta-examination (incorporating 66 randomized preliminaries) inspecting the impact of wholesome mixtures to treat malignancy or cardiovascular illnesses showed blended outcomes. Out of 66 preliminaries analyzed, 24 showed a positive result, 39 showed an adverse result, and three showed an unbiased result. The positive result was generally seen in members who were viewed as experiencing ailing health. These discoveries suggest that the situation with nourishment of patients will influence the viability of cancer prevention agent supplementation. It is viable for patients who do not have certain supplements that add to the cell reinforcement safeguard organization. Regardless of whether the extra admission will have beneficial effects in those patients who are not insufficient in these supplements are indistinct as of now.

8.1.2 Novel experimental antioxidant-based therapies

The second classes of mixtures are the inhibitors of oxidative pressure arrangement incorporate medications focusing on Xo, NOX, and MPO. The Xo inhibitor, allopurinol, is a promising remedial specialist. Albeit enormous scope planned investigations assessing allopurinol in cardiovascular sicknesses are as yet inadequate with regards to, little clinical examinations have shown a valuable impact of allopurinol in hypertension, I/R injury, and HF by restricting the oxidative pressure in endothelial cells [20]. A meta-investigation including 10 clinical examinations assessing the impact of allopurinol on circulatory strain additionally showed a minor decrease of pulse. Furthermore, the investigations assessing allopurinol in patients going through coronary conduit sidestep uniting or essential percutaneous coronary intercession showed diminished in-medical clinic mortality and heart complexities. Concerning the NOX inhibitors, GKT137831 is the principal NOX inhibitor in clinical turn of events [21]. The treatment utilizing GKT137831 brought about a significant enemy of atherosclerotic impact in apo-E KO mice. GKT137831 additionally protected cardiovascular capacity after I/R injury in mice. In a mice model of atherosclerosis, MPO inhibitors had the option to change the atherosclerotic sore piece and heart redesigning. The last group is the mixtures that permit utilitarian fix of ROS-prompted harm, which essentially centers around nitric oxide-cyclic guanosine monophosphate (NO-cGMP) flagging. ROS can meddle with NO-cGMP motioning through uncoupling NOS, artificially rummaging NO, or oxidatively harming the NO receptor (solvent guanylylcyclase, sGC). The medications in this group incorporate HNO givers, for example, CXL-1427, which is the second era of HNO contributor compounds [22]. A Phase 2a portion heightening investigation showed a great well-being profile and hemodynamic impacts in hospitalized patients with HFrEF. L-Citrulline and L-arginine are additionally remembered for this group since they can recouple the NOS. A meta-investigation of 11 randomized, twofold dazed, fake treatment controlled preliminaries showed that oral organization of L-arginine had the option to bring down the systolic circulatory strain by 5.39 mmHg and the diastolic pulse by 2.66 mmHg. A meta-examination of 15 randomized controlled preliminaries with 424 members showed that L-citrulline organization came about in 7.54 mmHg decrease in the systolic pulse and 3.77 mmHg decrease in the diastolic circulatory strain [23].

8.1.3 Antioxidant role of clinical drugs

There are drugs that are currently being used in a clinical setting that are known to have antioxidant effects, such as melatonin, proprotein convertase subtilisin/kexin type 9 (PCSK9) inhibitor, carvedilol, and metformin. Melatonin is a potent free radical scavenger because of its antioxidant properties and has been shown to have a protective role in myocardial I/R injury, HF, and atherosclerosis via targeting oxidative stress in animal studies [24]. PCSK9 inhibition decreases the ROS production in endothelial cells and smooth muscle cells by inhibiting lectin-like oxidized low-density lipoprotein receptor-1 (LOX-1) expression in mice atherosclerosis models [25]. Carvedilol is a combined β1-, β2-, and α1-adrenergic blocking agent that has antioxidant properties. It is important in the treatment of HF [26]. Metformin reduces ROS production by inhibiting the NOX pathway and increasing antioxidant genes [27]. Furthermore, because metformin is a structural analog of ADMA (asymmetric dimethylarginine), it can help regulate the balance between NO and ADMA. Both the animal studies and the clinical trials showed a protective role in CAD and HF.

8.2 miRNAs

miRNAs are arrangements of little noncoding RNAs credited to control quality articulation by focusing on courier RNAs [28]. Aggregating proof shows that miRNAs are engaged with oxidative pressure reaction and are basic in managing oxidative pressure. miRNA-210 is a notable hypoxia-prompted RNA which is essentially up-regulated during hypoxia. miRNA-210 is considered to be the main anoxic-related miRNA in the body; it improves cardiovascular capacity by hindering apoptosis and advancing angiogenesis [29]. The outflow of miRNA-210 is mostly initiated by HIF-1α. In the HIF-1α knockout mice, the miRNA-210 level was diminished contrasted with that of the wild-type mice. In the infarcted myocardial tissues of patients who died from intense myocardial localized necrosis, the miRNA-210 level was expanded contrasted and the heart tissues from the benchmark group [30]. In an animal model of myocardial localized necrosis, the heart work was improved by a direct myocardial infusion of miRNA-210. Furthermore, flowing miRNA-210 level showed a critical relationship with cardiovascular-related mortality in patients giving intense coronary disorder. miRNA-1 is the most plentiful miRNA communicated in cardiovascular muscles and has a vital role in separation and multiplication of muscle cells. H_2O_2 was found to expand miRNA-1 in rodent cardiomyocytes. Over-expression of miRNA-1 brought about expanded ROS and reduced creation of SOD. In a rodent model of myocardial localized necrosis, the measure of miRNA-1 was significantly connected with infarct size [31]. Likewise, organization of miRNA-1 after myocardial localized necrosis improved cardiovascular capacity in mice. Furthermore, serum levels of miRNA-1 in patients with intense coronary disorder were associated with the coursing troponin T. Other miRNAs, such as miRNA-132, miRNA-21, and miRNA-17, are additionally demonstrated to be up-regulated during hypoxia. Furthermore, numerous miRNAs are significant in the guideline of atherosclerotic plaque arrangement [32]. A cross-sectional observational examination with 100 subjects showed an expansion in miRNA-133 level in the patients with CAD contrasted and the benchmark group, and the miRNA-133 level was expanded in suggestive plaques. Furthermore, restraint of miRNA-133 can target NOS to forestall endothelial brokenness. miRNA-92a can direct the declaration of endothelial NOS to influence endothelial cells [33]. miRNA-92a has been shown to decrease plaque irritation and expanded the plaque strength by advancing endothelial cell expansion and angiogenesis. Nonetheless, in view of low solidness and bioavailability, much work is required to make miRNAs an achievable remedial alternative. For future clinical applications, various techniques dependent on inciting or subduing miRNA articulation, like the utilization of miRNA foes or imitates, are additionally being investigated [34].

8.3 Limitations

Although oxidative pressure assumes a significant part in different cardiovascular illnesses, utilization of cancer prevention agent treatment so far has been restricted in clinical settings. To begin with, in spite of the fact that concealment of oxidative pressure utilizing cancer prevention agents has appeared to be useful in numerous animal models, the useful impacts of these cell reinforcement treatments in human clinical examinations have been questionable. One reason might be because of vague concealment of ROS, which may not be appealing or successful on the grounds that it could upset significant ROS-interceded cell flagging. In this manner, directed concealment at the site of ROS overproduction (e.g., in the heart for myocardial dead tissue), for example, utilizing focused on nanoparticles, may offer more compelling cell reinforcement treatment. In addition, the unique character of the infection makes it essential to pick a proper chance to utilize the cell reinforcements. At various phases of infections, oxidative pressure may have various jobs. Consequently, it is essential to give antioxidative medicines at the suitable time. Ultimately, for novel mixtures, the vast majority of the investigations depend on the research facility tests. All the more enormous scope clinical preliminaries with different cardiovascular patients are required right now.

9 Conclusions

This chapter focused on the relationship between oxidative stress and cardiovascular diseases, and discussed the current status of different antioxidant drugs that are being used or being studied for cardiovascular diseases. Furthermore, we included novel strategies that have potential to be used in future therapies. Oxidative stress plays an important role in the development and the evolution of cardiovascular diseases. Various therapeutic strategies targeting oxidative stress have been developed. Although animal studies have shown beneficial effects of antioxidant therapy in various cardiovascular diseases, the clinical outcomes vary in human trials. A deeper understanding of oxidative stress in the cardiovascular diseases and development of better antioxidant therapies are needed to ensure more effective treatments of cardiovascular diseases.

References

[1] L. Abid, Z. Ben Ali, I. Gtif, N. Kharrat, S. Charfeddine, R. Hammami, et al., Oxidative stress and severity of coronary artery disease in patients with acute coronary syndrome, Arch. Cardiovasc. Dis. Suppl. 11 (2019), https://doi.org/10.1016/j.acvdsp.2019.01.011, e287.

[2] N.R. Madamanchi, A. Vendrov, M.S. Runge, Oxidative stress and vascular disease, Arterioscler. Thromb. Vasc. Biol. 25 (2005) 29–38.

[3] A. Bastani, S. Rajabi, A. Daliran, H. Saadat, F. Karimi-Busheri, Oxidant and antioxidant status in coronary artery disease, Biomed. Rep. 9 (2018) 327–332.

[4] T. Senoner, W. Dichtl, Oxidative stress in cardiovascular diseases: still a therapeutic target? Nutrients 11 (2019).

[5] T. Costacou, R.W. Evans, G.L. Schafer, T.J. Orchard, Oxidative stress and response in relation to coronary artery disease in type 1 diabetes, Diabetes Care 36 (2013) 3503–3509.

[6] Z.S. Nedeljkovic, N. Gokce, J. Loscalzo, Mechanisms of oxidative stress and vascular dysfunction, Postgrad. Med. J. 79 (2003) 195–200.

[7] K. Cervantes-Gracia, K. Raja, D. Llanas-Cornejo, J.N. Cobley, I.L. Megson, R. Chahwan, et al., Oxidative stress and inflammation in the development of cardiovascular disease and contrast induced nephropathy, Vessel Plus 2020 (2020).

[8] J.L. Greaney, E.F.H. Saunders, L. Santhanam, L.M. Alexander, Oxidative stress contributes to microvascular endothelial dysfunction in men and women with major depressive disorder, Circ. Res. 124 (2019) 564–574.

[9] M. Jerkic, M. Letarte, Contribution of oxidative stress to endothelial dysfunction in hereditary hemorrhagic telangiectasia, Front. Genet. 6 (2015) 1.

[10] B.R. Silva, L. Pernomian, L.M. Bendhack, Contribution of oxidative stress to endothelial dysfunction in hypertension, Front. Physiol. 3 (2012) 1–5.

[11] A. García-Sánchez, A.G. Miranda-Díaz, E.G. Cardona-Muñoz, The role of oxidative stress in physiopathology and pharmacological treatment with pro- and antioxidant properties in chronic diseases, Oxidative Med. Cell. Longev. 2020 (2020).
[12] W.K. Peng, L. Chen, B.O. Boehm, J. Han, T.P. Loh, Molecular phenotyping of oxidative stress in diabetes mellitus with point-of-care NMR system, NPJ Aging. Mech. Dis. 6 (2020) 1–12, https://doi.org/10.1038/s41514-020-00049-0.
[13] J.G. Scandalios, Oxidative stress: molecular perception and transduction of signals, Braz. J. Med. Biol. Res. 38 (2005) 995–1014.
[14] C.N. Bairey Merz, C.J. Pepine, H. Shimokawa, C. Berry, Treatment of coronary microvascular dysfunction, Cardiovasc. Res. 116 (2020) 856–870.
[15] P. Ong, B. Safdar, A. Seitz, A. Hubert, J.F. Beltrame, E. Prescott, Diagnosis of coronary microvascular dysfunction in the clinic, Cardiovasc. Res. 116 (2020) 841–855.
[16] F. Crea, G.A. Lanza, P.G. Camici, Coronary Microvascular Dysfunction, 2013, pp. 1–257.
[17] U. Kołodziej, M. Maciejczyk, A. Zalewska, An elimination systems for oxidized biomolecules oxidative stress-repair systems of oxidatively damaged biomolecules, Prog. Health Sci. 8 (2018) 2018–2145.
[18] K.A. Johnson, E.K. Moran, J.A. Becker, D. Blacker, A.J. Fischman, M.S. Albert, Differences in mild cognitive impairment, Dementia 64 (2007) 240–247.
[19] R.J. Pickering, Oxidative stress and inflammation in cardiovascular diseases, Antioxidants 10 (2021) 1–2.
[20] A.L. Levonen, E. Vähäkangas, J.K. Koponen, S. Ylä-Herttuala, Antioxidant gene therapy for cardiovascular disease: current status and future perspectives, Circulation 117 (2008) 2142–2150.
[21] Y. Xu, C. Wang, J. Klabnik, O' Donnell J., Novel therapeutic targets in depression and anxiety: antioxidants as a candidate treatment, Curr. Neuropharmacol. 12 (2014) 108–119.
[22] D. Sorriento, N. De Luca, B. Trimarco, G. Iaccarino, The antioxidant therapy: New insights in the treatment of hypertension, Front. Physiol. 9 (2018).
[23] J.G. Farías, V.M. Molina, R.A. Carrasco, A.B. Zepeda, E. Figueroa, P. Letelier, et al., Antioxidant therapeutic strategies for cardiovascular conditions associated with oxidative stress, Nutrients 9 (2017) 1–23.
[24] G. Ribaudo, M. Bortoli, C. Pavan, G. Zagotto, L. Orian, Antioxidant potential of psychotropic drugs: from clinical evidence to in vitro and in vivo assessment and toward a new challenge for in silico molecular design, Antioxidants 9 (2020) 1–22.
[25] A. Zampelas, R. Micha, Antioxidants in Health and Disease, 2015, pp. 1–302.
[26] V.T.V. Dao, A.I. Casas, G.J. Maghzal, T. Seredenina, N. Kaludercic, N. Robledinos-Anton, et al., Pharmacology and clinical drug candidates in redox medicine, Antioxid. Redox Signal. 23 (2015) 1113–1129.
[27] L. Fási, F. Di Meo, C.Y. Kuo, S. Stojkovic Buric, A. Martins, N. Kúsz, et al., Antioxidant-inspired drug discovery: antitumor metabolite is formed in situ from a Hydroxycinnamic acid derivative upon free-radical scavenging, J. Med. Chem. 62 (2019) 1657–1668.
[28] T. Carbonell, A.V. Gomes, MicroRNAs in the regulation of cellular redox status and its implications in myocardial ischemia-reperfusion injury, Redox Biol. 36 (2020) 101607, https://doi.org/10.1016/j.redox.2020.101607.
[29] J. Banerjee, S. Khanna, A. Bhattacharya, MicroRNA regulation of oxidative stress, Oxidative Med. Cell. Longev. 2017 (2017) 1–4.
[30] Z. Chen, T.P. Shentu, L. Wen, D.A. Johnson, J.Y.J. Shyy, Regulation of SIRT1 by oxidative stress-responsive miRNAs and a systematic approach to identify its role in the endothelium, Antioxid. Redox Signal. 19 (2013) 1522–1538.

[31] E. Klieser, C. Mayr, T. Kiesslich, T. Wissniowski, P. Di Fazio, D. Neureiter, et al., The crosstalk of miRNA and oxidative stress in the liver: from physiology to pathology and clinical implications, Int. J. Mol. Sci. 20 (2019).
[32] J. Konovalova, D. Gerasymchuk, I. Parkkinen, P. Chmielarz, A. Domanskyi, Interplay between MicroRNAs and oxidative stress in neurodegenerative diseases, Int. J. Mol. Sci. 20 (2019).
[33] S. Wang, C. Guo, M. Yu, X. Ning, B. Yan, J. Zhao, et al., Identification of H2O2 induced oxidative stress associated microRNAs in HLE-B3 cells and their clinical relevance to the progression of age-related nuclear cataract, BMC Ophthalmol. 18 (2018) 1–11.
[34] A. Robson, Oxidation of miRNAs by ROS leads to cardiac hypertrophy, Nat. Rev. Cardiol 17 (11) (2020) 678.

CHAPTER

8

Biosensors in the detection of oxidative stress using discovered biomarkers

Swaraj Mohanty and Yasmin Ahmad

Peptide and Proteomics Division, Defence Institute of Physiology & Allied Sciences (DIPAS), Defence Research & Development Organization (DRDO), Delhi, India

1 Introduction

The living biological system relies on a series of biochemical reactions for the survival of species over time. The by-products of complex biochemical reactions show a variable trend in their accumulation inside the intracellular and extracellular fluids before elimination. The by-products of this process generate a large number of ionic molecules and free radicals, which can be detoxified by the production of a large number of antioxidants, and this is commonly known as oxidative stress [1]. The free radicals like hydroxyl (OH) [2], singlet oxygen ($^1O_2$) [3], superoxide (O_2^-) [4], and hydrogen peroxide (H_2O_2) [5] are mainly responsible for the generation of reactive oxygen species (ROS). The duration of stress classifies the stress into two different types, acute and chronic stress, in different animal models [6]. Stress that arises due to nutrition, physical exercise, and environmental factors like temperature, radiation, and air quality, and that can be reversed after a short period of time, is known as acute stress. Prolonged exposure to acute stress may lead to complete immobilization of the species and to a disease state termed as chronic stress [7].

The impact of oxidant its type, production and localization decide the harmful effect on an animal. The antioxidant comes into the picture as it produces various scavenging enzymes which encounter the produced oxidants and help in the reduction of stress conditions [8]. The technological

Novel Therapeutic Approaches Targeting Oxidative Stress
https://doi.org/10.1016/B978-0-323-90905-1.00008-0

advancements in the field of electronics and microchips for sustainable healthcare system improvised us in the manufacture of different types of biosensors. The new era of biotechnology is flooded with the discovery of putative biomarkers for different disease conditions which can gave us accuracy and precision in the early detection of a disease by simply studying a few standard parameters. Due to the biomarker discovery, the healthcare system ease the access to pre-diagnostic kits which can detect a disease at the very start of its progression [9]. Hence with the globalization policy for healthcare access and the guidelines by the expert committee from different countries helping each other to invade any kind of disease and also able to minimize the death rate.

2 Biosensors

A device that is capable of detecting the biological signals generated from a complex system and converting these signals into electric signals by means of a series of electronics components can be termed a biosensor [10]. Biosensors are mainly comprised of two principal elements: biological components and physical components. The biological components include the antibody, bacteria, cells, enzymes, nucleic acid, organelle, and tissue from a complex system that interact with the analyte and produce a signal known as a biological signal [11]. The physical components include transducers and amplifiers which detect the biological signals and convert them into electrical signals.

2.1 Working principle

The desired biological component is immobilized in a small compartment that can directly interact with the analyte, and this reaction produces the biological response which is converted into electrical signals by the transducers attached to it [12]. Before transmitting the electric signal to the detector, an amplifier is attached to enhance the low signal intensity to a detectable range, in order to close the circuit a detector/signal processor unit is also connected with a display component [13]. A schematic representation of the working principle and the internal components in a typical biomarker is illustrated in Fig. 1.

2.2 Types of biosensors

Based on the type of biological samples used and the electronic components used for sensing these biosensors, they can be classified into various types.

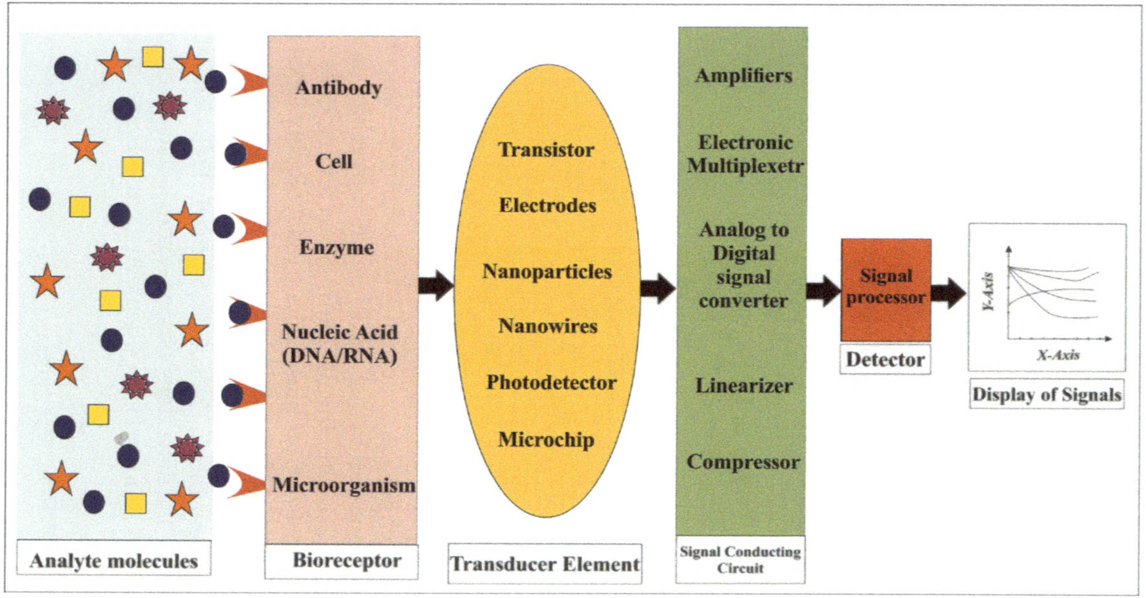

FIG. 1 Schematic diagram of internal components of a typical biosensor and its signal transduction process and detection.

2.2.1 *Electrochemical biosensors*

In the mid-19th century for the first time the initial type of biosensor was discovered for the detection of glucose levels in the blood in diabetic patients. The glucose oxidase enzyme was used for the detection, but due to the unstable nature of this enzyme, further calibration and improvements were necessary [14]. The use of biosensors is essential for early clinical detection and regular monitoring of various disease conditions. The measurement of antioxidant and reactive oxygen species (ROS) levels in order to study the physiology is a new invention in biosensor development. Different detection techniques like potentiometric [15], voltammetric [16], amperometric [17], and conductometric [18] are used for the development of biosensors depending upon the biological system and its output generated [19]. Advancements in three-dimensional bioprinting created a microenvironment for the live cell encapsulation, which forms a biosensor. Due to technological innovations in bio-fabrication, the electrochemical biosensors incorporate the nucleic acids of cells such as DNA and miRNAs and are helpful in the diagnosis of genetic disorders [20].

2.2.2 *Optical biosensors*

The immobilizing materials used for the development of optical biosensors are quartz, glass, and silica, which are believed to produce ultra-sensitive sensors. Progress in micro-fabrication techniques also helped to incorporate gold nanoparticles, silver nanoparticles, and quantum dots used for biosensors [21]. This fiber-optic biosensor has tremendous applications in clinical diagnosis and drug discovery. Silicon-based nanomaterials play a crucial role in the biosensor due to their biocompatibility and no cytotoxicity; coupled with gold nanoparticles, their conductivity increases and they are well known to produce ultra-sensitive biosensors [22,23]. Notably the 3D hydrogel technology has changed the sensing efficiency of optical biosensors by developing DNA-based sensors. The use of hydrogel not only increases the optical transparency but also protects the DNA from enzymatic degradation due to controlled release and entrapment inside the substrate matrix [24]. Optical biosensors are mainly used for the detection of blood sugar, lactate level, urinary tract infections, pO_2, pCO_2, and pH in patients in a critical care unit [25].

2.3 Other biosensors

There are several types of biosensors based on the mode of detection technologies used by the transducer elements of biosensors. Out of these, the most widely accepted in biomedical and healthcare systems are as follows.

2.3.1 *Piezoelectric biosensors*

Biosensors that can detect a sound wave through piezoelectric crystals are termed acoustic biosensors. The molecular propagation of acoustic wave on the surface of the crystals creates positive and negative frequencies which can be detected by electronic devices [26]. These types of biosensors have wide medical applications, such as measurement of pressure, vapor, humidity, and temperature, and can also be used as gas sensors [27].

2.3.2 *Calorimetric biosensors*

The production of heat is a very common phenomenon in biochemical reactions, and the sensors used to measure the heat energy of a process are known as thermal/calorimetric biosensors [28]. Tiny changes in the temperature can also be sensed and measured by the thermistors and give some signals. Thermal biosensors are mostly used in clinical data generation in patients for their glucose, cholesterol, urea, and uric acid levels, and also enzyme-linked immunoassay (ELISA) [29].

2.3.3 *Immuno-biosensors*

The specificity of this type of biosensors is based on the immunological techniques for its construction and sensing. The immobilized components in this biosensor can be antibody or antigen; accordingly the binding analyte can be chosen for the detection. Based on the labeled and unlabeled antigen used for detection, either thermometric or piezoelectric sensing modules are used for its construction [30]. These biosensors have a very wide range of applications in the agricultural, food manufacturing, and pharmaceutical industries; forensic laboratories also show potential detection during the management of epidemic diseases and biological warfare [31].

2.3.4 *Nano biosensors*

Marvelous developments in the sector of nanotechnology have brought conspicuous progress in biomedical science from the development of nano-drug delivery systems to the implantation of biomaterials made up of thin nanosheets. Likewise, the biosensors produced nowadays mainly use photonic crystals, carbon nanotubes, graphite, graphene, gold, copper, silver, and silicon nanoparticles [32,33]. Due to their very pure nature, having no toxicity and having stability to oxidation, gold nanoparticles are used in biosensors whereas there are limitations with the use of silver nanoparticles due to their toxicity in the internal environment of the body due to oxidation [34,35]. Nanoparticles are found to be very promising for the detection of a very low concentration and for single molecular detection; similarly the iron nanoparticles [36] and quantum dots [37] are widely used as nano-medicine and in targeted drug delivery.

2.3.5 Fluorescent biosensors

The tagging of fluorophore molecules with different small molecules and proteins for its tracking and detection in a complex system is the major goal of fluorescent biosensors. Mitochondrial physiology was previously studied by using the traditional genetically encoded biosensors for the oxidative stress, energy metabolism, and cAMP signaling [38]. To pave the way of biophysical discoveries fluorescence resonance energy transfer (FRET) technique came into picture in which the chromophore molecules are used for the live cell imaging and visualization of cGMP, cAMP, and Ca^{++} molecules inside cells [39]. Fluorescent biosensors coupled with fiber-optic biosensors have proven to be potential devices in terms of their sensitivity and applications in healthcare management, on an industrial scale, and in environmental science.

2.3.6 Whole cell biosensors

These are the trending biosensors nowadays which show great accuracy in the monitoring of different parameters due to their precise sensing ability. The whole cell of a particular microbial species or the recombinant protein molecules are used as the biological components which interact with an analyte and produce signals [40,41]. Different proteins and xenobiotic compounds have been identified for the detection of pesticide toxicity [42], bioremediation and biological oxygen demand [43], and the presence of heavy metals and radioactive isotopes [44].

3 Biomarkers

The study of biomarkers for any kind of disease is highly necessary for the development of therapeutic interventions and to study disease progression within a population. Biotechnological aid with the areas of different omics approaches simplifies the discovery of biomarkers and their use in clinical diagnosis. The World Health Organization (WHO) defines these biomarkers as the substances that are released from biochemical processes inside a cell or into body fluids before the onset of any kind of disease conditions in humans [45]. Different biomarkers reported so far act as the determinant indicator of a particular disease case for patients, which is detected by using patented laboratory techniques and biochemical analysis [46]. Depending upon the nature of the molecules considered as biomarkers for different diseases, the biomarkers have been broadly classified into four different groups: genomic biomarkers, transcriptomic biomarkers, proteomic biomarkers, and metabolomics biomarker [47] (Fig. 2).

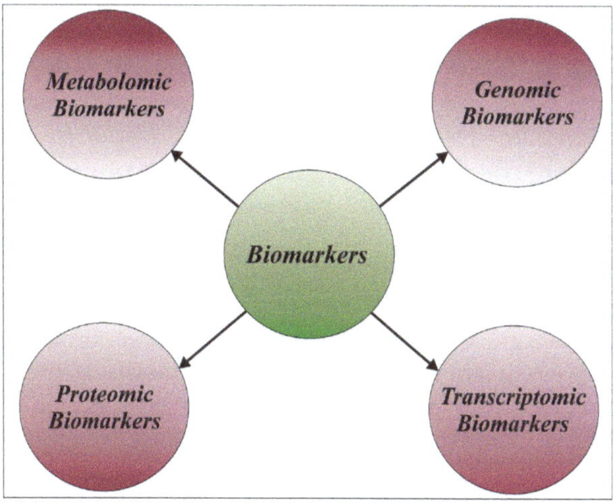

FIG. 2 Schematic diagram of different types of biomarkers based on its source of different omics approach.

3.1 Genomic biomarkers

The in-depth study of the human genome sequence opened up the way to understand the gene and its loci in the chromosomes. The variations in the DNA with a single or multiple nucleotide sequence encodes different DNA than its regular pattern in a somatic cell and these polymorphisms can be tracked and may be considered as biomarkers [48]. Few studies proven that the tumor cells which lead to destroy and release a certain amount of its DNA into the bloodstream and can be considered as biomarkers for the identification of the disease state [49]. The mutations of the nucleotides in the tumor suppressor genes like the adenomatous polyposis coil (APC) gene [50], retinoblastoma (RB_1) gene [51], cyclin dependent kinase inhibitor A($CDKN_2A$) [52], and oncogenic RAS [53] act as biomarkers for tumor prognosis.

3.2 Transcriptomic biomarkers

The technological advancement in the molecular biology techniques like northern blot, RNA sequencing [54], RT-PCR [55], microarray [56], and in situ hybridization [57] tremendously change the biomarker discovery at the transcriptional level. Apart from the very low stability of the RNAs under in vivo condition, scientists are able to identify and clinically use the biomarkers for different disease conditions. The RNA expression profile helps to identify a number of molecular subtypes which are associated with different cell types and regulate different disease conditions.

3.3 Proteomic biomarkers

The innovation in the omics technologies simplifies the study of human proteome analysis and enables the structure and function of a single protein to be explained from a complex network despite the mutation and posttranscriptional modification in proteins. These proteomic biomarkers help us in various aspects like the pathways that are affected, the susceptible individuals to a disease, how an individual will respond after the use of therapeutic intervention, and the possible side effects [48,58]. Due to the use of low sensitivity and specificity of the targets, high investment and advanced laboratory setup are required for detection using proteomic biomarkers. The clinical research and developments of these proteomic biomarkers are essential for the establishment of new strategies to lead the way regarding possible biomedical management of the diseases.

3.4 Metabolomic biomarkers

The complex biological systematic study of chemical fingerprinting can now be studied by targeting the metabolites of the cells and their expressions in different disease conditions. The biological databases are essential in understanding and analysis of the small molecules in different conditions. The discovered metabolic biomarkers give ideas on the fold change, activation, and deactivation of a target molecule, hence clinicians can easily trace back the biochemical pathways in a particular disease and the possible physiological conditions during these disease states [59,60]. The profiling of the metabolites has a larger sensitivity over classical and conventional detection methods as it uses high thoughtful techniques and in-depth analysis of those data by using advanced computing and bioinformatics tools.

4 Overview of discovered biomarkers for oxidative stress

Oxidative stress is generated due to high fluctuations in the concentration of ROS at the cellular level in almost all disease conditions. This assumption about the production of ROS was considered as a negative impact on animal physiology, but the real picture came into ray after the discovery of different antioxidant enzymes [61]. A study revealed that an alteration in the level of ROS is essential and helpful to maintain a balance of the reduction and oxidation state during progression of disease. The by-products of ROS can directly or indirectly affect different signaling pathways in the cell by interacting with different molecules of the cells [62]. Different diseases like chronic inflammation, cancer, glycemia, chronic obstructive pulmonary disease (COPD), metabolic disorders, and cardiovascular disorders arise due to oxidative stress only [63]. Periodic intracellular ROS production is the substantial cause of nucleic acid (DNA and RNA) damage [64], lipid peroxidation [65], and antioxidant imbalance in different cell types [66]. The measurement of different signaling molecules, enzymes, free radicals, and ions involved in the cellular process that helps in the detection of disease stages can be termed as a biomarker, and understanding with mechanistic interpretation of the data is very tedious work with traditional techniques. The elevation in the level of 8-hydroxyguanosine (8-OHdG) in a cell is responsible for the mutation in the DNA by promoting the transversion in the nucleotide sequence (i.e., adenine to cytosine and guanine to thymine). Hence 8-OHdG suppresses the DNA repair mechanism during the progression of disease which includes alteration in the genetic level [67]. The conversion of thymidine into thymidine glycol, due to the interaction of extra OH^- released by the oxidation process, is found in the body fluids of patients undergoing oxidative stress. This thymidine glycol is reported to deform the structure of DNA molecules by forming the double stranded breaks (DSBs), also known as lethal lesion [68]. From the last few decades it has been investigated that the oxidation of single stranded molecules (RNA) is due to the elimination of 7,8-dihydro-8-oxo-guanosine (8-oxoGsn) as the by-product in neurodegenerative disorders, viral infection, and autoimmune disease conditions. Lipid peroxidation is a common phenomenon in conditions like coronary artery disease, renal disorders, and cardiovascular disease [69]. An increase in the aldehyde compound results from the oxidation of membrane phospholipids, which can easily damage the DNA by modifying the nucleic acid bases. The level of aldehydes named crotonaldehyde [70], 4-hydroxy-2-nonenal [71], methylglyoxal [72], acrolein [73], and F_2-isoprosten [74] can be measured as a biomarker and they can be used as biomarkers for the clinical diagnosis of the abovementioned diseases. The harmful effect of free radicals generated from the biochemical processes can be balanced by the secretion of antioxidant molecules. The antioxidants can be broadly classified into two different groups: enzymatic (glutathione peroxidase, superoxide dismutase, and catalase) and low molecular weight molecules (albumin, ascorbic acid, α-tocopherol, bilirubin, β-carotene, ceruloplasmin, polyphenols, ferritin) [75–77]. Emerging biochemical technologies pave the way to measure the levels and can act as putative biomarkers.

5 Management of oxidative stress conditions

The management of oxidative stress conditions in humans varies from one disease state to another, and depends on the condition of patients. The management of oxidative stress depends upon various factors, and a few of them can be described as follows.

5.1 Habits and lifestyle

The ease of living with a lots of workload in professional and personal life has increased oxidative stress in any age group among society. Exposure to environmental conditions with electromagnetic and UV radiation, xenobiotic pollutants in the atmosphere, and psychological stress are responsible for DNA damage and variation in telomere length in chromosomes [78]. If we look closely into the daily routine of a person working in corporate life, they are frequently facing very stressful conditions in achieving their short- and long-term goals. In case of a healthy family different causes like emergency situation in family, domestic violence, and financial crises can elevate the levels of oxidative stress in family members [79,80]. Sedentary work, improvident social and official life, overtime to achieve goals in a very short period of time, and comparative psychological thoughts should be avoided to balance the stress levels in humans.

5.2 Food and beverage intake

A fit human requires a healthy diet. Due to modern and busy lifestyles, the food habitat is also drastically changed and human civilization often depends upon ready-to-eat food, junk food, and prepacked foods which mainly consist of chemical additives and preservatives [81,82]. The chemicals used for cultivation of vegetables, as flavor and preservation purpose acts like the stimulant of oxidative stress in humans and is harmful to their health [83]. The packaged drinking products, alcoholicism, and smoking are found to have very bad effects on the physiological system in humans and are known to cause critical disease. Thus, the intake of organic and proper nutritious food on a routine basis in the required quantities is part of the best strategy in minimizing oxidative stress conditions.

5.3 Supplements and antioxidants

Different supplementing products have been marketed to boost the strength and overcome the deficiency of nutrients, but a few manufacturing industries do not maintain the proper quality of supplements. The use of low-quality supplements directly impacts on human health as they contain various impurities like aluminum, lead, silicon, silver, manganese, and cobalt [84]. The rise in the level of these metal ions in the body is responsible for ionic influx inside the cytosol and leads to oxidative stress in different tissue types. Different types of herbal and pharmaceutical antioxidant capsules are also marketed and used as therapeutics in different disease conditions [85]. The use of natural vegetables and fruits as well as pharmaceutical products rich in antioxidant molecules has been proven to counter the produced free radical in stressful situations and hence balances oxidative stress in humans.

5.4 Yoga, exercise, and meditation

Over-exercise beyond the threshold capacity of any living animal is known to cause the production of free radicals and ions, leading to stress in tissue and organs [86]. The effect of this oxidative stress condition can mainly be healed by the ancestral and strategic habits of meditation, yoga, and exercise for different mental distractions as well as chronic disorders in patients with established naturopathy [87]. A recent study revealed that yoga and meditation-based lifestyle interventions are relatively helpful in lowering the level of biomarkers associated with cellular aging, ROS, and oxidative stress.

5.5 Healthcare situations and its management

Nowadays, deterioration of mental health conditions in different-aged groups is a major cause of oxidative stress. It is believed that emergency situations relating to family and relatives have an impact on mental health, and are a major cause of stress [88]. Although it is difficult to avoid these types of situations in daily life, proper understanding and psychological support are highly useful to avoid stress conditions [89]. The government is also working with the medical council to develop low-cost and strategic healthcare facilities for the financially more vulnerable in society.

6 Research trends and future challenges in development of biosensors

The development of different types of biosensors has opened a new door in the clinical diagnostic of critical disease conditions during its progression. Due to the electrochemical biosensors and bio-fabrication, electronic skins and tattoo are being used with the combination of nanotechnology for physiological and security detection of chemical ingredients [90]. New technological advancements in fuel cell electronics have completely changed biosensors, which traditionally used whole cell biomass and introduced biosensors that work on the principle of microbial fuel cell technology and can be used for a longer duration without replacement [91]. Graphene-based biosensors are now widely used for the fabrication of chip biosensors due to their biocompatibility and conductive nature when implanted inside the bodies of living beings.

Cost-effective sophisticated instrumentation facilities and the use of multiplex analytical procedures are now essential for building of efficient biosensors. Advancements in transducer elements by using the latest fabrication technology for 2D and 3D detection ability offer a futuristic approach in biosensor technology [92]. The use of minute quantities of analytes for the detection of changes in multiple clinical parameters by performing a single experiment is the trending research theme [93]. Developments in nano-electronics and the use of robust chip technologies enable noninvasive biosensors to be made for detection of different parameters [94]. Recent study delineates affinity-based biosensors uses the electrochemical method of analysis of metabolites and electrolytes with the help of wearable devices and has a wide use for the tracking of doping tests in sportspersons [95]. Different technical advancements are underway to achieve cost-effective, small molecular detection, toxicity-free, and noninvasive biosensors in the near future.

7 Conclusion

Cellular homeostasis is maintained by the balanced state of free radicals and molecular signaling for communication between cells. However, oxidative and redox stress in animals eventually creates an imbalanced state of free radical generation and ions, leading to ROS, which needs to be strategically managed at the earliest stage. Hence the role of biomarkers comes into play for early detection before evolution of disease, the main goal of biomarker discovery is to search for stable molecules that show a dynamic nature with the progression of disease condition. Since oxidative stress is the major cause of different diseases, its management is the new challenge in the healthcare sector. Moreover, the small electronics research and developments will lead to promising biosensors with better efficacy of measurements, higher sensitivity, full automation, and a higher sampling ratio. It is believed that understanding of molecules within the analyte and its interaction with the sensors needs in-depth analysis, and the improved fabricating materials process will be helpful in the production of new-generation biosensors.

Acknowledgments

Funding: S.M. is a recipient of the DRDO-JRF fellowship. The authors have not received any special funding for this work.
Conflict of interest: The authors state no conflict of interests.
Availability of data and materials: The figures are not adopted from any other sources.
Authors' contribution: Y.A. conceptualized the manuscript. S.M. designed and wrote the manuscript and provided the figures for convenient representation. Y.A. critically evaluated the manuscript.
Ethics approval: Not applicable.
Consent for publication: All authors agreed.

References

[1] A.M. Pisoschi, A. Pop, The role of antioxidants in the chemistry of oxidative stress: a review, Eur. J. Med. Chem. 97 (2015) 55–74.
[2] C.F. Vogel, et al., Physicochemical properties of iron oxide nanoparticles that contribute to cellular ROS-dependent signaling and acellular production of hydroxyl radical, Free Radic. Res. 50 (11) (2016) 1153–1164.

[3] E. Choe, D. Min, Chemistry and reactions of reactive oxygen species in foods, J. Food Sci. 70 (9) (2005) R142–R159.

[4] M. Tada, M. Kohno, Y. Niwano, Scavenging or quenching effect of melanin on superoxide anion and singlet oxygen, J. Clin. Biochem. Nutr. 46 (3) (2010) 224–228.

[5] C. Rampon, et al., Hydrogen peroxide and redox regulation of developments, Antioxidants (Basel) 7 (11) (2018) 159.

[6] X. Wang, et al., Metabonomics approach to understanding acute and chronic stress in rat models, J. Proteome Res. 8 (5) (2009) 2511–2518.

[7] B.M. Kudielka, S.J.S. Wüst, Human models in acute and chronic stress: assessing determinants of individual hypothalamus–pituitary–adrenal axis activity and reactivity, Stress 13 (1) (2010) 1–14.

[8] H.M. Ochs-Balcom, et al., Antioxidants, oxidative stress, and pulmonary function in individuals diagnosed with asthma or COPD, Eur. J. Clin. Nutr. 60 (8) (2006) 991–999.

[9] L.L. De Matos, et al., Immunohistochemistry as an important tool in biomarkers detection and clinical practice, Biomark. Insights 5 (2010). p. BMI. S2185.

[10] B.R. Eggins, Biosensors: An Introduction, Springer-Verlag, 2013.

[11] F.W. Scheller, et al., Research and development in biosensors, Curr. Opin. Biotechnol. 12 (1) (2001) 35–40.

[12] V. Perumal, U. Hashim, Advances in biosensors: principle, architecture and applications, J. Appl. Biomed. 12 (1) (2014) 1–15.

[13] Y.D. Han, H.J. Chun, H.C.J.B.J. Yoon, Low-cost point-of-care biosensors using common electronic components as transducers, BioChip J. 14 (1) (2020) 32–47.

[14] M. Keusgen, Biosensors: new approaches in drug discovery, Naturwissenschaften 89 (10) (2002) 433–444.

[15] R. Koncki, Recent developments in potentiometric biosensors for biomedical analysis, Anal. Chim. Acta 599 (1) (2007) 7–15.

[16] V. Vyskočil, et al., Voltammetric detection of damage to DNA caused by nitro derivatives of fluorene using an electrochemical DNA biosensor, Anal. Bioanal. Chem. 397 (1) (2010) 233–241.

[17] P.E. Erden, E. Kilic, A review of enzymatic uric acid biosensors based on amperometric detection, Talanta 107 (2013) 312–323.

[18] T.M. Anh, et al., Conductometric tyrosinase biosensor for the detection of diuron, atrazine and its main metabolites, Talanta 63 (2) (2004) 365–370.

[19] M. Pohanka, P. Skládal, Electrochemical biosensors—principles and applications, J. Appl. Biomed. 6 (2) (2008).

[20] Y. Du, S. Dong, Nucleic acid biosensors: recent advances and perspectives, Anal. Chem. 89 (1) (2017) 189–215.

[21] M. Velasco-Garcia, Optical biosensors for probing at the cellular level: a review of recent progress and future prospects, in: Seminars in Cell & Developmental Biology, Elsevier, 2009.

[22] Z. Liao, et al., Microfluidic chip coupled with optical biosensors for simultaneous detection of multiple analytes: a review, Biosens. Bioelectron. 126 (2019) 697–706.

[23] F. Long, A. Zhu, H. Shi, Recent advances in optical biosensors for environmental monitoring and early warning, Sensors (Basel) 13 (10) (2013) 13928–13948.

[24] I. Khimji, et al., Visual optical biosensors based on DNA-functionalized polyacrylamide hydrogels, Methods 64 (3) (2013) 292–298.

[25] M.A. Cooper, Optical biosensors in drug discovery, Nat. Rev. Drug Discov. 1 (7) (2002) 515–528.

[26] P. Skládal, Piezoelectric biosensors, TrAC Trends Anal. Chem. 79 (2016) 127–133.

[27] M.T. Chorsi, et al., Piezoelectric biomaterials for sensors and actuators, Adv. Mater. 31 (1) (2019), e1802084.

[28] C. Karunakaran, R. Rajkumar, K. Bhargava, Introduction to biosensors, in: Biosensors and Bioelectronics, Elsevier, 2015, pp. 1–68.

[29] K. Ramanathan, B. Danielsson, Principles and applications of thermal biosensors, Biosens. Bioelectron. 16 (6) (2001) 417–423.
[30] S. Sang, et al., Progress of new label-free techniques for biosensors: a review, Crit. Rev. Biotechnol. 36 (3) (2016) 465–481.
[31] X. Zhang, et al., Electrochemical functionalization of gold and silicon surfaces by a maleimide group as a biosensor for immunological application, Acta Biomater. 9 (3) (2013) 5838–5844.
[32] W. Zhang, Y. Du, M.L. Wang, Noninvasive glucose monitoring using saliva nano-biosensor, Sensing Bio-Sensing Research 4 (2015) 23–29.
[33] S. Priyadarsini, et al., Graphene and graphene oxide as nanomaterials for medicine and biology application, J. Nanostructure. Chem. 8 (2) (2018) 123–137.
[34] A. Loiseau, et al., Silver-based plasmonic nanoparticles for and their use in biosensing, Biosensors 9 (2) (2019) 78.
[35] Y. Xing, et al., A sensitive and selective electrochemical biosensor for the determination of beta-amyloid oligomer by inhibiting the peptide-triggered in situ assembly of silver nanoparticles, Int. J. Nanomedicine 12 (2017) 3171–3179.
[36] S. Kumar, et al., Electrochemical paper based cancer biosensor using iron oxide nanoparticles decorated PEDOT: PSS, Anal. Chim. Acta 1056 (2019) 135–145.
[37] F. Faridbod, A.L.J.C.A.C. Sanati, Graphene quantum dots in electrochemical sensors/biosensors, Curr. Anal. Chem. 15 (2) (2019) 103–123.
[38] S.K. Krishnan, et al., A review on graphene-based nanocomposites for electrochemical and fluorescent biosensors, RSC Adv. 9 (16) (2019) 8778–8881.
[39] M. Polito, et al., The NO/cGMP pathway inhibits transient cAMP signals through the activation of PDE2 in striatal neurons, Front. Cell. Neurosci. 7 (2013) 211.
[40] L.T. Bereza-Malcolm, G. Mann, A.E. Franks, Environmental sensing of heavy metals through whole cell microbial biosensors: a synthetic biology approach, ACS Synth. Biol. 4 (5) (2015) 535–546.
[41] E. Michelini, et al., Field-deployable whole-cell bioluminescent biosensors: so near and yet so far, Anal. Bioanal. Chem. 405 (19) (2013) 6155–6163.
[42] B. Bucur, et al., Advances in enzyme-based biosensors for pesticide detection, Biosensors (Basel) 8 (2) (2018) 27.
[43] B.K. Behera, et al., Polycyclic aromatic hydrocarbons (PAHs) in inland aquatic ecosystems: perils and remedies through biosensors and bioremediation, Environ. Pollut. 241 (2018) 212–233.
[44] G. GuanJun, et al., Engineering Deinococcus radiodurans into biosensor to monitor radioactivity and genotoxicity in environment, Chin. Sci. Bull. 53 (11) (2008) 1675–1681.
[45] S. O'Neill, et al., Blood-based biomarkers for metabolic syndrome, Treat. Endocrinol. 27 (6) (2016) 363–374.
[46] E. MacLean, et al., A systematic review of biomarkers to detect active tuberculosis, Nat. Microbiol. 4 (5) (2019) 748–758.
[47] S. de Franciscis, L. Metzinger, R. Serra, The discovery of novel genomic, transcriptomic, and proteomic biomarkers in cardiovascular and peripheral vascular disease: the state of the art, Biomed. Res. Int. 2016 (2016) 7829174.
[48] F.D. Shah, et al., A review on salivary genomics and proteomics biomarkers in oral cancer, Indian J. Clin. Biochem. 26 (4) (2011) 326–334.
[49] T. Mikeska, J.M. Craig, DNA methylation biomarkers: cancer and beyond, Genes (Basel) 5 (3) (2014) 821–864.
[50] M.K. Choong, G. Tsafnat, Genetic and epigenetic biomarkers of colorectal cancer, Clin. Gastroenterol. Hepatol. 10 (1) (2012) 9–15.
[51] H. Tsuyoshi, Y. Yoshida, Molecular biomarkers for uterine leiomyosarcoma and endometrial stromal sarcoma, Cancer Sci. 109 (6) (2018) 1743–1752.
[52] P.C. Galipeau, et al., NSAIDs modulate CDKN2A, TP53, and DNA content risk for progression to esophageal adenocarcinoma, PLoS Med. 4 (2) (2007), e67.

[53] T. Wang, et al., Gene essentiality profiling reveals gene networks and synthetic lethal interactions with oncogenic Ras, Cell 168 (5) (2017) 890–903 e15.
[54] X. Qian, et al., RNA-Seq technology and its application in fish transcriptomics, OMICS 18 (2) (2014) 98–110.
[55] D.M. Pedrotty, M.P. Morley, T.P. Cappola, Transcriptomic biomarkers of cardiovascular disease, Prog. Cardiovasc. Dis. 55 (1) (2012) 64–69.
[56] B.B. Raddatz, et al., Microarray-based gene expression analysis for veterinary pathologists: a review, Vet. Pathol. 54 (5) (2017) 734–755.
[57] S. Kiflemariam, et al., Scalable in situ hybridization on tissue arrays for validation of novel cancer and tissue-specific biomarkers, PLoS One 7 (3) (2012), e32927.
[58] S.K. Al-Tarawneh, et al., Defining salivary biomarkers using mass spectrometry-based proteomics: a systematic review, OMICS 15 (6) (2011) 353–361.
[59] B. Li, et al., Novel applications of metabolomics in personalized medicine: a mini-review, Molecules 22 (7) (2017) 1173.
[60] R.D. Beger, A review of applications of metabolomics in cancer, Meta 3 (3) (2013) 552–574.
[61] A.A. Alfadda, R.M. Sallam, Reactive oxygen species in health and disease, J. Biomed. Biotechnol. 2012 (2012), 936486.
[62] G. Poli, et al., Oxidative stress and cell signalling, Curr. Med. Chem. 11 (9) (2004) 1163–1182.
[63] A. Spector, Review: oxidative stress and disease, J. Ocul. Pharmacol. Ther. 16 (2) (2000) 193–201.
[64] X. Liu, et al., Elevated levels of urinary markers of oxidative DNA and RNA damage in type 2 diabetes with complications, Oxidative Med. Cell. Longev. 2016 (2016) 4323198.
[65] U. Asmat, K. Abad, K. Ismail, Diabetes mellitus and oxidative stress—a concise review, Saudi Pharm. J. 24 (5) (2016) 547–553.
[66] R.K. Gupta, et al., Oxidative stress and antioxidants in disease and cancer: a review, Asian Pac. J. Cancer Prev. 15 (11) (2014) 4405–4409.
[67] P. Katakwar, et al., Oxidative stress marker in oral cancer: a review, J. Cancer Res. Ther. 12 (2) (2016) 438–446.
[68] L. Rambabu, I.L. Megson, M.J.C.T. Eddleston, Does oxidative stress contribute to toxicity in acute organophosphorus poisoning?–a systematic review of the evidence, Clin. Toxicol. 58 (6) (2020) 437–452.
[69] S. Jiang, et al., A potential urinary biomarker to determine frailty status among older adults, Arch. Gerontol. 88 (2020), 104038.
[70] M. Gowtham, R.J.F. Prakash, Control of regulated and unregulated emissions from a CI engine using reformulated nano fuel emulsions, Fuel 271 (2020), 117596.
[71] D. Dham, et al., 4-hydroxy-2-nonenal, a lipid peroxidation product, as a biomarker in diabetes and its complications: challenges and opportunities, Free Radic. Res. (2020) 1–50.
[72] Y. Toya, H. Shimizu, Flux controlling technology for central carbon metabolism for efficient microbial bio-production, Curr. Opin. Biotechnol. 64 (2020) 169–174.
[73] J. Jeevanandam, M.K. Danquah, Nanosensors for better diagnosis of health, in: Nanofabrication for Smart Nanosensor Applications, Elsevier, 2020, pp. 187–228.
[74] Y.M. Hosakote, K. Rayavara, Respiratory syncytial virus-induced oxidative stress in lung pathogenesis, in: Oxidative Stress in Lung Diseases, 2020, pp. 297–330.
[75] S. Ghimenti, et al., Salivary lactate and 8-isoprostaglandin F 2α as potential noninvasive biomarkers for monitoring heart failure: a pilot study, Sci. Rep. 10 (1) (2020) 1–11.
[76] E. Pedone, et al., Enzymatic antioxidant signatures in hyperthermophilic archaea, Antioxidants (Basel) 9 (8) (2020) 703.
[77] I. Gulcin, Antioxidants and antioxidant methods: an updated overview, Arch. Toxicol. 94 (3) (2020) 651–715.
[78] M. Sharifi-Rad, et al., Lifestyle, oxidative stress, and antioxidants: back and forth in the pathophysiology of chronic diseases, Front. Physiol. 11 (2020) 694.

[79] B. Fang, et al., Association between elder abuse and telomere shortening in older adults: a 2-year prospective study, Int. J. Geriatr. Psychiatry 36 (1) (2021) 54–63.
[80] S. Schiavone, et al., Severe life stress and oxidative stress in the brain: from animal models to human pathology, Antioxid. Redox Signal. 18 (12) (2013) 1475–1490.
[81] H. Zhang, R. Tsao, Dietary polyphenols, oxidative stress and antioxidant and anti-inflammatory effects. Current opinion in food, Science 8 (2016) 33–42.
[82] E.I. Medina-Reyes, et al., Food additives containing nanoparticles induce gastrotoxicity, hepatotoxicity and alterations in animal behavior: the unknown role of oxidative stress, Food Chem. Toxicol. 146 (2020), 111814.
[83] H.N. Wang, et al., Protective effects of a freeze-dried extract of vegetables and fruits on the hydroxyl radical-mediated oxidative damage of DNA and decrease of erythrocytes deformability, Appl. Biochem. Biotechnol. 141 (2–3) (2007) 241–249.
[84] M.C. Gomez-Cabrera, et al., Oxidative stress in marathon runners: interest of antioxidant supplementation, Br. J. Nutr. 96 (Suppl 1(S1)) (2006) S31–S33.
[85] G. Bjelakovic, C. Gluud, Surviving antioxidant supplements, J. Natl. Cancer Inst. 99 (10) (2007) 742–743.
[86] S.K. Powers, W.B. Nelson, M.B. Hudson, Exercise-induced oxidative stress in humans: cause and consequences, Free Radic. Biol. Med. 51 (5) (2011) 942–950.
[87] M. Sengar, et al., Cancer treatment: role of yoga, naturopathy and prayer, Health Adm. 17 (1) (2012) 151–157.
[88] S.R. Horn, et al., Childhood adversity, mental health, and oxidative stress: a pilot study, PLoS One 14 (4) (2019), e0215085.
[89] M. Jamilian, et al., Carnitine and chromium co-supplementation affects mental health, hormonal, inflammatory, genetic, and oxidative stress parameters in women with polycystic ovary syndrome, J. Psychosom. Obstet. Gynaecol. (2019) 1–9.
[90] B.W. An, et al., Smart sensor systems for wearable electronic devices, Polymers (Basel) 9 (8) (2017) 303.
[91] S.S. Mishra, et al., Based microbial fuel cell, in: Microbial Fuel Cells: Materials Applications, vol. 46, 2019, pp. 101–124.
[92] A. Hasan, et al., Recent advances in application of biosensors in tissue engineering, Biomed. Res. Int. 2014 (2014), 307519.
[93] N. Gupta, et al., Cell-based biosensors: recent trends, challenges and future perspectives, Biosens. Bioelectron. 141 (2019), 111435.
[94] S. Campuzano, M. Pedrero, J.M.J.S. Pingarrón, Noninvasive breast cancer diagnosis through electrochemical biosensing at different molecular levels, Sensors 17 (9) (2017) 1993.
[95] S. Herranz, A.F. Gavela, L.M. Lechuga, Label-free biosensors based on bimodal waveguide (BiMW) interferometers, in: Biosensors and Biodetection, Springer, 2017, pp. 161–185.

CHAPTER

9

Biochemical approaches/methods in the detection of free radicals

Imteyaz Qamar and Uditi

School of Biotechnology, Gautam Buddha University, Greater Noida, UP, India

1 Introduction

Free radicals and reactive oxygen species have been known to play different roles in mediating cytotoxicity by the means of alterations in proteins, lipids, and nucleic acid structure and functions which results in disruptions of cellular homeostatic mechanisms [1]. Free radicals, which can be defined as an atom or molecule having one or more unpaired electron in the orbit and is capable of independent existence, have a central role in various physiological conditions and have implications in a diverse range of diseases [2]. At higher concentrations, these free radicals are hazardous for cellular systems and are capable of causing damage to all major cellular constituents; at moderate concentrations, they play a crucial role as a regulatory mediator in various different signaling processes [3]. There exists a highly regulated equilibrium between the production of free radicals and their counter defense mechanism by scavenging systems and repair enzymes for the elimination of degraded by-products. An imbalance in this regulated equilibrium results in serious oxidative stress, which further initiates a series of pathological processes [4,5]. Though the detection is extremely important but the measurement or detection of these highly reactive and short-lived free radicals is a challenging task due to their aspecificity and sophisticated behaviors inside the living systems [6]. Increasing evidence of the involvement of these free radicals in the pathogenesis of various human diseases indicates the explicit need for some techniques to detect these free radicals in different clinical situations [7,8] (Fig. 1).

https://doi.org/10.1016/B978-0-323-90905-1.00009-2

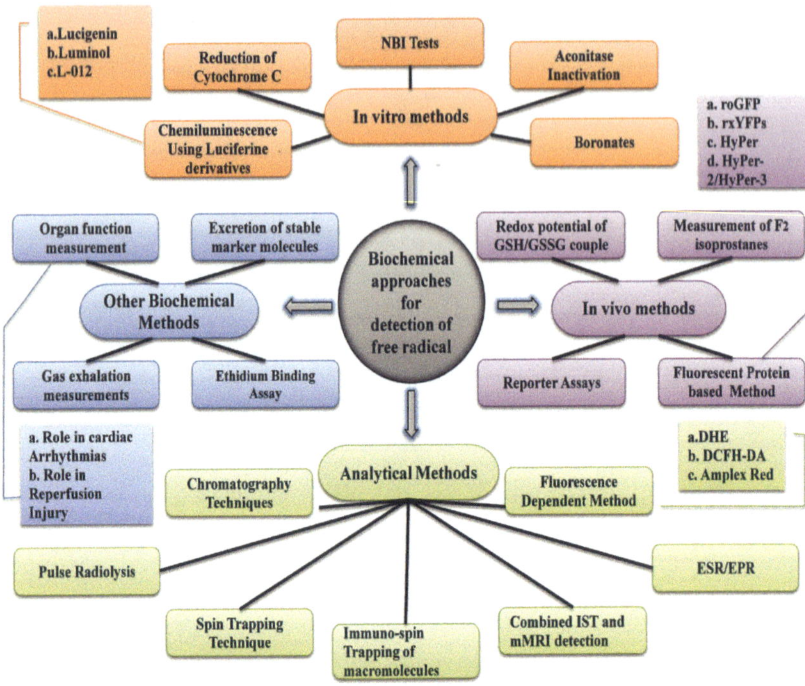

FIG. 1 Summarized view of all the techniques of free radical detection described in this chapter.

This chapter discusses the various methods of quantification and detection of free radicals on the basis of the approach chosen to design them. Different methods have been formulated according to the emerging needs of the experimental studies for such detection of free radicals or oxidative stress in the cellular systems and hence the objectives of the methods and approaches developed also varied. This chapter attempts to provide an insight to various conventional, traditional, and also newly developed methods of free radical detection in a comprehensive way.

2 In vitro methods

For in vitro free radical detection and quantification, various different types of assays are employed which have been used individually and combined with others over a period of time for accuracy and in search of suitable and exact detection of free radical species [9]. A common limitation of all these approaches is that the results of these assays are aspecific and can only be applied to detect the radicals in extracellular space—for example, the free radicals generated during the time of phagocytosis or generated in body fluids during ischemia/reperfusion episodes [10].

The main working principle for these approaches is based on the reaction radicals and redox substances, and the difference in the absorbance of which under different wavelengths between substrates and products leads to the semi-quantification of free radicals [11]. For the determination of intracellular free radical reactions, reporter or in vivo methods are considered more suitable.

3 Chemiluminescence-derived methods using luciferine derivatives

Though bearing immense similarities to fluorescent assays, chemiluminescence methods are more sensitive toward the free radicals with an additional advantage of being easy to operate with the help of chemiluminescent probes [12]. To be captured by the photometer or scintillation counter, these probes react with the free radicals and form photons without any additional requirement of excitation light source. The probe radical intermediates also play a critical role in such a mechanism. Like fluorescent probes, luminescence has similar limitations [9,11].

(a) *Lucigenin*: bis-*N*-methylacridinium nitrate is a probe which is known for its good membrane permeability and is used in the measurement of free radicals produced by the macrophages, neutrophils, as well as the extracellular isolated enzymes [13]. Lucigenin method when employed in superoxide detection in various enzymatic and cellular systems, such as the xanthine oxidase/xanthine, the lipoamide dehydrogenase/NADH, the isolated or intracellular mitochondria, as well as the phagocytic NADPH oxidase, proving the lucigenin method to be a valid probe for O_2 (•–) for the mentioned conditions [14]. For the relevant reaction, LC2 + is reduced by flavoprotein reductase to LC•+, which is able to react with O_2 to generate O_2 (•–). And since superoxide is not the specific radical that could react with lucigenin, it could also induce the luminescence of this probe, through generation of the radical LC•+ [15]. In addition, as per the prior knowledge reductases are crucial for generation of LC•+ which makes this assay extremely sensitive to any change in the reductase activity. Considering the low reactivity between lucigenin and superoxide, the probe becomes somewhat incapable in detecting superoxide radicals at a low level [12] (Fig. 2).

(b) *Luminol*: 5-Amino-2,3-dihydroxy-1, 4-phthalayineidone (luminol) is known for specific detection of superoxide among other free radicals like O_2 (•–), hydroxyl radical, hydrogen peroxide, and peroxynitrite [11]. The luminol reaction takes place through a series of sequential steps that lead to the final luminescent product. Firstly luminol (LH–) is converted into intermediate radical LH• which is not specific to superoxide but instead oxidants like •OH and CO_3 (•–) are proved

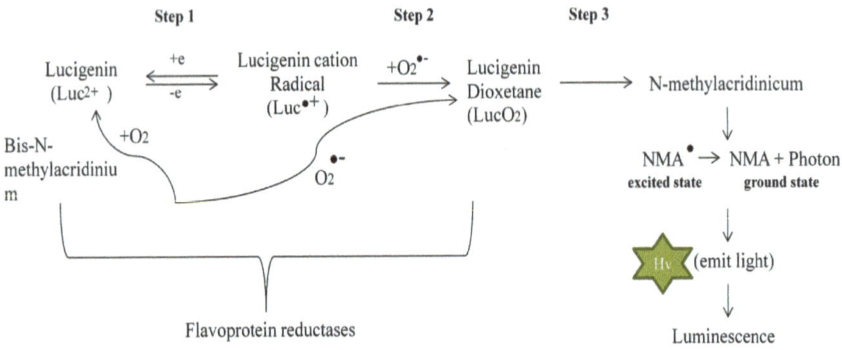

FIG. 2 Mechanism of chemiluminescence-based free radical detection using the lucigenin probe.

to be more competent for such a sensitive reaction [9]. Then the LH• radical reacts to O_2 to produce O_2 (•– which may obstruct the detection of superoxide. While it is observed that the antioxidants in cells could compete with radicals that react with LH•, or even directly react to LH•, which makes this probe capable for testing and detection of antioxidants as well [11]. For luminol and its analogues, the light emitting wavelength is 400 nm where auto-absorption of biomolecules in cells can also easily take place [16]. One of the limitations or drawbacks posed by this luminescence reaction is that luminol could not react with free radicals efficiently under the neutral pH conditions, which speaks for the fact that the factors influencing the chemiluminescence of luminol are complex and hard to control [17]. Hence, though this reaction could be used for the detection of O_2 (•–) without the interference of other free radicals, the radical generation is still not advised to be detected with this probe only, which makes the approach a little less reliable for cellular systems [18] (Fig. 3).

(c) *L-012*: 8-Amino-5-chloro-2,3-dihydro-7- phenyl-pyrido [3,4-d] pyridazine sodium salt which is a luminol analogue has been widely used to analyze free radical production like superoxide in many cellular systems and its individual components like human whole blood or isolated leukocytes and mitochondria, the neutrophils and vascular cells, as well as the chronic inflammation systems, where it shows high sensitivity recognition ability toward superoxide [19]. This confirms why L-012 was originally used for superoxide detection though it does not react with the superoxide radical directly and it is often used with the fluorescent dye hydroethidine (HE) [20]. As per the mechanism, the relative rates of oxidation of L-012 and HE in the presence of O_2 (•–) alone and in combination with horseradish peroxidase (HRP) are monitored and further analyzed by the help of HPLC. L-012 is such a chemical analogue, which gives rise to significantly higher

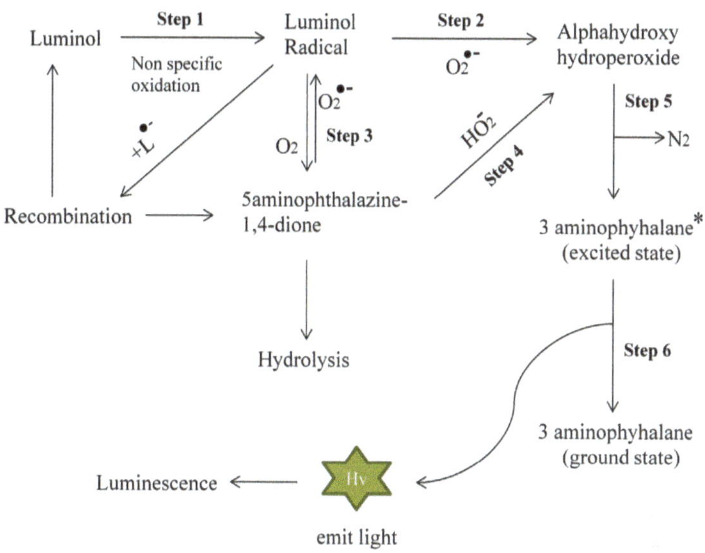

FIG. 3 Mechanism of chemiluminescence-based free radical detection using the luminol probe.

luminescence yield and increased sensitivity than other CL probes, lucigenin, and MCLA [21]. In the methodology for L-012, a non-specific intermediate radical LumH• is generated which reacts with O_2 to generate O_2 (•–) and L-012 quinone (q-Lum). Now L-012 quinone (q-Lum) forms a small part of LumH-OOH under the elimination of H_2O_2, which results in false positive results of O_2 (•–) [22]. Though L-012 (luminol-based chemiluminescent (CL) probe) is widely used for detection of NADPH oxidase (Nox)-derived superoxide O_2 (•–) and identification of Nox inhibitors, the exact understanding of the free radical detection mechanism of the L-012 probe is still unknown, which makes this assay less ideal for specific detection and identification of O_2 (•–) and NOX inhibitors [19] (Fig. 4).

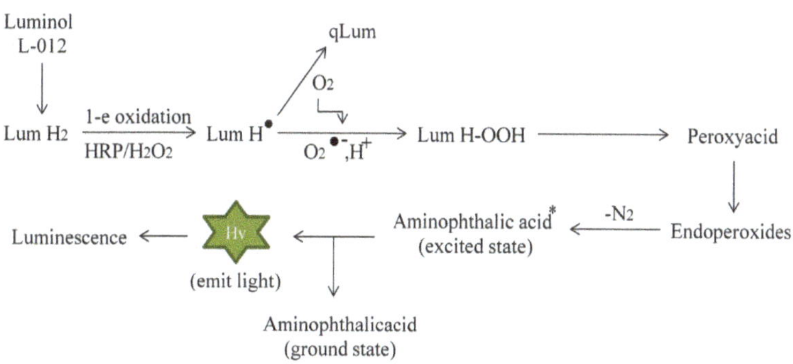

FIG. 4 Mechanism of chemiluminescence-based free radical detection using the L-012 probe.

4 Reduction of cytochrome *c*

The reduction of horse heart ferricytochrome *c* by the superoxide anion radical has been used for some time to detect the production of free radicals by different enzyme systems under various studies [23,24]. The reduction of ferricytochrome *c* to ferrocytochrome *c* measures the rates of formation of superoxide by numerous enzymes, whole cells, and vascular tissue at pH 8.5 and room temperature [25]. This spectrophotometric reaction has been optimized at 550nm in most cases, while several precautions need to be undertaken for detection of superoxide by this mechanism as reduction of cytochrome *c* is not absolutely specific for superoxide [26]. Various cellular reductants which are present in high concentrations within tissue extracts are natural cellular reductases which are capable of carrying out cytochrome *c* reduction enzymatically such as ascorbate, glutathione, etc. [27]. Various other enzymes such as xanthine oxidase are also capable of reducing redox-active dyes and cytochrome *c*, which makes this method non-exclusive of superoxide detection [27,28].

An important limitation posed by this method of cytochrome *c* reduction is that reduced cytochrome *c* can be reoxidized by cytochrome oxidases, cellular peroxidases, and oxidants, including hydrogen peroxide (H_2O_2) and peroxynitrite (ONOO–). Therefore, to avoid such a drawback, various enzyme inhibitors or scavengers of reactive species can be added to the reaction mixture, or to enhance the specificity of the assay, acetylation of cytochrome *c* can done [29]. Lysine residues present within ferricytochrome *c* are acetylated to decrease the direct electron transfer by mitochondrial and microsomal reductases to cytochrome *c* and from cytochrome *c* by cytochrome oxidase, this acetylation is done throughout which superoxide's ability of to reduce cytochrome *c* is cautiously maintained [24,25].

To decrease the reduction of cytochrome *c* by NADPH-cytochrome P-450 reductase or cytochrome *b*5 and oxidation by cytochrome *c* oxidase, succinoylation of cytochrome *c* can also be done, which in turn also decreases the rate constant of the reaction and this higher concentration of succinoylated cytochrome *c* is then required [30]. The reason for reduction of cytochrome *c* being an in vitro method of free radical reduction is because of the size and charge considerations and having intracellular access, hence it cannot be a preferred method for free radical detection within intact cells, as opposed to cellular or tissue extracts [31,32]. Thus, this method of cytochrome *c* reduction shows tremendously positive results with in vitro assays and activated leukocytes but still has its limitations in terms of in vivo applications like insensitivity to detect lower rates of free radical formation as well as interference by endogenous reductants [24].

5 NBT tests

Nitro blue tetrazolium (2,2′-bis(4-nitrophenyl)-5,5′-diphenyl-3,3′-(3,3′-dimethoxy-4,4′-diphenylene) ditetrazolium chloride) is the most commonly used probe for the detection of free radicals out of all the tetrazolium compounds, as it is a substrate of dehydrogenase and peroxidase [33]. A water-insoluble dye diformazan (which can be monitored microscopically or spectrophotometrically) is generated when NBT reacts with O_2 (•–) and its DMSO lysates absorbance under 620nm is in coordination of the superoxide or free radical levels, so this achieves the semi- quantification of O_2 (•–) [34]. Specifically O_2 (•–) produced for monocytes and macrophages is detected by the use of NBT assay, although the limitation of this assay is its incapability to detect low levels of radicals in the system [35]. According to the principle of the NBT test, target cells are incubated in the NBT solution and subsequently take up the tetrazolium salt into their cytoplasm, where it is converted by the activity of superoxide anions to water-insoluble purple-blue formazan crystals which are trapped intracellularly, and hence they can be visualized within the target cells using an optical microscope [10]. In addition, formazan crystals can be released using a solubilization reagent and subsequently quantified by measuring the absorbance of the resulting purple-blue suspension [36].

This NBT assay poses application in the diagnosis of chronic granulomatous disease (CGD), a defect of peripheral blood granulocytes in killing certain bacteria due to its ability to recognize the superoxide. The measurement of superoxide for diagnosis of CGD is generally used along with other probes like ferricytochrome *c* reduction, chemiluminescent dyes, or dihydrorhodamine oxidation (DHR), out of which DHR is preferable for its ease, sensitivity, and identification to different types of CGDs [11]. Despite NBT being both a traditional and conventional method, new dyes are being researched and introduced which challenge its efficiency and solve any drawback that this conventional method may have [10].

6 Aconitase inactivation

Both cytoplasm and mitochondria contain aconitase which has activity in catalyzing the inter-conversion between citrate and isocitrate which could be inactivated by oxidation of superoxide through the reversible loss of Fe, resulting in its absorbance change at 240nm and 340nm [11,37]. Cell and tissue aconitase activity is thus determined by the relative rates of inactivation by superoxide or other oxidants, and the rate of reactivation by reduction and restoration of iron to the cluster [38,39]. This generated superoxide could cause the ongoing conversion between active and inactive aconitase, which is largely influenced by the rate of O_2 (•–) production, making it possible to determine the relative change of superoxide level [40]. For further confirmation, it has been established that the spectrophotometric detection of A340nm and A240nm of both isocitrate and cis-aconitate can measure the generation of superoxide under NADPH presence [41]. This reaction between superoxide and aconitase is rapid and sensitive, which explains its prevalent use in cells like cultured neuronal cells, macrophages, fibroblasts, and tissues like liver, heart, and brain [42].

Other free radicals such as NO, oxygen, and H_2O_2 when subjected to this approach can also cause the inactivation of aconitase at a much lower rate than superoxide causing interference to specific detection to O_2 (•–) [43]. Thus, other oxidants, in addition to superoxide, can modify aconitase activity. Hence influences for these other free radicals should be avoided so that the accuracy of superoxide detection is not compromised [40].

7 Boronates

Boronates are synthetic organic compounds that react with peroxynitrite radicals and H_2O_2 and form a stable phenolic product. Boronates have been applied as one of the most effective methods for the monitoring and quantifying of ONOO – and H_2O_2 in cells [44]. The conventional probe is improvised by attaching with a fluorophore that generates a fluorescent product when it reacts with H_2O_2, which makes the detection even more selective and sensitive than before [45]. Based on this improvisation, various highly selective boronate-based fluorescent probes such as Peroxyfluor-1 (PF1), Peroxy green 1 (PG1), and Peroxy Crimson 1 (PC1) have been used for the fluorescence imaging of H_2O_2 in many tissues and cells [46].

These probes have the advantages of both the methods combined, which makes this assay much more suitable for the detection and quantification of H_2O_2 and provides possibilities to create innovative multifunctional probes in a similar manner for exact and more accurate targeting and detection of free radicals in the living systems or cells [47].

These are some widely accepted and utilized approaches for free radical detection for in vitro systems. There are various other employed methods for in vitro free radical detection, such as production of nitrite from hydroxylamine, determination of ethylene formation, degradation of deoxyribose, production of sulfinic acid from DMSO, bleaching of carotenoids, cleavage of methionine or keto-methyl butyric acid to ethylene, decarboxylation of 7–14C-benzoic acid, oxidation of epinephrine, and electrochemical detection, that have been utilized for free radical detection and quantification. Each has its advantages and limitations which keeps the door open for improving the conventional and traditional methods [9,11].

8 In vivo methods

Detection of free radicals for the in vivo systems is challenging compared to in vitro detection, the reason being that most methods have limitations in membrane permeability issues and interference due to sub-cellular components, which makes in vivo detection for free radicals a tedious task. However, some new approaches are being developed constantly based upon the intracellular mechanisms, such as intracellular redox potential. Various methods are used in combination with in vitro and analytical techniques for better and more accurate results.

9 Redox potential of the GSH/GSSG couple

It is well known that the prerequisite requirement of all cellular systems is the detoxification of reactive oxygen species. Addressing the prior knowledge that an imbalance caused between the defense networks at multiple levels of enzymatic and non-enzymatic defenses leads to the oxidative stress [48]. Reduced glutathione (GSH) is one of the most widely known non-enzymatic oxidant defenses of the body which exists in very large amounts in the cells and performs crucial functions such as detoxifying peroxides and maintaining other physiologically important antioxidants (e.g., α-tocopherol, ascorbic acid) in their reduced forms [49]. This reduced glutathione is generated from GSSG (its oxidized form) under the action of NADPH-dependent reductase continuously as the rates of synthesis of GSH, export of GSH and GSSG from the intracellular compartment to the extracellular space, and formation of protein-bound GSH mixed disulfides are slow relative to the rates of reduction of GSSG and oxidation of GSH. For in vivo systems, this balance of GSH and GSSG proves to be an indicator of oxidative stress in the system [24]. These alterations in the redox environment of the tissues have been known to be reflected in the alterations taking place in cell growth, differentiation, and other important cellular mechanisms such as apoptosis which calls for the quantification of this GSH/GSSG pool in the tissues and plasma as a prerequisite for getting an exact picture of the amount of oxidative stress borne by the system or the amount of free radical concentration [10]. This quantification of the GSH/GSSG pool could also be done biochemically or analytically by employing various techniques such as HPLC, which gives a dynamic indication of oxidative stress in the system [50].

A serious limitation posed by this method is that there is a possibility of artifactual overestimation or underestimation of plasma GSH and GSSG; hence it is advised to avoid hemolysis, limit oxidation, and inhibit the GSH degrading enzyme α-glutamyltranspeptidase [49].

10 Measurement of F2-isoprostanes as indicators of lipid per-oxidation

One of the widely used indicators of free radical generation in vivo is lipid per-oxidation [51]. However, these methods of lipid peroxidation detection rely on the detection of thiobarbituric acid (TBA)-reactive substances such as malondialdehyde (MDA) or other reactive aldehydes generated from the in vivo decomposition of lipid peroxidation products in blood plasma or tissue, which makes these methods non-specific [52]. It has been established through various studies that non-enzymatic, free radical-induced lipid peroxidation produces F2-like prostanoid derivatives of arachidonic acid, and as these oxidized lipids are cyclooxygenase-generated prostaglandin F2α isomers, they are commonly known as IsoP (F2-isoprostanes) [53]. Various potent oxidants such as hydroxyl radicals, peroxyl radicals, nitrogen dioxide, peroxynitrite, and higher oxidation states of heme and hemoproteins (ferryl heme) initiate the lipid *per*-oxidation of polyunsaturated fatty acids, and these agents induce the IsoP formation in an in vivo system [54]. Initially IsoPs are formed as esterified fatty acids attached to phospholipids and then are released to their free form by the action of phospholipases, which is why measurement of IsoP makes this approach one of the most reliable attractive markers for assessing oxidative stress in vivo [55]. The most accurate advantage provided by this method is that these compounds are stable and not produced by any other known enzymatic pathways using arachidonate such as the cyclooxygenase or lipoxygenase pathways; in addition, IsoP levels can be quantified in extracellular fluids such as plasma and urine, making this relatively non-invasive approach particularly useful for human and whole animal studies, which justifies this approach being a widely accepted formulation for determination of free radical generation/oxidative stress [56]. These oxidized lipids are known to increase in response to already known free radical generators such as diquat or CCl_4 administration [24].

A major drawback of this method lies in the detection system that is responsible for quantifying these oxidized lipids. As the most sensitive, reliable, and quantitative method for the detection of IsoPs is by mass spectrometry, it makes the detection of IsoPs expensive and time-consuming [57,58]. In addition, a considerable amount of interference from the substances found in biological fluids and tissue extracts has also been reported during the quantification of IsoPs [59].

11 Reporter assays

The results of various in vitro methods and some in vivo methods have major limitations of either being aspecific or only being able to detect the free radicals present in the extracellular space when secreted during phagocytosis and those which are generated in body fluids during ischemia/reperfusion processes but fail to detect or quantify intracellular free radicals. To solve this problem, various reporter assays on the basis of chemical and biological end points have been formulated [9,10] (Table 1).

12 Fluorescent protein-based methods

For the specific in vivo free radical detection, fluorescent protein-based redox probes are designed which are combinations of fluorescent proteins (FPs) and the prokaryotic redox-sensitive proteins where the recombinant protein is introduced into cells by the means of plasmids or adenovirus methods with an aim to target sub-cellular organelles and hence report the redox states [60]. The mechanism lying under the structural changes in the disulfide bonds under the oxidized conditions leads to the spectral changes of the recombinant proteins which give redox dependent fluorescence [11].

An advantage of such probes over the others is that they do not have any permeability of the target cells requirement because these provide the dynamic detection of the redox reversible reaction which is mediated by

TABLE 1 Different reporter assays based on the chemical and biological end points which have been attempted to quantify and detect the free radical presence in various different studies.

Determination based on the chemical end points (general indicator molecules are used)	Determination based on the biological end points (assays based on the cellular level)
Hyperoxides (determined by gas chromatography (GC), mass spectrometry (MS), high performance liquid chromatography (HPLC), chemiluminescence, fluorescence	Viability measurements (survival curves, plating efficiency) cellular transformation, chromosome damage
Aldehydes (determined by thiobarbituric acid reactive material (TBARS)	Flow cytometric of fluorescence microscopic techniques
Gaseous breakdown products (ethane, pentane)	Membrane potential measurements
Cholesterol epoxides and hydroperoxides	Membrane fluidity measurements (electrode techniques, spin or fluorescence labeling)
Oxidized proteins	Oncogene activation
DNA degradation products	Gene amplification
DNA fragments or "waste" products of DNA repair	Specific mRNA production

Though any single or specific method is not considered the exact go to method for such detection.

the free radicals [60]. These proteins can reside in the cells for long-term free radical detection, and their combination and interactions with the specific targeting signal peptides or retention sequences help then to target different organelles, which enables the localized free radical detection [61,62].

a. *Redox-sensitive green fluorescent protein (roGFP)*: This is formed by the insertion of redox reactive cysteines to enhance GFP, leading to disulfide bond formation under a reducing environment [63]. This enables its response to intra- and extracellular systems to be accompanied by reversible ratio-metric changes in fluorescence intensity. Various different probes have been designed on the basis of this strategy (Table 2).

b. *rxYFPs*: These proteins are engineered from yellow shifted GFP-derived protein (YFP) by introducing pairs of cysteines near the chromophore region, and reversible disulfide bonds are formed leading to the spatial conformational change in YFP under oxidative condition [73]. This fluorescence intensity is reduced and can be used

TABLE 2 Various different fluorescent protein based methods along with their characteristic features are mentioned which are being applied for oxidative stress detection among different model systems in a variety of studies.

roGFP probe	Features
roGFP1/2 [64] [65]	• Monitors "dynamic redox change" rather than directing "ROS" • Able to reflect conversion of dithiol/disulfide induced by accumulation of ROS • More suitable for reducing environments, such as cytoplasm, mitochondrial, nucleus, and peroxisomes
roGFP1-iX [66] [67]	• Developed based on roGFP1 • Lower sensitivity to pH than roGFP1 • More competent for redox monitor in oxidizing organelles like ER
roGFP1-iL and roGFP1-RX [68]	• Relatively high level of reduction potential for monitoring the change of redox status • More sensitive to convention of 2GSH/GSSG • Higher reaction rate and dynamic range than roGFP1
Grx1-roGFP2-iL/ roGFP2-Orp1 [69] [70,71] [72]	• Broader midpoint potentials, thus more suitable for usage both in cytosol and ER, and is more specific than roGFP2 for the measurement of glutathione redox potential • Developed through combination of the yeast peroxidase Orp1 with roGFP2

Improved versions of the roGFP1 are constantly being developed to make the monitoring of redox free radical reactions more competent.

to visualize the redox status in vivo. Due to its appropriate midpoint redox potential, it is applied to detect the dynamic variations of GSH/GSSG and thiol/disulfide in reducing compartments (like nucleus, cytosol, and mitochondria) [74]. These probes are pH-sensitive, and the pKa of their chromophore varies according to the pH of the system [75].

c. *cpYFP-OxyR recombinant protein (HyPer)*: Circularly yellow fluorescent protein (cpYFP) is inserted into the regulatory domain (RD) of prokaryotic H_2O_2-sensing protein OxyR, leading to a disulfide bridge formation between the key reactive cysteines [76]. This results in the conformational change in the confirmation of OxyR, leading to alterations in both the conformational and fluorescence intensity of the whole protein, hence this probe is specifically designed for H_2O_2 detection [77]. This method has been used for many kinds of cell systems and live animals, such as *Arabidopsis*, the *Caenorhabditis elegans*, yeast, and mice, as a powerful tool for detecting intracellular dynamic change of hydrogen peroxide [78,79].
d. *HyPer-2/HyPer-3*: Improved mutations based on HyPer are more stable and show improved performance in response to time and speed toward the reaction with H_2O_2 [80]. Both HyPer-2 and HyPer-3 have a broader dynamic range compared to original HyPer. These two probes are successfully used in the fluorescent lifetime imaging of H_2O_2 in vivo [81].

The abovementioned approaches are used for the determination of generalized oxidative stress in vivo and no specific method claims to be the appropriate method for the free radical detection and quantification. Each approach offers different advantages and have its limitations, and these are routinely applicable methods but still remain aspecific to some extent.

13 Analytical methods

Analytical methods are the most widely used and direct method of free radical estimation. Compared to the other methods of the same, analytical methods are considered to be the most accurate as far as the extent of accuracy in terms of methods of free radical detection goes. This is probably due to the fact that analytical methods are more technique-oriented and do not need to be constrained under the filters of fulfilling some additional and specific requirements such as analysis under in vivo or in vitro conditions. These are classical methods having their share of applicability and limitations.

14 Fluorescence-dependent methods

Due to their sensitivity toward reactive oxygen species by being non-fluorescent before being oxidized by oxygen species, fluorescent probes are a good fit for free radical detection [9]. Dihydroethidium, dichloro-dihydro fluorescein, and Amplex Red are widely used probes which basically provide information regarding free radical production under a stimulus and reflect the oxidative state of different cellular compartments being analyzed. These probes are oxidized by a one-electron free radical mechanism which yields a radical intermediate and ultimately results in fluorescent products [11].

(a) *Dihydroethidium (DHE) staining*: Dihydroethidium utilizes its ability to get into the cells and get oxidized by the intracellular oxygen species to create a red fluorescent product called 2-hydroxyethidium (2-OH-E^+), or it could be oxidized by other oxidants present in the cell to generate a non-specific red fluorescent product called ethidium (E^+). Both of these products can be captured and analyzed by fluorescence-based techniques. These two fluorescent products can be identified by the application of analytical techniques such as HPLC and LC–MS assays, which confirm the presence of free radicals in the cellular compartments [82].

A limitation of this method of detection is that more clarity in the exact mechanism and accurate stoichiometry is required for the better localized detection. It is also reported that the presence of other oxidative radicals interferes with this reaction and hence prevents accurate quantification of free radical concentration [11,82].

(b) *Dichlorodihydro fluorescein diacetate (DCFH-DA)*: 2,7-Dichlorodihydro fluorescein diacetate (DCFH-DA) is a probe which is generally used for the directly measuring of intracellular redox states upon its oxidation [11]. This probe enters the cell and gets hydrolyzed by the intracellular esterase which generates membrane-impermeable DCFH, which is diffused inside and oxidized by non-specific oxygen radicals [83]. This finally yields a fluorescent DCF, through a non-fluorescent intermediate free radical DCFH (•–). This fluorescence intensity of DCF is correlated with levels of oxidative stress in cells. It is essential to study the behavior of DCFH-DA/DCFH in specific cells beforehand because distinct cells show different permeability to this particular probe, which varies the concentration inside and outside of cells; in addition, the activity of esterase may also vary a lot among different cells which could be a factor to consider before opting this method for detection [82].

A considerable number of drawbacks are associated with the application of this method; for example, the fluorescence intensity of DCF is not directly correlated with any certain free radical and also the different factors like cytochrome *c*, hematin, peroxidases, ferrous ion, and horseradish peroxidase (HRP) could also promote DCFH oxidation even in the absence of any specific or target free radical, which makes DCFH-DA unable to detect a specific free radical in the cells while being perfectly capable of reflecting the changes of cellular oxidative stress under intra- or extracellular stimulus [82].

(c) *Amplex Red*: Amplex Red is a probe which works upon one electron mechanism under catalysis of HRP to fluorescent resorufin which reflects the concentration of hydrogen peroxide by its intensity, hence allowing detection under both oxidative and reductive conditions [82]. This ability of the probes makes it widely applicable under different ranges such as to detect the role of mitochondria-sourced hydrogen peroxide in the process of mitochondrial gene damage and ATP synthesis. This method also poses a variety of limitations including the need for specific inhibitors to maintain selectivity, HRP being able to react with different electron donors such as NADPH, and being reduced to glutathione which results in production of additional resorufin, which could also pose problems in identification due to being reduced to a colorless, non-fluorescent compound under the action of NADPH-CYP450 reductases [83].

This probe also shows a low affinity to the cell membrane which does not allow the intracellular free radicals to be reported, hence due to interference caused by other radicals and enzymes, the Amplex Red method loses its appeal among other methods for free radical detection but can still be used in cell-free systems [11].

15 Electron spin resonance (ESR) or electron paramagnetic resonance (EPR) spectroscopy

Electron paramagnetic resonance (EPR) is a widely accepted versatile, non-destructive analytical technique that can directly detect the presence of free radicals qualitatively and quantitatively [84]. It is an analogous technique to nuclear magnetic resonance (NMR); the difference is the excitation of electron spins instead of the spins of atomic nuclei, so instead of nuclear transitions, transitions of unpaired electrons under an applied magnetic field are detected by EPR [85]. Free radicals are short-lived and play a crucial role in different processes; they are directly and accurately detected by this technique without interference from the sample properties, including its phase which could be any solid, liquid, or gas. This approach reports the magnetic properties of the unpaired electrons and their molecular environments, and the energy differences in these states correspond to the microwave region which can be detected by the electron spin resonance (ESR) spectrum. The ESR spectrum can detect both the presence of free radicals and can also determine their concentrations; furthermore, these ESR peaks can also determine the structure of the free radicals as well [86].

EPR uses phase-sensitive detection at the modulation frequency to observe the EPR spectrum, which is detected directly and scans the magnetic field through resonance [85].

The most prominent limitation of the ESR technique is that it requires spin traps that reduce specificity and stability for the measurement which require low temperatures and long recording times for detection in physiological conditions [84,87].

16 Spin trapping technique

In the case of failure of the ESR spectroscopy technique, another technique is used called spin trapping for detection of the presence of any free radical and its involvement [85]. This technique involves a compound that is added to the reaction medium which contributes to reactive free radical which in turns more persistent radical and this can be observed by the ESR spectrum. The free radicals are trapped by the trapping agent used, and then are detected by the ESR spectrum. Factors of ESR characteristics such as g-value, hyperfine coupling constant (hfcc), and spin concentration are observed by the ESR spectrum of spin adducts [85].

$$\text{Trapping Agent} + \text{Free Radical Reactions} \rightarrow \text{ESR Detector}$$

It is well known that these oxygen radicals are short-lived; specific spin traps are required to stabilize down, either by incorporating the radicals into their structures or by oxidizing them to a relatively stable radical, which makes them easy to capture using electronic spin resonators [85].

A five-membered ring nitron derivative, 5,5-dimethyl-1-pyrroline-N-oxide (DMPO), is the most widely accepted trapping agent used in this technique. The most desirable characteristic of a spin adduct to be used is for the agent being stable for a relatively a long time so that the spectrum can be easily analyzed. ESR spin trapping can also be used for kinetic analyses and determine the formation and elimination velocities of a free radical [87].

An important limitation of this technique is that if a free radical reacts immediately with a molecule other than the spin-trapping agent, the spin adduct will not be generated, hence it cannot be detected by ESR [87].

17 Immuno-spin trapping of macromolecules

While electron spin resonance (ESR) and electron paramagnetic resonance (EPR) spectroscopy are widely used as conventional methods for the free radical detection, a significant drawback of these techniques is that they are used after being coupled with the spin trapping methods and these spin adducts are short-lived due to biological reductive and/or oxidative processes [85,86]. To cope with such a problem, Immuno-spin trapping works as a solution because it involves the use of an antibody which recognizes macromolecular DMPO spin adducts and also the corresponding anti-DMPO antibody, regardless of the oxidative or reductive state of the trapped radical adducts [88]. This corresponds to the IST methodology which has been employed in combination with various different techniques as well, according to the particular requirements of the detection experiment such as with molecular magnetic resonance imaging (MRI), which has served many applications in different experimental oxidative stress-associated animal disease models. High-field MRI is considered for this approach due to the superior image resolution (e.g., $<50\times50\,\mu m^2$ for mice and $<100\times100\,\mu m^2$ for rats) of these imaging systems [61,62].

Herein an anti-DMPO antibody needs to be attached to the fluorescent dye, which in turn allows the detection of the radical adducts which could be either as the free radical or reduced/oxidized products with the help of fluorescence microscopy. And the fact that they are EPR detectable or not does not hold any significance is the additional advantages that this technique provides [61,62,88] (Fig. 5).

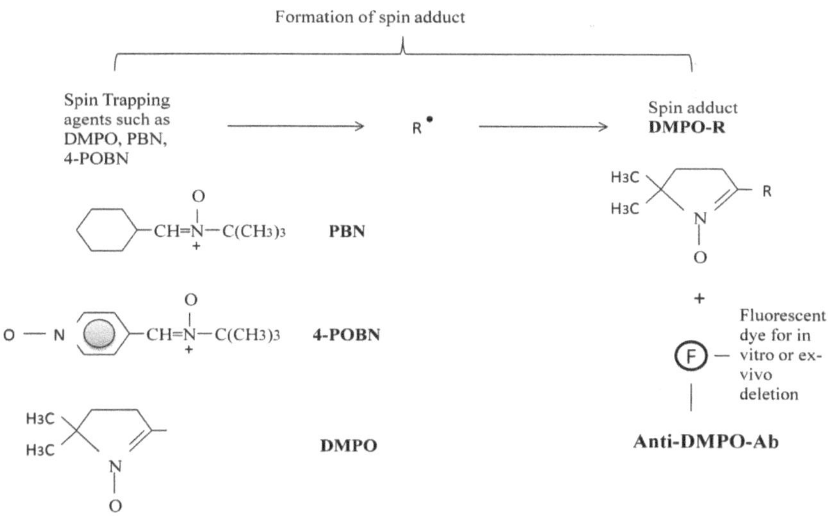

FIG. 5 Mechanism of Immuno-spin trapping by the formulations of different spin adducts (DMPO, PBN, 4-POBN).

18 Combined IST and mMRI detection

The combined IST–mMRI approach has been used in several rodent disease models, including diabetes, amyotrophic lateral sclerosis (ALS), gliomas, and septic encephalopathy. The advantage of this approach is that heterogeneous levels of trapped free radicals can be detected directly in vivo and in situ to pinpoint where free radicals are formed in different tissues [61,62].

This method extends the use of fluorescence in the approach, which involves the use of IST in conjunction with mMRI. The methodology involves the use of a spin trapping agent called DMPO, which is used to trap the free radicals in a disease model and then an mMRI probe is administered, which is known as an anti-DMPO probe [89]. This aims to combine an antibody against DMPO-radical adducts and an MRI contrast agent as well, which would be responsible for targeted free radical adduct mMRI [90]. Different contrast agents can be used according to the requirements; the one most common and that is proposed in many studies is albumin-gadolinium diethylene tri-amine penta-acetic acid (Gd-DTPA)-biotin construct, where the anti-DMPO antibody is covalently linked to the cysteine residues of albumin, forming an anti-DMPO adduct antibody-albumin-Gd-DTPA-biotin entity, which acts as a MRI signaling component and thereby increases MRI signal intensity. These MRI parameters can be used to assess the presence of the anti-DMPO probe, and this moiety helps in ex vivo validation of the presence of the anti-DMPO probe in concerned tissues, by using streptavidin fluorescence (e.g., Cy3) or streptavidin-horseradish peroxidase (HRP) to tag the biotin in the anti-DMPO probe [61,62,89].

The main identified drawbacks of this method are the limited access to preclinical MRI systems, availability of the anti-DMPO antibody, and further identifying the radical source that is being trapped [90]. This method has shown tremendous potential in non-neurological studies in numerous pathological/toxicological models, while in neurological studies the approach is limited in terms of permeability, and also possibility of DMPO/spin adduct to access the target tissues as well, which could serve as a limitation and could cause the anti-DMPO probe strategy to become inefficient [61,62] (Fig. 6).

19 Pulse radiolysis

Pulse radiolysis is considered as the most original and one of the most effective techniques for characterization of radical reactions detection [9]. This method involves the generation of radicals by the ionizing radiation under the controlled conditions and their corresponding absorptions are observed by UV–vis spectroscopy. This technique employs time-resolved

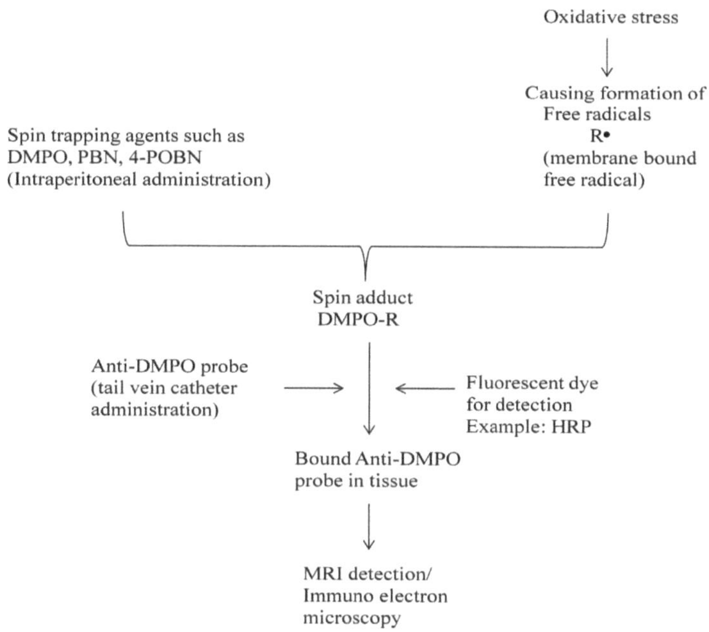

FIG. 6 Mechanism of combined IST and mMRI detection using fluorescent dye (horseradish peroxidase) for detection and immuno electron microscopy to confirm the presence of an anti-DMPO probe in plasma and nuclear membrane).

kinetic spectroscopy which is in the time range of micro- to milliseconds, which is the real-time scale for diffusion controlled reactions, and allows determination of the pertinent reaction parameters. Alkoxyl RO″ and peroxyl ROO″ are some examples of the radical intermediates of oxidative chain reactions, and these can be monitored by direct observations [91].

The methodology involves a short pulse of high-energy electrons delivered to a cell filled with a solution to be investigated, which yields kinetic data which become available from radiation-chemical studies [92]. Many different techniques are now available to detect transients generated by these methods, such as UV–vis spectroscopy, conductometry, light-scattering, polarography, Raman spectroscopy, electron spin resonance (ESR) spectroscopy, fluorescence-detected magnetic resonance (FDMR) spectroscopy, and chemically induced nuclear polarization (CIDNP) spectroscopy. Hence pulse radiolysis is considered as a very versatile instrument for detection of free radical reactions [93].

An important and fundamental limitation of this method comes to light when several species absorb in the same region as the compound of interest, which can obscure its spectrum and hence interferes with the results and makes the findings unreliable [93]. For the rectification of such an error, competition measurements are introduced to the original methodology such as addition of reactants to produce absorbing intermediates which are more easily accessible in the spectral regions, and hence narrow sense pulse radiolysis is also an indirect method which can be used for radical determination under such circumstances [9].

20 Chromatography-based techniques

Separation and identification of hydroxyl radicals and their products are the main reasons for the employment of chromatographic techniques in detection of free radicals. The free radicals generated in the system are reacted with some specific reagents to generate stable compounds detectable by the appropriate chromatographic assays. Out of all the other liquid chromatography methods, its combination with mass spectrometry are the most often used approaches for the concerned purpose [11].

Benzoic acid, salicylic acid, and DMSO are some of the reagents that are used for stabilization of free radicals. The 2,3-dihydroxybenzoic acid (2,3-DHBA) and 2,5-dihydroxybenzoic acid (2,5-DHBA) generated by the reaction between salicylic acid and hydroxyl radical are quantified by HPLC and electrochemical detector, proving this method to be useful to measure hydroxyl radical in vivo. HPLC-based methods have been applied for detection of hydroxyl radicals or activity of antioxidants in many reaction systems and tissues [94].

The efficiency and sensitivity of the chromatography methods in the detection of hydroxyl radicals are not considered very reliable, as the process of such reaction and its products are complicated and the treatment for the pre-detection is also very complex, which is why this method is not widely used or chosen over other methods [95].

Studies have shown that GC/MS analysis can be used to detect a novel product of the reaction between the CH_3 and the spin trapping agent 4-Oxo-TEMPO and this proved to be an accurate means of measuring hydroxyl radicals. The spin-trapping agent 4-Oxo-TEMPO could be used for the trapping of methyl radicals, and the other free radical adducts could be detected by GC/MS analysis as well [96].

A specific and sensitive gas chromatography/mass spectrometry (GC/MS) method for detection of hydroxyl and hydroxyethyl free radicals can be used to compensate for the fact that although detection of hydroxyl free radicals is frequently performed by electron spin resonance (ESR) following spin trapping of the radical using 5,5-dimethylpyrroline N-oxide (DMPO) to generate a stable free radical having a characteristic ESR spectrum, the ESR equipment required is too expensive and GC/MS can prove to be an excellent alternative [97]. Studies showed that in microsomal suspensions, DMPO proved to be better for detection of OH radicals and PBN was more suitable for detection of 1-hydroxyethyl radicals. The procedure is specific, sensitive, and potentially as useful as ESR [98].

Any one of these methods could provide the relative level of radicals, but in order to gain credible results, it is recommended that reasonable usage of each method based on experimental requirements should be made. Comparatively addressing fluorescent and luminescent probes meets the qualitative and semi-quantitative detection of most kinds of free radicals, as these are sensitive, flexible, and simple in data collection, enabling them to be used in many kinds of cells, hence being suitable for detection of short-term free radical flash in cells [11]. All these different methods have proved to be very appealing in terms of detection and can be used as appropriate for each study or analysis.

21 Other biochemical methods

Apart from the specific and system-targeted approaches for free radical detection, there are different organ level-based methods which are developed with aims of their specific importance and correlation in various disease model studies. The use of such methods is more study-oriented compared to the other conventional methods that are process-oriented, hence these methods are employed in conditions of their suitability with the particular experimental studies being performed.

Organ function measurements: Various assays have been formulated for the detection of free radicals which do not use the conventional approach and target the organ-specific strategy.

a) *Role in cardiac arrhythmias*: These methods employ different objectives such as determining whether reactive oxygen intermediates can induce electrophysiological disturbances leading to the genesis of arrhythmia, even in the absence of ischemia and reperfusion [99]. The underlying concept is that oxygen-derived free radicals produced during ischemia and reperfusion can cause injury to the membrane proteins, leading to ionic imbalance and electrical instability, and these free radical generating systems can make the heart more susceptible to arrhythmias [100].
b) *Role in reperfusion injury*: The studies evaluated in various different organs demonstrated that toxic oxygen metabolites participate in organ injury caused by reperfusion [101].

Superoxide dismutase (SOD), the highly specific scavenger of the superoxide radical, has been proved to prevent the injury if introduced at the end of the ischemic period and before reperfusion, which was the first demonstration of free radical-mediated reperfusion injury in any organ. This preventive effect was also shown by the hydrogen peroxide with catalase or the hydroxyl radical (or activated iron oxygen complexes that act like hydroxy radicals) with dimethyl sulfoxide [102].

Excretion of stable marker molecules (8-hydroxy-2′-deoxyguanosine (oh8dG): A highly sensitive and selective method for detecting the oxidatively modified DNA residue oh8dG by the means of electrochemical detection with HPLC is a powerful approach for the assessment of in vivo oxidative damage [103]. The detection of this oh8dG from DNA permits the quantification of the steady-state levels of this oxidatively modified deoxynucleoside and overcomes the detection problems associated with the extremely low levels present in DNA. The detection and quantification of this marker molecule in urine give an estimation of oh8dG in the DNA and hence enable estimating the oxidative DNA damage in the body, proving oh8dG to be a sensitive marker of oxidative DNA damage [104].

Gas exhalation measurements: A recently developed non-invasive method for monitoring in vivo lipid peroxidation and hence free radical concentration is the measurement of exhaled hydrocarbons [105,106]. The underlying mechanism of the approach is peroxidation of polyunsaturated fatty acids, i.e., linoleic acid and linolenic acid, which are both cellular membrane compounds, induce ethane and pentane formation in pulmonary air exhalation, and these two volatile hydrocarbons are considered potential lipoperoxidation markers [107]. Studies showed that exhalation of ethane demonstrated CCl_4-induced lipid peroxidation. The advantage of this method of indirect free radical detection is that this approach is non-invasive, as it does not require extraction of tissue or internal probing [108,109].

Ethidium binding assay: This takes advantage of the extreme sensitivity of supercoiled covalently closed circular (CCC) DNA to the oxidative lesions, which detects the generation of free radicals when they are used in conjunction with various enzymes that play an important role in cellular protection, growth, and repair. Ethidium bromide is responsible for binding intercalatively with little apparent sequence selectivity to duplex DNA and shows an enhancement in the fluorescence intensity of – 25-fold, which results in proton exchange. Under normal conditions, native negatively super coiled CCC DNA binds a certain amount of ethidium and shows fluorescence under the action of reactive free radicals. Nicking of the strands of the CCC DNA is caused, leading to the open circular (OC) form, which can consequently accept more ethidium. This results in a 30–38% increase in the fluorescence reading of the DNA ethidium complex depending on the super helix density [110].

An important advantage served by this assay is that by addition of some selective enzymes and chemical inhibitors, individual reactive free radicals can be identified by observing the inhibitory effect on the DNA scission process. These enzymes include superoxide dismutase for (O•–) and catalase for (H_2O_2) or glutathione peroxidase [10].

Apart from these, several other methods have been developed with respect to the requirements of the experimental studies and they are advised to be used accordingly, such as methods based on the perfused organ chemiluminescence and oxystat techniques [16,111]. However, these techniques are not considered a one-stop shop for free radical detection, but more to achieve estimation of the generalized oxidative stress load of the system, and they are thus advised to be employed under such requirements.

22 Summary

Production of free radicals in cellular systems is a pervasive process because of the continuous exposure of reactive oxygen species as a consequence of biochemical reactions and external factors. As these radicals are capable of initiating pathological cascades at all times, the counteracting enzyme activity and scavenging systems are maintained to counter these. The aim of developing these methods for free radical detection or quantification of oxidative stress/damage is to elucidate the exact mechanism of action of free radicals in different diseases, especially in genetic, degenerative, and carcinogenic disorders.

For in vitro analysis, the free radicals measured are only the ones present in the extracellular space by samples such as blood, urine, and breath, while for in vivo analysis, the laws of chemical kinetics fail to prevail for the scavenging reactions, so the techniques become non-specific and semi-quantitative under in vivo conditions. Although they have their share of limitations and drawbacks, analytical methods have been proved to offer the most direct detection and promising quantification by fundamentally establishing the role of free radicals in the pathogenesis of various diseases.

Every method yet discovered/designed offers some advantage over the others and has its own limitations, as discussed in this chapter, and it is obvious that each can provide a relative estimation of free radicals in the system. However, for more credible and qualitative results, reasonable usage of each method based on experimental requirements and combined usage of more than one approach are recommended. New strategies and approaches are constantly being developed with the expectations of availability for such a method that eliminates the limitations of all the discovered methods and provides the most credible results according to the experimental requirements.

References

[1] M. Valko, et al., Free radicals, metals and antioxidants in oxidative stress-induced cancer, Chem. Biol. Interact. 160 (1) (2006) 1–40.

[2] V. Sakac, M. Sakac, Free oxygen radicals and kidney diseases—part I, Med. Pregl. 53 (9–10) (2000) 463–474.

[3] M. Valko, et al., Free radicals and antioxidants in normal physiological functions and human disease, Int. J. Biochem. Cell Biol. 39 (1) (2007) 44–84.
[4] F.-C. Cheng, J.-F. Jen, T.-H. Tsai, Hydroxyl radical in living systems and its separation methods, J. Chromatogr. B 781 (1–2) (2002) 481–496.
[5] E. Tvrdá, P. Massanyi, N. Lukáč, Physiological and pathological roles of free radicals in male reproduction, in: Spermatozoa-Facts and Perspectives, IntechOpen, 2017.
[6] B. D'Autréaux, M.B. Toledano, ROS as signalling molecules: mechanisms that generate specificity in ROS homeostasis, Nat. Rev. Mol. Cell Biol. 8 (10) (2007) 813–824.
[7] Z. Cheng, et al., Study on the multiple mechanisms underlying the reaction between hydroxyl radical and phenolic compounds by qualitative structure and activity relationship, Bioorg. Med. Chem. 10 (12) (2002) 4067–4073.
[8] H. Togashi, et al., Analysis of hepatic oxidative stress status by electron spin resonance spectroscopy and imaging, Free Radic. Biol. Med. 28 (6) (2000) 846–853.
[9] M. Saran, W. Bors, Direct and indirect measurements of oxygen radicals, Klin. Wochenschr. 69 (1991) 957–964, https://doi.org/10.1007/BF01645139.
[10] S. Hollán, Methods for measurements of increased release of free oxygen radicals, Complement. Med. Res. 2 (5) (1995) 254–258.
[11] Y. Zhang, M. Dai, Z. Yuan, Methods for the detection of reactive oxygen species, Anal. Methods 10 (38) (2018) 4625–4638.
[12] Y.A. Vladimirov, E.V. Proskurnina, D.Y. Izmailov, Chemiluminescence as a method for detection and study of free radicals in biological systems, Bull. Exp. Biol. Med. 144 (3) (2007) 390–396.
[13] A.S. Yalçın, et al., Chemiluminescence measurements for the detection of free radical species, in: Free Radicals, Oxidative Stress, and Antioxidants, Springer, Boston, MA, 1998, pp. 385–390.
[14] F.R. Ochsendorf, et al., Lucigenin chemiluminescence in human seminal plasma, Free Radic. Res. 34 (2) (2001) 153–165.
[15] I.B. Afanas'ev, Lucigenin chemiluminescence assay for superoxide detection, Circ. Res. 89 (11) (2001) e46.
[16] M.E. Murphy, H. Sies, Range low-level chemiluminescence in biological systems, in: L. Packer, A.N. Glazer (Eds.), Visible Oxygen Radicals in Biological Systems, Oxygen Radicals and Antioxidants, 1993.
[17] R.P. Brandes, M. Janiszewski, Direct detection of reactive oxygen species ex vivo, Kidney Int. 67 (5) (2005) 1662–1664.
[18] Y. Yamamoto, B. Frei, B.N. Ames, Assay of lipid hydroperoxides using high-performance liquid chromatography with isoluminal chemiluminescence detection, Methods Enzymol. 186 (1990) 371–380.
[19] J. Zielonka, J. David Lambeth, B. Kalyanaraman, On the use of L-012, a luminol-based chemiluminescent probe, for detecting superoxide and identifying inhibitors of NADPH oxidase: a reevaluation, Free Radic. Biol. Med. 65 (2013) 1310–1314.
[20] H.-Y. Sohn, et al., Sensitive superoxide detection in vascular cells by the new chemiluminescence dye L-012, J. Vasc. Res. 36 (6) (1999) 456–464.
[21] Y. Nishinaka, et al., A new sensitive chemiluminescence probe, L-012, for measuring the production of superoxide anion by cells, Biochem. Biophys. Res. Commun. 193 (2) (1993) 554–559.
[22] J.-S. Kim, et al., A quantitative method to measure low levels of ROS in nonphagocytic cells by using a chemiluminescent imaging system, Oxid. Med. Cell. Longev. 2019 (2019).
[23] A.Y. Michelle, et al., Radical formation in cytochrome c oxidase, Biochim. Biophys. Acta (BBA) Bioenerget. 1807 (10) (2011) 1295–1304.
[24] M.M. Tarpey, D.A. Wink, M.B. Grisham, Methods for detection of reactive metabolites of oxygen and nitrogen: in vitro and in vivo considerations, Am. J. Physiol. Regul. Integr. Comp. Physiol. 286 (3) (2004) R431–R444.

[25] A. Azzi, C. Montecucco, C. Richter, The use of acetylated ferricytochrome c for the detection of superoxide radicals produced in biological membranes, Biochem. Biophys. Res. Commun. 65 (2) (1975) 597–603.
[26] V. Massey, The microestimation of succinate and the extinction coefficient of cytochrome c, Biochim. Biophys. Acta 34 (1959) 255–256.
[27] I. Yamazaki, The reduction of cytochrome c by enzyme-generated ascorbic free radical, J. Biol. Chem. 237 (1) (1962) 224–229.
[28] I. Fridovich, Cytochrome c, Handb. Methods Oxygen Radic. Res. (1985) 121–122.
[29] L. Thomson, et al., Kinetics of cytochrome C2+ oxidation by peroxynitrite: implications for superoxide measurements in nitric oxide-producing biological-systems, Arch. Biochem. Biophys. 319 (2) (1995) 491–497.
[30] H. Kuthan, V. Ullrich, R.W. Estabrook, A quantitative test for superoxide radicals produced in biological systems, Biochem. J. 203 (3) (1982) 551–558.
[31] Y.-R. Chen, R.P. Mason, Mechanism in the reaction of cytochrome c oxidase with organic hydroperoxides: an ESR spin-trapping investigation, Biochem. J. 365 (2) (2002) 461–469.
[32] U. Landmesser, et al., Oxidation of tetrahydrobiopterin leads to uncoupling of endothelial cell nitric oxide synthase in hypertension, J. Clin. Invest. 111 (8) (2003) 1201–1209.
[33] C.K. Sen, L. Packer, Redox Cell Biology and Genetics, Part B, Elsevier, 2002.
[34] O. Tunc, J. Thompson, K. Tremellen, Development of the NBT assay as a marker of sperm oxidative stress, Int. J. Androl. 33 (1) (2010) 13–21.
[35] P.K. Javvaji, et al., An efficient nitroblue tetrazolium staining and bright-field microscopy based method for detecting and quantifying intracellular reactive oxygen species in oocytes, cumulus cells and embryos, Front. Cell Dev. Biol. 8 (2020) 764.
[36] J. Gosalvez, E. Tvrda, A. Agarwal, Free radical and superoxide reactivity detection in semen quality assessment: past, present, and future, J. Assist. Reprod. Genet. 34 (6) (2017) 697–707.
[37] P.R. Gardner, I. Fridovich, Inactivation-reactivation of aconitase in Escherichia coli. A sensitive measure of superoxide radical, J. Biol. Chem. 267 (13) (1992) 8757–8763.
[38] D.H. Flint, J.F. Tuminello, M.H. Emptage, The inactivation of Fe-S cluster containing hydro-lyases by superoxide, J. Biol. Chem. 268 (30) (1993) 22369–22376.
[39] L. Castro, M. Rodriguez, R. Radi, Aconitase is readily inactivated by peroxynitrite, but not by its precursor, nitric oxide, J. Biol. Chem. 269 (47) (1994) 29409–29415.
[40] A. Hausladen, I. Fridovich, Measuring nitric oxide and superoxide: rate constants for aconitase reactivity, Methods Enzymol. 269 (1996) 37–41.
[41] P.R. Gardner, et al., Superoxide radical and iron modulate aconitase activity in mammalian cells, J. Biol. Chem. 270 (22) (1995) 13399–13405.
[42] M. Patel, et al., Requirement for superoxide in excitotoxic cell death, Neuron 16 (2) (1996) 345–355.
[43] C. Frejaville, et al., 5-(Diethoxyphosphoryl)-5-methyl-1-pyrroline N-oxide: a new efficient phosphorylated nitrone for the in vitro and in vivo spin trapping of oxygen-centered radicals, J. Med. Chem. 38 (2) (1995) 258–265.
[44] A.E. Albers, et al., A red-emitting naphthofluorescein-based fluorescent probe for selective detection of hydrogen peroxide in living cells, Bioorg. Med. Chem. Lett. 18 (22) (2008) 5948–5950.
[45] J. Zielonka, et al., Boronate probes as diagnostic tools for real time monitoring of peroxynitrite and hydroperoxides, Chem. Res. Toxicol. 25 (9) (2012) 1793–1799.
[46] A. Sikora, et al., Direct oxidation of boronates by peroxynitrite: mechanism and implications in fluorescence imaging of peroxynitrite, Free Radic. Biol. Med. 47 (10) (2009) 1401–1407.
[47] B.C. Dickinson, C.J. Chang, A targetable fluorescent probe for imaging hydrogen peroxide in the mitochondria of living cells, J. Am. Chem. Soc. 130 (30) (2008) 9638–9639.

[48] D.P. Jones, Redox potential of GSH/GSSG couple: assay and biological significance, Methods Enzymol. 348 (2002) 93–112.
[49] C. Berndt, C.H. Lillig, L. Flohé, Redox regulation by glutathione needs enzymes, Front. Pharmacol. 5 (2014) 168.
[50] D.P. Jones, Y. Liang, Measuring the poise of thiol/disulfide couples in vivo, Free Radic. Biol. Med. 47 (10) (2009) 1329–1338.
[51] D. Milatovic, M. Aschner, Measurement of isoprostanes as markers of oxidative stress in neuronal tissue, Curr. Protoc. Toxicol. 39 (1) (2009) 12–14.
[52] W.A. Pryor, J.P. Stanley, E. Blair, Autoxidation of polyunsaturated fatty acids: II. A suggested mechanism for the formation of TBA-reactive materials from prostaglandin-like endoperoxides, Lipids 11 (5) (1976) 370–379.
[53] B. Halliwell, J.M.C. Gutteridge, Free Radicals in Biology and Medicine, Oxford University Press, USA, 2015.
[54] M.B. Grisham, Oxidants and free radicals in inflammatory bowel disease, Lancet 344 (8926) (1994) 859–861.
[55] J.D. Morrow, et al., A series of prostaglandin F2-like compounds are produced in vivo in humans by a non-cyclooxygenase, free radical-catalyzed mechanism, Proc. Natl. Acad. Sci. 87 (23) (1990) 9383–9387.
[56] L.J. Roberts II, J.D. Morrow, Measurement of F2-isoprostanes as an index of oxidative stress in vivo, Free Radic. Biol. Med. 28 (4) (2000) 505–513.
[57] D. Milatovic, T.J. Montine, M. Aschner, Measurement of isoprostanes as markers of oxidative stress, in: Vitro Neurotoxicology, Humana Press, Totowa, NJ, 2011, pp. 195–204.
[58] C. Soffler, V.L. Campbell, D.M. Hassel, Measurement of urinary F2-isoprostanes as markers of in vivo lipid peroxidation: a comparison of enzyme immunoassays with gas chromatography–mass spectrometry in domestic animal species, J. Vet. Diagn. Invest. 22 (2) (2010) 200–209.
[59] T.J. van't Erve, et al., Classifying oxidative stress by F2-isoprostane levels across human diseases: a meta-analysis, Redox Biol. 12 (2017) 582–599.
[60] C.T. Dooley, et al., Imaging dynamic redox changes in mammalian cells with green fluorescent protein indicators, J. Biol. Chem. 279 (21) (2004) 22284–22293.
[61] R.A. Towner, N. Smith, Detecting in vivo free radicals in various disease models, in: Free Radicals, Antioxidants and Diseases, 2018.
[62] R.A. Towner, N. Smith, In vivo and in situ detection of macromolecular free radicals using immuno-spin trapping and molecular magnetic resonance imaging, Antioxid. Redox Signal. 28 (15) (2018) 1404–1415.
[63] G.T. Hanson, et al., Investigating mitochondrial redox potential with redox-sensitive green fluorescent protein indicators, J. Biol. Chem. 279 (13) (2004) 13044–13053.
[64] C.R. Arias-Barreiro, et al., A bacterial biosensor for oxidative stress using the constitutively expressed redox-sensitive protein roGFP2, Sensors 10 (7) (2010) 6290–6306.
[65] T. Yano, et al., A novel fluorescent sensor protein for visualization of redox states in the cytoplasm and in peroxisomes, Mol. Cell. Biol. 30 (15) (2010) 3758–3766.
[66] J.R. Lohman, S. James Remington, Development of a family of redox-sensitive green fluorescent protein indicators for use in relatively oxidizing subcellular environments, Biochemistry 47 (33) (2008) 8678–8688.
[67] J. Birk, et al., Endoplasmic reticulum: reduced and oxidized glutathione revisited, J. Cell Sci. 126 (7) (2013) 1604–1617.
[68] M.B. Cannon, S. James Remington, Re-engineering redox-sensitive green fluorescent protein for improved response rate, Protein Sci. 15 (1) (2006) 45–57.
[69] I. Aller, N. Rouhier, A.J. Meyer, Development of roGFP2-derived redox probes for measurement of the glutathione redox potential in the cytosol of severely glutathione-deficient rml1 seedlings, Front. Plant Sci. 4 (2013) 506.
[70] M. Gutscher, et al., Proximity-based protein thiol oxidation by H2O2-scavenging peroxidases, J. Biol. Chem. 284 (46) (2009) 31532–31540.

[71] B. Morgan, M.C. Sobotta, T.P. Dick, Measuring EGSH and H2O2 with roGFP2-based redox probes, Free Radic. Biol. Med. 51 (11) (2011) 1943–1951.
[72] A. Müller, et al., Systematic in vitro assessment of responses of roGFP2-based probes to physiologically relevant oxidant species, Free Radic. Biol. Med. 106 (2017) 329–338.
[73] G. Maulucci, et al., High-resolution imaging of redox signaling in live cells through an oxidation-sensitive yellow fluorescent protein, Sci. Signal. 1 (43) (2008) pl3.
[74] G.S. Baird, D.A. Zacharias, R.Y. Tsien, Circular permutation and receptor insertion within green fluorescent proteins, Proc. Natl. Acad. Sci. 96 (20) (1999) 11241–11246.
[75] M. Dardalhon, et al., Redox-sensitive YFP sensors monitor dynamic nuclear and cytosolic glutathione redox changes, Free Radic. Biol. Med. 52 (11–12) (2012) 2254–2265.
[76] A. Hernandez-Barrera, et al., Using hyper as a molecular probe to visualize hydrogen peroxide in living plant cells: a method with virtually unlimited potential in plant biology, Methods Enzymol. 527 (2013) 275–290.
[77] V.V. Belousov, et al., Genetically encoded fluorescent indicator for intracellular hydrogen peroxide, Nat. Methods 3 (4) (2006) 281–286.
[78] C.F. Labuschagne, A.B. Brenkman, Current methods in quantifying ROS and oxidative damage in Caenorhabditis elegans and other model organism of aging, Ageing Res. Rev. 12 (4) (2013) 918–930.
[79] K.A. Lukyanov, V.V. Belousov, Genetically encoded fluorescent redox sensors, Biochim. Biophys. Acta (BBA) Gen. Subjects 1840 (2) (2014) 745–756.
[80] K.N. Markvicheva, et al., A genetically encoded sensor for H2O2 with expanded dynamic range, Bioorg. Med. Chem. 19 (3) (2011) 1079–1084.
[81] D.S. Bilan, et al., HyPer-3: a genetically encoded H2O2 probe with improved performance for ratiometric and fluorescence lifetime imaging, ACS Chem. Biol. 8 (3) (2013) 535–542.
[82] M. Katerji, M. Filippova, P. Duerksen-Hughes, Approaches and methods to measure oxidative stress in clinical samples: research applications in the cancer field, Oxid. Med. Cell. Longev. 2019 (2019).
[83] E.H. Verbon, J.A. Post, J. Boonstra, The influence of reactive oxygen species on cell cycle progression in mammalian cells, Gene 511 (1) (2012) 1–6, https://doi.org/10.1016/j.gene.2012.08.038. Epub 2012 Aug 31. PMID: 22981713.
[84] M. Kohno, Applications of electron spin resonance spectrometry for reactive oxygen species and reactive nitrogen species research, J. Clin. Biochem. Nutr. (2010) 1006170036.
[85] S. Suzen, H. Gurer-Orhan, L. Saso, Detection of reactive oxygen and nitrogen species by electron paramagnetic resonance (EPR) technique, Molecules 22 (1) (2017) 181.
[86] O. Augusto, J. Schreiber, R.P. Mason, Direct ESR detection of a free radical intermediate during the peroxidase-catalyzed oxidation of the antimalarial drug primaquine, Biochem. Pharmacol. 37 (14) (1988) 2791–2797, https://doi.org/10.1016/0006-2952(88)90042-1. Erratum in: Biochem Pharmacol 1988 Oct 15;37(20):4033. PMID: 2840077.
[87] E.G. Janzen, J.L. Gerlock, Detection of gas phase free radicals by electron spin resonance spin trapping techniques, Nature 222 (5196) (1969) 867–868.
[88] R.P. Mason, D. Ganini, Immuno-spin trapping of macromolecules free radicals in vitro and in vivo—one stop shopping for free radical detection, Free Radic. Biol. Med. 131 (2019) 318–331, https://doi.org/10.1016/j.freeradbiomed.2018.11.009. Epub 2018 Dec 12. PMID: 30552998.
[89] R.A. Towner, et al., Combined molecular MRI and immuno-spin-trapping for in vivo detection of free radicals in orthotopic mouse GL261 gliomas, Biochim. Biophys. Acta (BBA) Mol. Basis Dis. 1832 (12) (2013) 2153–2161.
[90] B. Ahn, et al., Using MRI to measure in vivo free radical production and perfusion dynamics in a mouse model of elevated oxidative stress and neurogenic atrophy, Redox Biol. 26 (2019), 101308.

[91] P.E.T.E.R. Wardman, Application of pulse radiolysis methods to study the reactions and structure of biomolecules, Rep. Prog. Phys. 41 (2) (1978) 259.
[92] K.-D. Asmus, Pulse radiolysis methodology, Methods Enzymol. 105 (1984) 167–178.
[93] C. Houée-Levin, K. Bobrowski, The use of the methods of radiolysis to explore the mechanisms of free radical modifications in proteins, J. Proteomics 92 (2013) 51–62.
[94] P. Piszcz, K. Żurawski, B.K. Głód, Application of HPLC to study the reaction of free radicals with antioxidants and/or toxins, J. Chem. 2014 (2014).
[95] K. Makino, F. Moriya, H. Hatano, Separation of free radicals by high- performance liquid chromatography with electron spin resonance detection, J. Chromatogr. A 332 (1985) 71–106.
[96] F. Moriya, K. Makino, HPLC-ESR detection of free radicals, in: Advances in Liquid Chromatography: 35 Years of Column Liquid Chromatography in Japan 1, 1996, p. 75.
[97] S.K. Putta, B.K. Talapula, Detection of free radicals using GC/MS trapped by Tempo, Res. J. Pharm. Technol. 4 (9) (2011) 1367–1372.
[98] G.D. Castro, A.M.A. Delgado, de Layño, and J. A. Castro., Hydroxyl and 1-hydroxyethyl free radical detection using spin traps followed by derivatization and gas chromatography—mass spectrometry, Redox Rep. 3 (5–6) (1997) 343–347.
[99] D.J. Hearse, Y. Kusama, M. Bernier, Rapid electrophysiological changes leading to arrhythmias in the aerobic rat heart. Photosensitization studies with rose bengal-derived reactive oxygen intermediates, Circ. Res. 65 (1) (1989) 146–153.
[100] S.L. Shuter, et al., Studies on the effects of antioxidants and inhibitors of radical generation on free radical production in the reperfused rat heart using electron spin resonance spectroscopy, Free Radic. Res. Commun. 9 (3–6) (1990) 223–232.
[101] S.R.J. Maxwell, G.Y.H. Lip, Reperfusion injury: a review of the pathophysiology, clinical manifestations and therapeutic options, Int. J. Cardiol. 58 (2) (1997) 95–117.
[102] M.S. Sussman, G.B. Bulkley, Oxygen-derived free radicals in reperfusion injury, Methods Enzymol. 186 (1990) 711–723.
[103] M.K. Shigenaga, et al., In vivo oxidative DNA damage: measurement of 8-hydroxy-2′-deoxyguanosine in DNA and urine by high-performance liquid chromatography with electrochemical detection, Methods Enzymol. 186 (1990) 521–530.
[104] M.K. Shigenaga, B.N. Ames, Assays for 8-hydroxy-2′-deoxyguanosine: a biomarker of in vivo oxidative DNA damage, Free Radic. Biol. Med. 10 (3–4) (1991) 211–216.
[105] P. Hotz, et al., Development of a method to monitor low molecular mass hydrocarbons in exhaled breath of man: preliminary evaluation of its interest for detecting a lipoperoxidation process in vivo, Clin. Chim. Acta 162 (3) (1987) 303–310.
[106] O. Ortolani, et al., Free radical lipid peroxidation through expired ethane and pentane. An improved method, Boll. Soc. Ital. Biol. Sper. 62 (3) (1986) 383–389.
[107] L. Massias, et al., Thermal desorption-gas chromatographic determination of ethane and pentane in breath as potential markers of lipid peroxidation, Biomed. Chromatogr. 7 (4) (1993) 200–203.
[108] G.D. Lawrence, G. Cohen, [36] concentrating ethane from breath to monitor lipid peroxidation in vivo, Methods Enzymol. 105 (1984) 305–311.
[109] G.D. Lawrence, G. Cohen, Ethane exhalation as an index of in vivo lipid peroxidation: concentrating ethane from a breath collection chamber, Anal. Biochem. 122 (2) (1982) 283–290.
[110] J.W. Lown, [70] ethidium binding assay for reactive oxygen species generated from reductively activated adriamycin (doxorubicin), Methods Enzymol. 105 (1984) 532–539.
[111] M.E. Murphy, H. Sies, Visible-Range Low-Level Chemiluminescence in Biological Systems, In'Oxygen Radicals in Biological Systems, Oxygen Radicals and antioxidants', Eds. L. Packer and AN Glazer, 1993.

CHAPTER

10

Novel therapeutic approaches targeting oxidative stress in breast and lung cancer

Pooja Kaushik, Medha Kaushik, and Suhel Parvez

Department of Toxicology, School of Chemical and Life Sciences, Jamia Hamdard, New Delhi, India

List of abbreviations

ACS	American Cancer Society
ADCs	antibody-drug conjugation systems
ADEPT	antibody directed enzyme prodrug therapy
ALND	axillary lymph node dissection
BET	bromodomain and extra-terminal motif
BRAC	breast cancer genes
COPD	chronic obstructive pulmonary disease
COX-2	cyclooxygenase 2
CSCs	cancer stem cells
CT	computed tomography
CTLA4	cytotoxic T-lymphocyte-associated protein 4
EGFR	epidermal growth factor receptor
EMT	epithelial-mesenchymal transition
ERBB2	erythroblastic oncogene B2
ERK1/2	extracellular regulated kinase 1/2 activation
ERRα	estrogen-related receptor alpha
ETC	electron transport chain
GLUT1	glucose transporter
GPx	glutathione peroxidase
HDI	human development index
HIF-1α	hypoxia inducible factor 1 alpha
HR	human receptor
IARC	International Agency for Research on Cancer
IBC	invasive breast cancer
ICIs	immune checkpoint inhibitors
IR	ionizing radiations
JNK	c-Jun N-terminal kinase

https://doi.org/10.1016/B978-0-323-90905-1.00010-9

keap1	Kelch-like protein 19
LCC	large cell carcinoma
LMICs	low- and middle-income countries
MAPK	mitogen-activated protein kinases
NIBC	noninvasive breast cancer
NLC	nanostructured lipid carriers
Nrf2	nuclear factor erythroid 2-related factor 2
NSCLC	nonsmall cell lung carcinoma
OS	oxidative stress
OXPHOS	oxidative phosphorylation
PD-1	programmed cell death protein 1
PROTAC	proteolysis targeting chimera
PRX	peroxiredoxins
ROS	reactive oxygen species
Sags	superantigens
SCC	squamous cell carcinoma
SCLC	small cell lung carcinoma
SLN	sentinel lymph node
SOD	superoxide dismutase
TGF-β	tumor growth factor-β
TKI	tyrosine kinase inhibitor
Treg	regulatory T cells
TTF-1	thyroid transcription factor-1
VEGF	vascular endothelial growth factor
WHO	World Health Organization

1 Introduction

Cancer is one of the leading causes of death and disability, and is a significant block on growing life expectancy worldwide. According to a report by the World Health Organization (WHO), in 2020, 19.3 million new cases and 9.4 million cancer-related deaths were encountered worldwide. It is expected that by 2040, the number of new cases of cancer will rise to 30.2 million and cancer-associated deaths to 16.4 million. It has been come across that the countries having high life expectancy, education system, and high living standards have higher incidence of cancer. However, for some type of cancers, like cervical cancer, the trend is reversed and the rate of incidence is highest in countries in which these measures are low (see https://www.who.int/news-room/fact-sheets/detail/cancer). Breast and lung cancer have high prevalence rates worldwide, contributing to large numbers of deaths. Breast cancer is the most common malignant and metastatic cancer, largely metastasizing to the lungs, thus increasing the risk of cancer development in other body regions as well [1,2]. Contrary to common belief, breast cancer has a moderate prevalence rate among males as well, which is a major concern due to lack of awareness. Increase in the number of both male and female smokers, along with environmental, genetic and occupational risk factors, largely contribute to increased risk of lung cancer development. Additionally, other respiratory disorders such as chronic obstructive pulmonary disease (COPD) and SARS-CoV-2 have contributed to a large number of indirect deaths among lung cancer patients, as well as greatly enhanced the risk of cancer development [3].

The available treatments for cancer are known to exert significant side effects, such as cytotoxicity, to healthy cells. Thus, there is a prevailing gap in the research, which needs to be filled to develop better therapeutic approaches. In order to develop effective and novel therapies for the treatment and prevention of cancer, novel targets must be identified. The role of oxidative stress has been well established in cancer pathogenesis; however, its role in cancer treatment has not been exploited in conventional therapies. Recent developments have explored oxidative stress-based therapies for their efficacy in the treatment of cancer. Cancer cells thrive by elevating the reactive oxygen species (ROS) levels in the cellular microenvironment; however, an intense and sudden increase in ROS could also lead to cancer cell death. This phenomenon is now being exploited to its full potential in novel therapeutic approaches. Moreover, the center of ROS generation, the mitochondria, is also being explored as a target to treat various types of cancer [4].

In this chapter, we discuss the epidemiology and classify lung and breast cancer in detail. The role of oxidative stress in cancer pathophysiology is also discussed and the conventional therapies used for the treatment of these cancer types briefly described. Additionally, we have provided detailed review of novel therapeutic approaches in cancer treatment and oxidative stress-based approaches including targeting mitochondrial markers for cancer treatment.

2 Epidemiology and risk factors

With proper and precise understanding of cancer epidemiology, it is conceivable to provide and develop appropriate clinical interventions aimed at establishing competent policies for screening, diagnosis, and ultimately, prevention.

The International Agency for Research on Cancer (IARC) stated that the incidence rate is 1 in 5 people throughout their lifetime and the mortality rate is 1 in 8 men and 1 in 11 women. These figures suggest that in the coming 5 years, 50 million people will be living with cancer. Aging is the leading risk factor that contributes to its increase and will create a huge socio-economic burden. Globally, in women, breast cancer represents one in four of all cancers diagnosed, while colorectal, lung, cervical, and thyroid cancers are also prevalent. For the first time, breast cancer has been reported as the most common diagnosed cancer—replacing lung cancer—in low- and middle-income countries (LMICs); however, lung cancer still persists as the leading cause in most high-income countries (see https://www.uicc.org/news/globocan-2020-new-global-cancer-data).

In 2020, the most common cancers were breast cancer (2.26 million cases), lung cancer (2.21 million cases), colorectal cancer (1.93 million cases), prostate cancer (1.41 million cases), skin cancer (1.20 million cases), and stomach cancer (1.09 million cases). The mortality rates of cancers in 2020 were as follows: lung, 1.80 million deaths; colorectal, 935,000 deaths; liver, 830,000 deaths; stomach, 769,000 deaths; and breast, 685,000 deaths.

The burden of cancer in the United States is increasing gradually, the number of new cases in 2020 was estimated at around 1,806,590, while 606,520 died of cancer. Prostate, lung, and colorectal cancers were prevalent in men, at 43%, while breast, lung, and colorectal were prevalent in women, at 50% of new cases diagnosed. It has been shown that 442.4 per 100,000 persons per year have been diagnosed, and 158.3 per 100,000 person per year have died with cancer. Among children and adolescents, the diagnosed cases were 16,850, out of whom 1730 died of the disease. Over the years, the mortality rates have declined in comparison to incidence rates, suggesting availability of improved treatment options (Figs. 1 and 2).

2.1 Breast cancer epidemiology

In 2020, breast cancer in women was the largest cause of global cancer incidence, with approximately 2.3 million new cases, accounting for 11.7% of all cancer cases. With 685,000 deaths worldwide, it is the fifth largest cause of cancer mortality. One in four cancer cases and one in six cancer deaths is due to breast cancer, placing it at the top for incidence in 159 out of 185 countries and accounting for mortality in 100 countries. Lifestyle risk factors such as alcohol intake, excess body weight, and physical inactivity, and reproductive and hormonal risk factors such as early age at menarche, later age at menopause, advanced age at first birth, fewer children, less breastfeeding, menopausal hormone therapy, oral contraceptives, and augmented detection through mammographic screening is the cause of the increased incidence rates of breast cancer in higher human development index (HDI) countries. There is a high prevalence of mutations in breast cancer genes (BRCA), such as *BRCA1* and *BRCA2*, in women.

Using cancer registry data supplemented with tumor maker information and findings from studies in the United States, Denmark, and Ireland has shown that the increase in incidence is limited to estrogen receptor-positive cancer, while it is reduced in estrogen receptor-negative cancer. This can be explained through the obesity epidemic, wherein a stronger relation is found between estrogen receptor-positive cancer and excess body weight, along with mammographic screening, which especially detects slow-growing estrogen receptor-positive cancers.

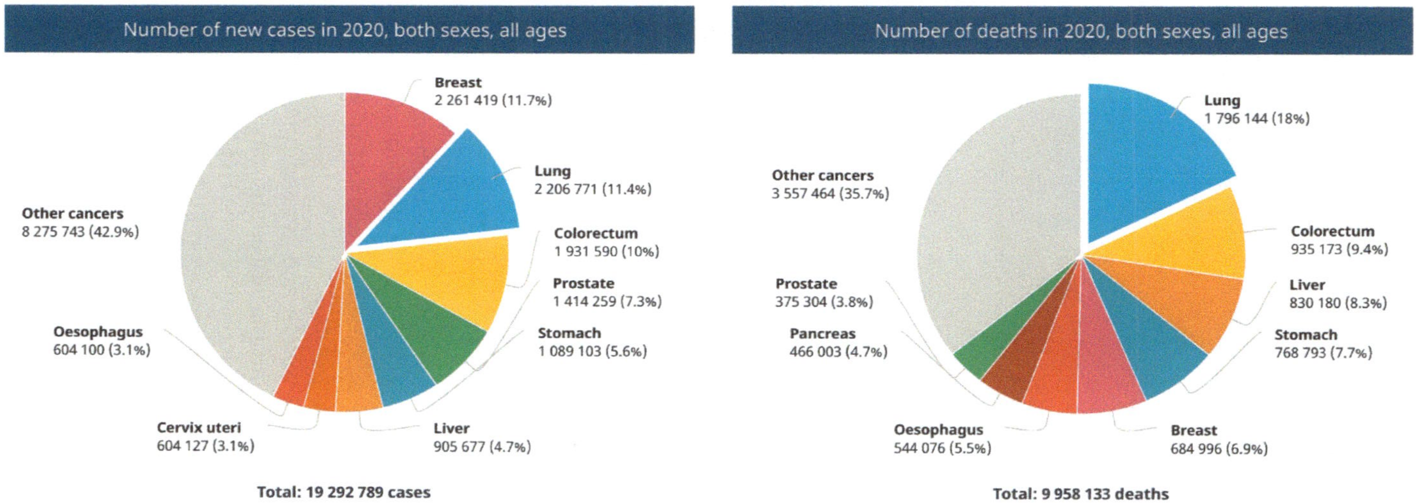

FIG. 1 All cancer statistics data. *Source: Globocan 2020.*

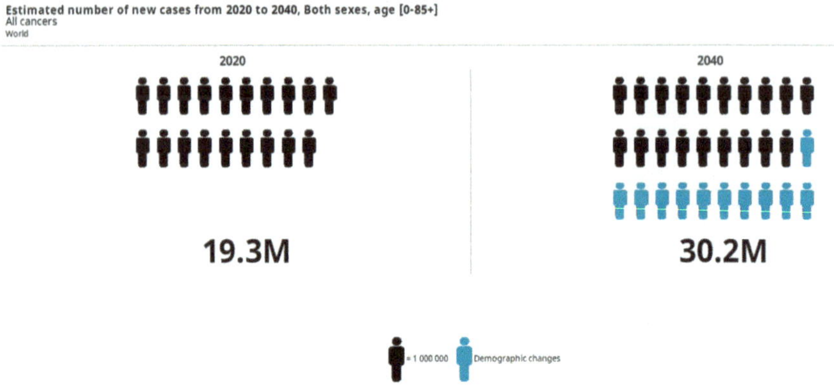

FIG. 2 Data depicting estimated number of new cancer cases in both sexes from 2020 to 2024. *Source: IARC.*

Transitioning countries in South America, Africa, and Asia are showing high incidence rates of breast cancer. Changes in lifestyle, socio-cultural factors, and built environments that are the consequences of economic growth and increased participation of women in the industrial workforce have led to the increase in prevalence of breast cancer risk factors. This can be attributed to a weak health system and consequently poor survival outcomes. Another cause can be late-stage presentation. A report consisting of 83 studies conducted in 17 sub-Saharan African countries concluded that 77% of the all the diagnosed cases were stage III/IV (see https://gco.iarc.fr/today/data/factsheets/cancers/20-Breast-fact-sheet.pdf) (Fig. 3).

Early detection through improved breast cancer awareness and clinical breast examination by skilled health workers followed by appropriate treatment can improve survival. This is essential in low resource settings where large and organized population-based mammography screening is not possible. The Breast Health Global Initiative has created a series of evidence-based, resource-stratified guidelines that support phased implementation into real-world practice.

Reducing excess body weight and consumption of alcohol, and promoting physical activity and breastfeeding may help in eradicating the incidence of breast cancer globally. The WHO has recommended an organized, population-based mammography screening of women at an average risk (aged 50–69 years) every 2 years. The American Cancer Society (ACS) recommends the following:

- Women aged 45–54 years should screen annually.
- Women aged 40–44 years should have the opportunity to screen annually.

Incidence, both sexes

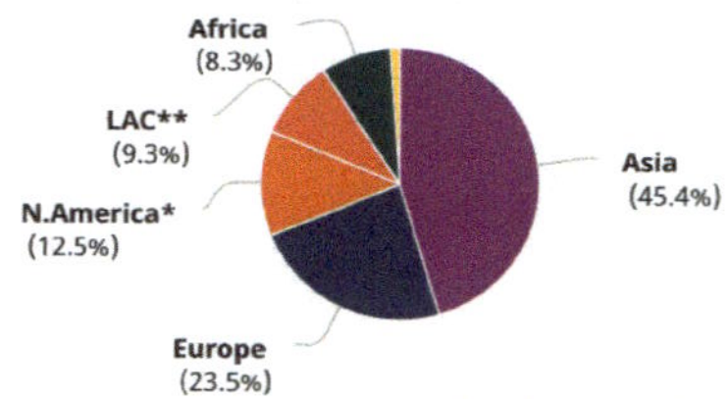

	Population	Number
	Asia	1 026 171
	Europe	531 086
	*Northern America	281 591
	**Latin America and the Caribbean	210 100
	Africa	186 598
	Oceania	25 873
	Total	2 261 419

Mortality, both sexes

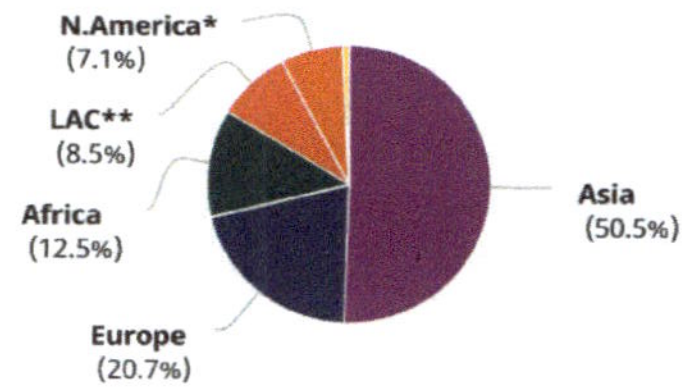

	Population	Number
	Asia	346 009
	Europe	141 765
	Africa	85 787
	**Latin America and the Caribbean	57 984
	*Northern America	48 407
	Oceania	5 044
	Total	684 996

5-year prevalence, both sexes

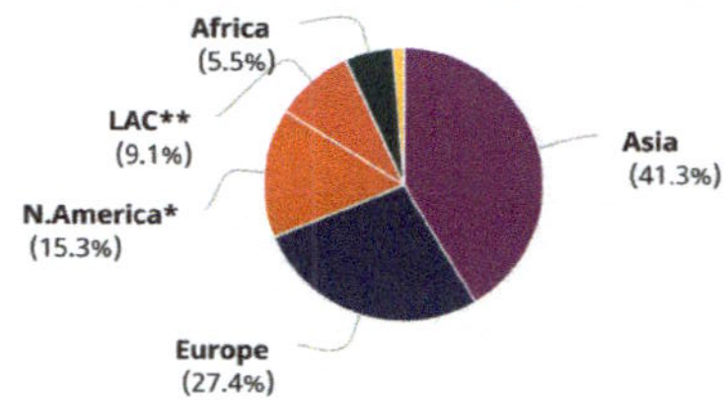

	Population	Number
	Asia	3 218 496
	Europe	2 138 117
	*Northern America	1 189 111
	**Latin America and the Caribbean	710 039
	Africa	429 220
	Oceania	105 734
	Total	7 790 717

FIG. 3 Epidemiological data of breast cancer prevalence in different populations. *Source: Globocan 2020.*

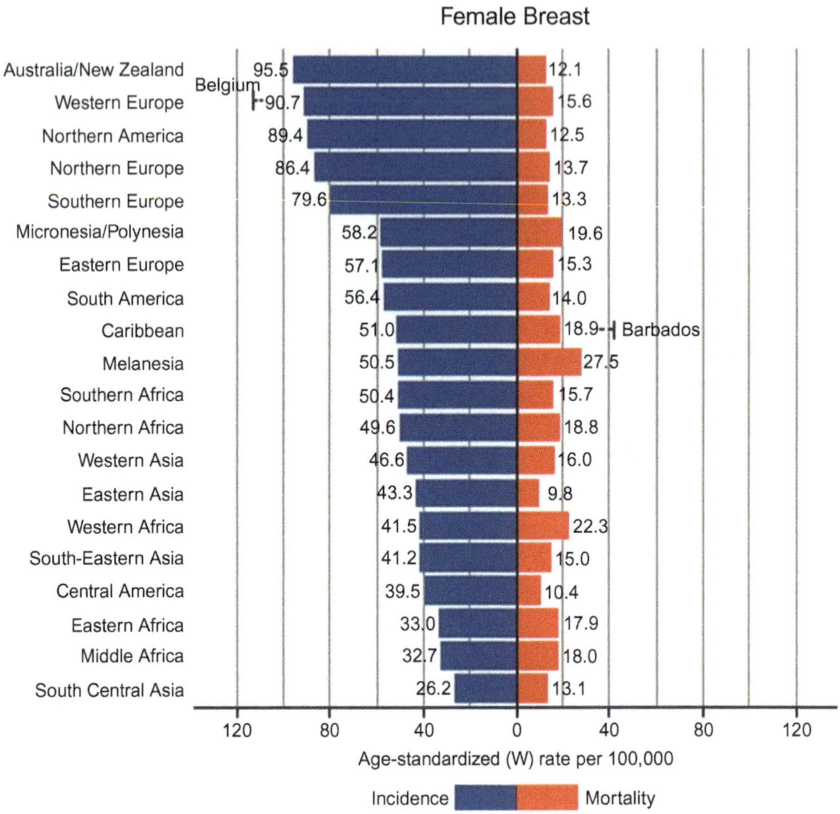

FIG. 4 Incidence and mortality rates for female breast cancer in different regions worldwide. *Source: Globocan 2020.*

- Women older than 55 years should have biennial screening.
- Women should continue to screen as long as their overall health is good and they have a life expectancy of ≥10 years.

Overdiagnosis and overtreatment are among the limitations of mammographic screening (Fig. 4).

2.2 Lung cancer epidemiology

Lung cancer leads to cancer deaths and was the second most diagnosed cancer in 2020, with 2.2 million new cases and 1.8 million deaths. In men, it is the foremost cause of cancer morbidity and mortality; while in women, it is the third largest cause of mortality and morbidity after breast and colorectal cancer. Mortality and incidence rates in men are approximately two times higher than in women and this relation varies across different geographical regions—for instance, it ranges from 5.6 in Northern Africa to 1.2 in North America.

In men, lung cancer is the most diagnosed cancer in 36 countries and the main cause of cancer death in 93 countries, with Micronesia/Polynesia, Eastern and Southern Europe, Eastern Asia, and Western Asia reporting the greatest incidence, and Turkey leading the list. There is intermediate to high reporting of incidence rates in Southern and Northern Africa, although the overall incidence rates remain low (see https://gco.iarc.fr/today/data/factsheets/cancers/15-Lung-fact-sheet.pdf).

In women, lung cancer is the main cause of the cancer death in 25 countries of North America, Oceania, and parts of Europe, with the greatest rates reported in North America, Northern and Western Europe, Micronesia/Polynesia, and Australia/New Zealand, and with Hungary leading the list. In Eastern China, high outdoor air pollution, and indoor pollution due to burning of solid fuels are major causes of the high burden of lung cancer among Chinese women (Fig. 5).

Globally, outdoor air pollution owing to fine particulate matter is the major cause of lung cancer deaths, with 14% in 2017, which ranges from 4.7% in the United States to 20.5% in China. Another major cause of lung cancer is the tobacco intake. A sharp increase in lung cancer was found in high-income countries such as the United Kingdom, United States, Finland, Australia, New Zealand, Netherlands, Singapore, Germany, and Uruguay where smoking was first established. The causalities of tobacco intake and lung cancer are still less defined among women, with a few populations, including the United States and Switzerland, showing signs of peak and stabilization, although at a very reduced rate when compared with men. More studies have found that the female to male incidence ratio is either equal or is higher in countries like the United States, Canada, Denmark, Germany, New Zealand, and the Netherlands due to increased incidence rates in women and a sharp decline in men. Countries like China, Indonesia, and a few African countries are expected to continue to show increased incidence of lung cancer, as the tobacco intake is at the nascent stage in these parts of the world.

Because smoking is the causal factor of two-thirds of lung cancer deaths worldwide, it can be prevented with interventions and effective tobacco-control policies and regulations. The WHO Framework Convention on Tobacco Control introduced the MPOWER package as a country-level implementation that aims to reduce the demand for tobacco. It includes six policy intervention strategies:

- Monitor tobacco use and prevention policies.
- Protect people from tobacco smoke.
- Offer help to quit tobacco use.
- Warn about the dangers of tobacco.

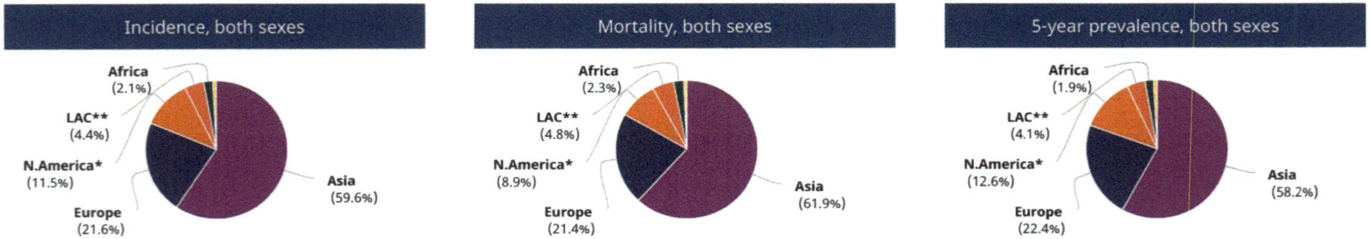

Population	Number
Asia	1 315 136
Europe	477 534
*Northern America	253 537
**Latin America and the Caribbean	97 601
Africa	45 988
Oceania	16 975
Total	2 206 771

Population	Number
Asia	1 112 517
Europe	384 176
*Northern America	159 641
**Latin America and the Caribbean	86 627
Africa	41 171
Oceania	12 012
Total	1 796 144

Population	Number
Asia	1 515 321
Europe	582 924
*Northern America	328 224
**Latin America and the Caribbean	106 459
Africa	50 186
Oceania	21 677
Total	2 604 791

FIG. 5 Epidemiological data of lung cancer prevalence in different populations. *Source: Globocan 2020.*

- Enforce bans on tobacco advertising, promotion, and sponsorship.
- Raise taxes on tobacco.

In 2018, it was reported for the first time that tobacco usage in men had declined significantly, even though the population continues to increase, while female tobacco consumers have continued to show a decline since 2000. Forty-nine LMICs countries are yet to adopt a single MPOWER, making the global progress unequal. Survival rate is 10%–20% after diagnosis of patients with lung cancer at 5 years with exceptions in a few countries, like Japan showing 33%, Israel with 27%, and the Republic of Korea with 25%.

Using low-dose computed tomography (CT) to screen high-risk individuals can help in detecting and diagnosing cancer early. Numerous independent, international, randomized controlled clinical trials have shown that using annual low-dose CT screening helps in reducing lung cancer mortality. One of the recent studies, a Dutch–Belgian lung cancer screening trial, showed the reduction of lung cancer mortality by 24% in men and 33% in women in a 10-year follow-up by introducing the grow-rate assessment as an imaging biomarker for intermediate tests in CT screening. The general population, however, is yet to receive the benefits of this, creating a hindrance for lung cancer screening as part of a global plan to reduce the disease burden (Fig. 6).

3 Role of oxidative stress in cancer pathophysiology

Cancer pathophysiology is known to alter the cellular microenvironment along with the disturbance to homeostasis of several cellular organelles like mitochondria and endoplasmic reticulum. Mitochondria is the powerhouse of the cell, actively involved in the process of oxidative metabolism. During respiration, the majority of consumed oxygen is completely reduced to form water by cytochrome *c* oxidase; however, a small amount of oxygen is partially reduced. This partial reduction leads to the generation of ROS. Enzymatic reactions that generate ROS consist of enzymes such as xanthine oxidase, NADPH oxidase, cytochrome P450 enzymes, lipoxygenase, etc. [5]. ROS consists of unstable molecules capable of initiating oxidation via unpaired electrons such as superoxide anions and hydroxyl radicals [6]. The production of ROS is a causal effect of exogenous stimuli as well as endogenous release, due to inflammatory responses from cells. ROS is responsible for nuclear/mtDNA damage and RNA damage, along with interrupting several inter- and intracellular signaling cascades. The damage caused by ROS is dependent not only on its concentration in the cell, but also the amount of endogenous antioxidants available to remove the toxic ROS from the cellular microenvironment. Some of the natural antioxidants present in the cells include superoxide

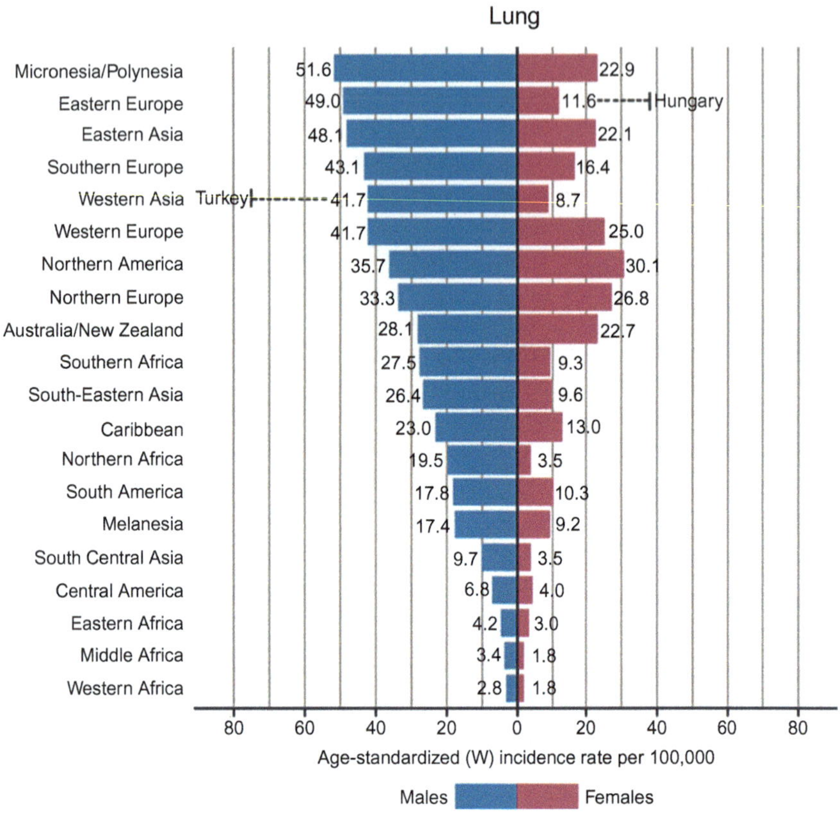

FIG. 6 Incidence and mortality rates for lung cancer in different regions worldwide. *Source: Globocan 2020.*

dismutase (SOD), catalase, glutathione peroxidase (GPx), peroxiredoxins (PRX), melatonin, glutathione, etc. Thus, the equilibrium between ROS and antioxidants is constantly maintained by the healthy cells, an imbalance of which results in oxidative stress (OS) [7].

OS is crucial for the pathophysiology of a wide variety of diseases, such as neurodegenerative diseases, cardiovascular diseases, inflammatory diseases, diabetes, immune dysfunction, aging, and cancer. Tumor development and progression are directly linked to OS in the cellular microenvironment because of the involvement of ROS in several aspects of its mechanism (Fig. 7). Four major stages of cancer development consist of: (a) cellular proliferation involving extracellular regulated kinase 1/2 activation (ERK1/2); (b) anoikis or apoptotic evasion via NF-κB and PI3K/Akt activation; (c) metastasis through extracellular matrix metalloproteinases, tumor growth factor-β (TGF-β), and other signaling pathways;

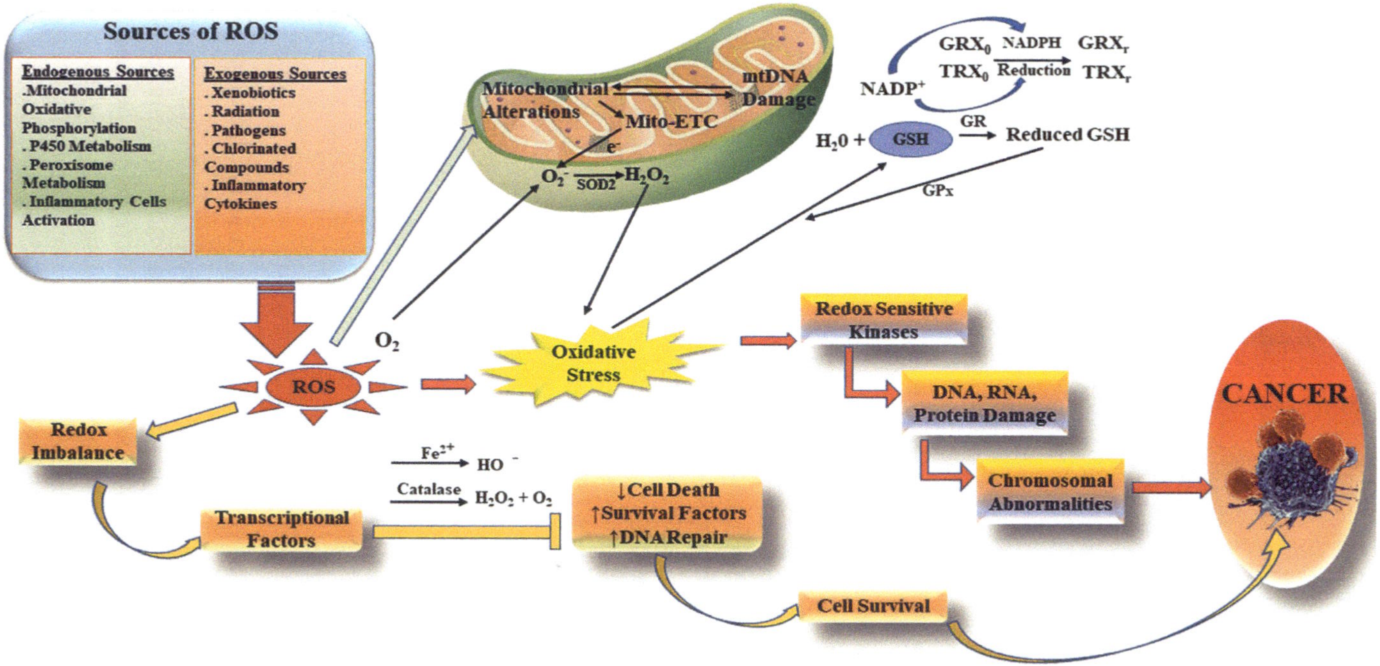

FIG. 7 A schematic diagram showing oxidative stress involving signaling pathways of cancer progression. *GPx*: glutathione peroxidase; *GR*: glutathione reductase; *GRXo*: glutaredoxin (oxidized); *GRXr*: glutaredoxin (reduced); *GSH*: glutathione; *GSHr*: glutathione (reduced); *Mito-ETC*: mitochondrial electron transport chain; *SOD*: superoxide dismutases; *TRXo*: thioredoxin (oxidized); *TRXr*: thioredoxin (reduced).

and (d) angiogenesis with the help of vascular endothelial growth factor (VEGF) release and angiopoietin [7].

3.1 Tumor cellular proliferation and oxidative stress

OS is known to interrupt several biochemical pathways actively involved in cellular growth such as epidermal growth factor receptor (EGFR) and mTOR signaling. Several key signaling proteins, namely nuclear factor erythroid 2-related factor 2 (Nrf2), kelch-like protein 19 (keap1), Ras, Raf, mitogen activated protein kinases (MAPK), p53, p38α, c-Jun N-terminal kinase (JNK), etc., are easy targets for ROS. Interrupting these pathways and signaling proteins suppresses the antioxidant and tumor suppressor responses of the cells along with their sensitivity to detect OS [8].

3.2 Metastasis and oxidative stress

Metastasis is the crucial stage of cancer pathogenesis, which influences OS to a great extent. Metastasis-related processes such as epithelial-mesenchymal transition (EMT), angiogenesis, and hypoxia are responsible for generation of large concentrations of ROS in the cellular matrix. ROS-associated signaling pathways are highly implicated in the EMT-related process. Numerous proteins like Smad, E-cadherin, β-catenin, matrix metalloproteinases, and Ets-1 are activated in a ROS-dependent manner [9]. Furthermore, ROS regulates metastasis with the help of NF-κB along with hypoxia inducible factor 1 alpha (HIF-1α) and cyclooxygenase 2 (COX-2). Other cellular pathways targeted by ROS include Wnt/TCF, MAPK, protein kinase C and p21-activated kinase 1 signaling pathways. Aberrant activation of these pathways promotes tumorigenesis among several cancer types, mainly breast cancer. Moreover, tumors produce pro-angiogenic factors such as VEGF and angiopoetin-1. Hypoxia and ROS production are directly related to upregulation of VEGF in tumor cells, thus enhancing the process of angiogenesis and tumor survival. NADPH oxidases such as Nox activate certain redox signaling pathways, resulting in generation of ROS and ultimately leading to angiogenesis [7].

3.3 Oncogenes, tumor suppressor genes, and oxidative stress

Cancer cells have a high metabolic rate, which is attributed to increased ROS production. ROS are generally less deleterious for cancer cells because of the variety of mutations caused in several oncogenes and tumor suppressor genes. Nrf2 and its repressor protein keap1 tightly regulate the ROS levels in normal cells. However, somatic mutations have recently been found in cancer cells, which disrupt the Nrf2-keap1 interaction and thus aggravate the ROS levels [10]. Similarly, the tumor suppressor gene in breast cancer BRCA1 has been found to be mutated in approximately 50%

of hereditary breast cancers and absent/expressed at low levels among 40% of sporadic breast cancers [11].

OS and cancer are strongly interconnected with the Ras pathway. The Ras gene family is actively involved in cell signaling, thus activation of Ras in tumors is achieved via point mutations [12]. The mutant Ras is known to increase the ROS levels, causing extensive DNA damage. Moreover, it positively regulates the NADPH oxidase system, producing peroxide radicals in the cellular microenvironment. Hence, oncogenic Ras enhances mitochondrial mass and accumulation of ROS in cancer cells. Similarly, overexpression of oncogenic proteins, such as Raf, hTERT, and MEK, and silencing of tumor suppressor genes p53 and p21 also result in the increase of ROS, leading to senescence. Apart from ROS elevation, these oncogenes are also actively involved in the stimulation of glycolytic metabolism and mitochondrial oxidative phosphorylation in the cancer cells [7].

3.3.1 *MicroRNAs regulated by ROS in cancer*

MicroRNAs regulate gene expressions and are known to be actively involved in different pathological conditions, including cancer. They are sensitive to OS and thus are easily altered by the free radicals and other ROS-inducing alterations in microRNA expression patterns. miR-210 is overexpressed in breast cancers and is associated with invasions and poor clinical outcomes. It is strongly mediated by hypoxic conditions and promotes metastasis. Similarly, miR-141 and miR200a target the sensor of OS, p38α, making cells nonresponsive toward elevating OS levels and ROS accumulation [13].

4 Classification of breast and lung cancer

4.1 Breast cancer

Invasive and noninvasive breast cancer are two types of breast cancer, classified on the basis of site [14].

4.1.1 *Noninvasive breast cancer (NIBC)*

In this type, the cancer is restricted to the lobule or ducts where it originated. Ductal carcinoma in situ and lobular carcinoma are examples of NIBC. Ductal carcinoma in situ arises with the development of atypical cells within the milk ducts without spreading in the tissue in close proximity as its name suggest: "in situ," which means "in place." Although these atypical cells do not invade the tissues, they continue to grow, which leads to invasive breast cancer. Breast-preserving treatment is considered the best therapy in the management of the ductal carcinoma in situ and to avoid this illness. Lobular carcinoma in situ originates in the breast lobules and the cancer does not penetrate to the exterior, or from the lobules into the breast tissue, and is hence classified as NIBC (Fig. 8).

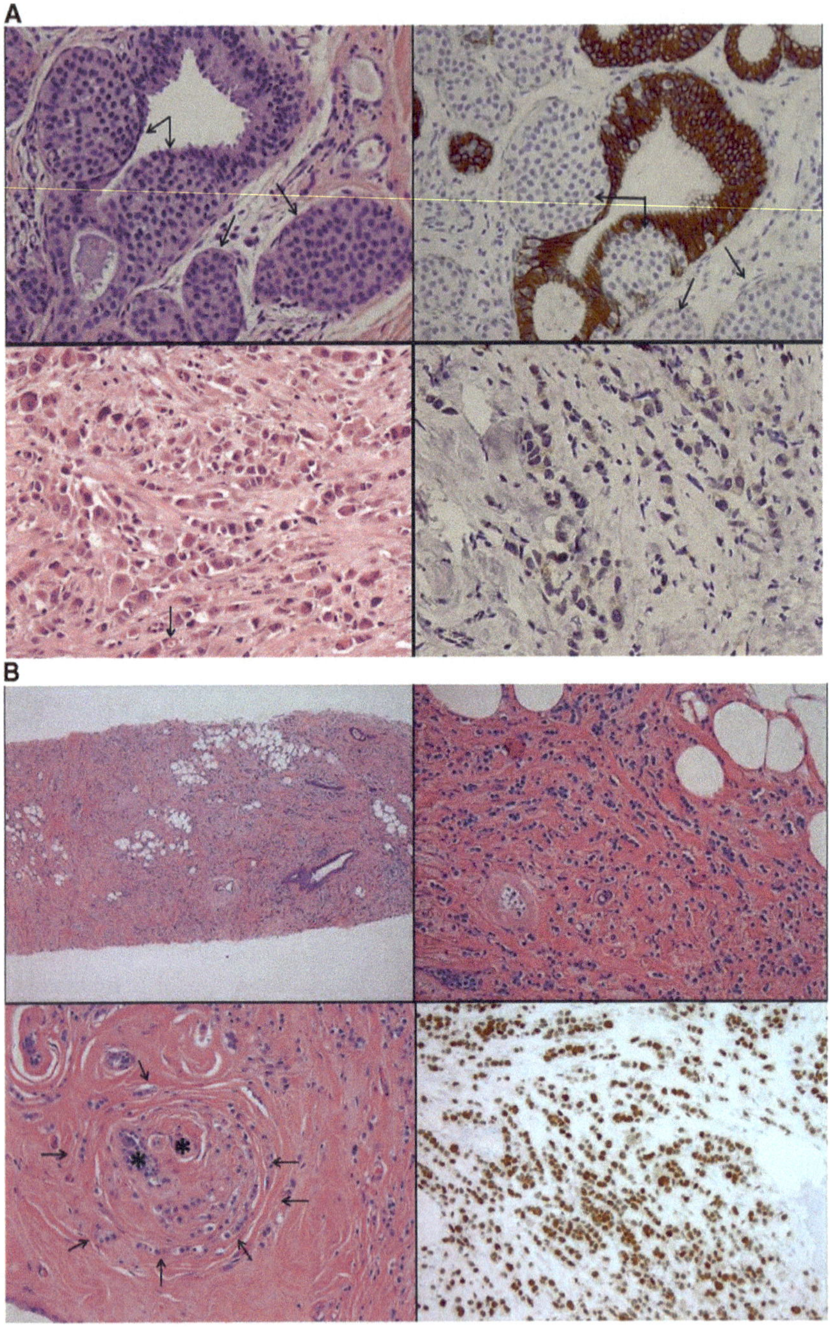

FIG. 8 Histopathological features of invasive lobular carcinoma [15].

4.1.2 Invasive breast cancer (IBC)

In IBC, the cancerous or abnormal cells in the lobules or milk ducts spread to the adjacent breast tissues. The cells from the breast can spread to different parts of the body, facilitated by the systemic circulation or immune system. Movement of the cancerous cells may occur when the tumor is in the nascent stage or when it has become a huge, invasive breast cancer. Approximately 6% of females (younger than 75 years of age) in Australia and Europe are under increased threat of this type of cancer. Prevalence increases with the increase in age. When IBC spreads to other organs, it is then known as metastatic breast cancer. The liver, brain, bones, and lungs are common areas to which these cancerous cells extend. These cells can spread sporadically, forming new cancers. However, these cells forming new cancers in other parts of the body would be still considered as BC. Infiltrating lobular carcinoma (develops in the milk glands but regularly spreads to other parts of the body), infiltrating ductal carcinoma (develops in the milk ducts and can spread to the duct wall, penetrating the fatty tissues of the breast, and eventually to other parts of the body), medullary carcinoma (makes a separate margin of normal tissue and medullary tissue), mucinous carcinoma (develops from the mucus-forming cancer cells and is a rare type of breast cancer), and tubular carcinoma are classified as IBC.

4.1.3 Inflammatory breast cancer

Inflammatory breast cancer is characterized by the presence of swollen breasts, which are red and warm. Additional characteristics include the appearance of broad ridges, which is due to the blockage of lymph vessels in the skin of the breast by the cancerous cells. It is a rare type of cancer but can grow very quickly. Cautious usage of all the strategies, such as radiation therapy, chemotherapy, surgery, and imaging, is included in the treatment options. Neoadjuvant chemotherapy has proven to be a successful treatment in improving survival of inflammatory breast cancer patients, and can be used to replace locoregional treatments such as radiation and surgery.

4.1.4 Paget's disease of the breast

Paget's disease of the breast is characterized by the presence of observable changes in the nipple. It is a rare type of cancer. Symptoms of this disease are itchy rashes that are reddish in color around the nipple area, which might spread to other areas of the skin. A resemblance to eczema or psoriasis is common; however, Paget's disease of the breast affects only one breast and it develops at the nipple, while in skin conditions, both the breasts are affected and it develops from the areola. Approximately 1%–3% of breast cancers are Paget's disease of the breast. Both men and women are affected by this disease. Pathogenesis is still unclear; however, there are theories that establish its pathogenesis. Bleeding and discharge from the nipple, inversion of the nipple, and a lump found in the breast are a few warning signs. Punch biopsy is used to diagnose it. Prediction or prognosis can be done effectively if the disease is limited to the nipple or the ducts of the breast.

4.1.5 Phyllodes tumor

Phyllodes tumors grow in the connective tissues of the breast. Treatment mainly involves removing the tumor through surgery. It is one of the rarest cancers, with fewer than 10 females dying of it annually in the United States. These tumors can be either benign or malignant.

4.2 Lung cancer

Lung cancer is classified into two major histological classes based on its spread and growth pattern: small cell lung carcinoma (SCLC) and nonsmall cell lung carcinoma (NSCLC) [16,17].

4.2.1 Small cell lung carcinoma

SCLC comprises 10%–15% of all lung cancers. It is characterized by a rapid growth rate and metabolism. SCLCs are derived from the lung's hormonal cells and are the most dedifferentiated cancers with central mediastinal tumors. These cancerous cells aggressively disseminate to local lymph nodes and distant sites. As a result, SCLCs are highly sensitive to chemotherapy and radiation therapy (Fig. 9).

4.3 Nonsmall cell lung carcinoma

NSCLC comprises 85%–90% of all lung cancers. This group is further subclassified into adenocarcinoma (approx. 40% of lung cancers),

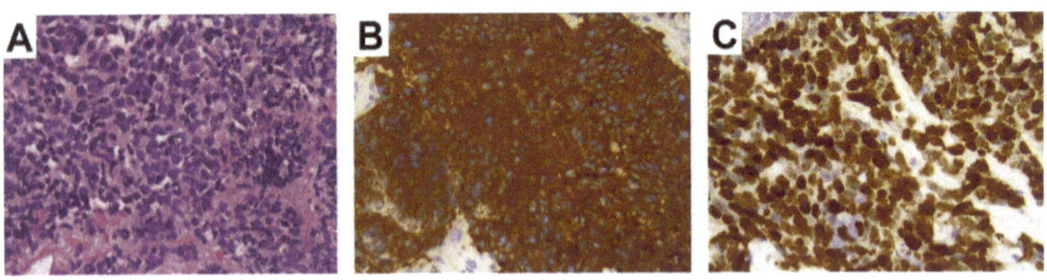

FIG. 9 Histopathological identification of SCLC and its characteristic features. (A) SCLC with small-sized nuclei and scanty cytoplasm (B) SCLC with positive staining for synaptophysin (C) SCLC with greater than 80% Ki67 labeling index [17].

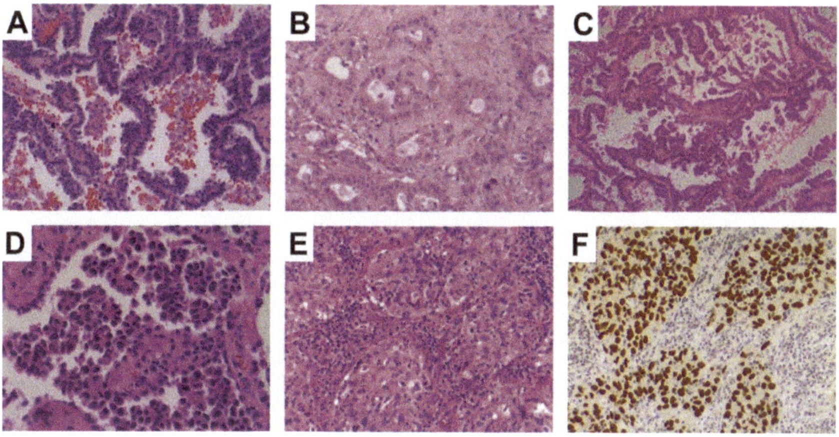

FIG. 10 Histopathological identification of types of adenocarcinomas. (A) Lepidic adenocarcinoma; (B) acinar adenocarcinoma; (C) papillary adenocarcinoma; (D) micropapillary adenocarcinoma; (E) solid adenocarcinoma; (F) solid adenocarcinoma [17].

squamous cell carcinoma (approx. 25%–30% of lung cancers), and large cell carcinoma (approx. 10%–15% of lung cancers).

Adenocarcinoma is the most common type of NSCLC, defined by malignant epithelium neoplasm and glandular differentiation or mucin production. Morphologically, it is known to form a peripheral mass characterized by central fibrosis and pleural puckering. Histological patterns include acinar, lepidic, papillary, micropapillary, and solid patterns. Pneumocytic markers are additional characteristic features of lung adenocarcinoma cells along with thyroid transcription factor-1 (TTF-1) and Napsin A (Fig. 10).

Squamous cell carcinoma (SCC) is found in the major airways of the lungs located centrally and results in the formation of cavities on achieving a large size. SCC is characterized by keratinization, intercellular bridges, and solid nested growth patterns. It does not possess the glandular structure and lacks mucin production. Apart from morphological markers, SCC is characterized by certain selective markers such as p63, p40, CK5/6, CK5, and negative TTF-1 (Fig. 11).

Large cell carcinomas (LCCs) account for very few cases of lung cancer. They are undifferentiated NSCLCs and lack morphological and histological features of adenocarcinoma, SCC, and null immunophenotype. LCCs are located in the peripheral lung regions and have necrotic appearance along with large, polygonal-shaped tumor cells with pleomorphic and vesicular nuclei. LCC is known to form patternless solid sheets.

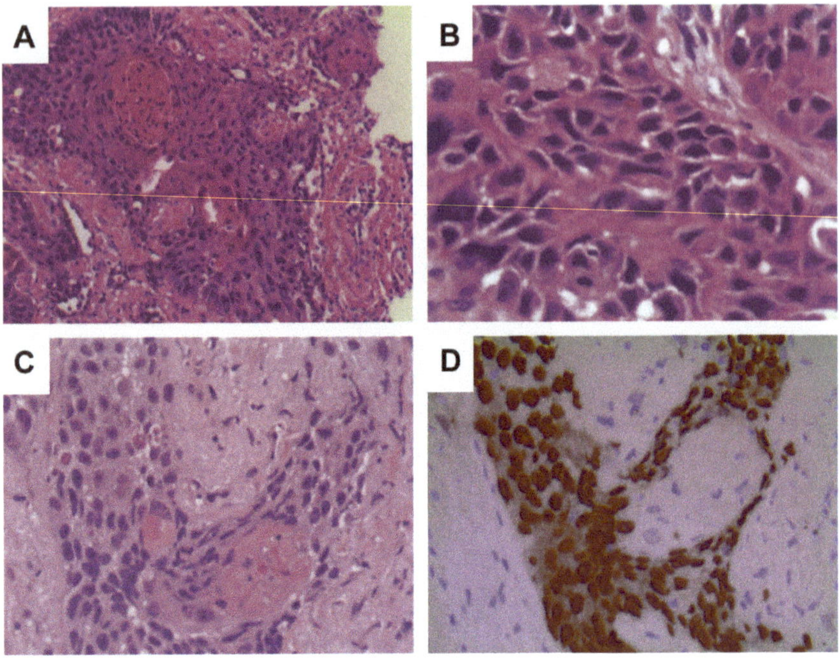

FIG. 11 Histopathological identification of squamous cell carcinoma and its characteristic features. (A) Keratin pearl formation; (B) intercellular bridge formation; (C) no keratinization or discernible intercellular bridges; (D) tumor cells stain positive for p40 (nuclear stain) [17].

5 Conventional therapies for cancer treatment

5.1 Conventional therapies for breast cancer treatment

5.1.1 Surgery

Surgical treatment of breast cancer has significantly advanced in the past decades, and is focused on reducing the long-term cosmetic and functional ramifications of conventional therapy. As per the standard protocols, established based on decades of research, the two approaches are either an excision followed by radiation or a total mastectomy, assuming that clear margins can be achieved. Both these approaches are known to be equivalent in promoting relapse-free overall survival. Conservative surgery is contraindicated by:

(1) diffuse suspicious microcalcifications present on breast imaging;
(2) the presence of positive pathologic margins post lumpectomy;
(3) disease that cannot be addressed by excision of a single breast tissue region with satisfactory cosmetic results, except in highly select patients;
(4) certain collagen-vascular diseases, such as scleroderma; and.
(5) history of radiotherapy availed for the involved breast.

Surgical treatment of axillary lymph nodes must be considered distinct from surgical therapy of the breast. The purpose for removal of lymph nodes can be both diagnostic (to help determine the anatomic extent of the breast cancer) and therapeutic (to help remove cancerous cells). The evident involvement of axillary lymph nodes at diagnosis and administration of neoadjuvant systemic therapy are crucial for surgical decision-making [18].

The standard of care in any patient having clinically evident axillary involvement at diagnosis who undergoes surgery as the initial treatment is axillary lymph node dissection (ALND). This was the universal approach to the axilla until it was demonstrated in clinical trials that there was no significant difference in regional recurrence or survival outcomes in women diagnosed with clinically node-negative (cN0) breast cancer who underwent full ALND versus women who underwent sentinel lymph node (SLN) biopsy, with conversion to ALND only if the SLN was positive. The surgical treatment of patients who receive neoadjuvant systemic therapy is evolving. Neoadjuvant chemotherapy was found to increase the patient's eligibility for breast-conserving therapy with minimal effects on long-term outcomes, as is evidenced by several prospective randomized trials and a recent meta-analysis [19].

Evidence also suggests that the false-negative results of SLN biopsy are similar in patients who are found to be clinically node-positive (cN1) and then convert to cN0 after neoadjuvant therapy compared to patients undergoing surgery upfront, given that appropriate measures are taken to maximize SLN identification (e.g., using dual SLN mapping methodology, placing a clip in the biopsy-proven node and retrieving the clipped node at the time of SLN dissection, and retrieving at least three SLNs). As per recent guidelines and current clinical practice, the use of SLN biopsy in this setting (cN1 converted to cN0) is promoted with omission of ALND as a stringent criterion in patients who are SLN negative. However, optimal treatment for disease affecting the lymph node following neoadjuvant therapy is still an area undergoing investigation [20].

5.1.2 Radiation therapy

Radiation therapy in breast cancer may be employed to a portion of the breast (after lumpectomy), the chest wall (after mastectomy), the whole breast, or the regional lymph nodes. Whole-breast radiation post lumpectomy is a standard component in breast-conserving therapy. A meta-analysis study demonstrated that administration of radiation following lumpectomy was associated with reduced recurrences in breast cancer.

Several prospective studies have explored the efficacy of a short course of radiation therapy post lumpectomy in order to identify patients who can benefit from an escalated dosing regimen. Radiation therapy to the partial breast following lumpectomy, as opposed to the whole breast, is an approach that has predominantly been studied in low-risk patients aged

50 years and older. Although some trials suggested that radiation therapy for partial breast was associated with a slightly higher rate of local recurrence and a somewhat worse cosmesis, data from emerging studies contradict this, and consensus guidelines promote nonintraoperative partial breast radiation for patients at a lower risk. Improvement in local control is seen with the administration of a boost of radiation particularly to the tumor bed; however, there is no improvement in overall survival and hence it should only be considered in patients at a higher risk. Furthermore, prospective randomized trials have demonstrated that there is no significant improvement in the overall survival or on the distant recurrence with radiation therapy for whole breast following lumpectomy in women who are 65 years or older and are diagnosed with a low-risk HR+ or *ERBB2* breast cancer. Individualized discussions are recommended for older women who fit the profile and plan to undergo whole-breast radiation. Radiation therapy post mastectomy is given to the chest wall, sometimes with integration of a boost to the scar due to mastectomy and/or radiation to regional nodes.

A significantly improved disease-free survival was observed with the addition of regional nodal radiation (covering the axillary, paraclavicular, and/or internal mammary nodes) either following mastectomy or lumpectomy, which was also associated with an increase in radiation toxicities such as pneumonitis and lymphedema. The benefits of regional nodal radiation were observed in women who had undergone ALND. However, it was not associated with overall survival, hence it is not administered universally even in node-positive patients, but it can be considered for patients with a high-risk biology or a higher nodal disease burden [21].

5.1.3 Systemic therapy

HR+/*ERBB2*− subtype

The primary systemic therapy for HR+/*ERBB2*− breast cancers is endocrine therapy, which acts by counteracting estrogen-promoted tumor growth. Standard endocrine therapy options vary according to the menopausal status of the patient, but mostly consist of oral antiestrogen medication taken daily for a duration of 5 years. Tamoxifen is a selective estrogen receptor modulator that is effective in both pre- and postmenopausal women. It acts by competitively inhibiting estrogen's binding to ER. Aromatase inhibitors such as anastrozole, exemestane, and letrozole are effective only in postmenopausal women (including those who are postmenopausal because of medical ovarian suppression or oophorectomy) and work by decreasing the circulating estrogen levels by inhibiting conversion of androgens to estrogens. The absolute benefit of tamoxifen is proportional to the risk associated with a given tumor and it has been observed that 5 years of therapy with tamoxifen is more effective than 5 years of therapy with aromatase inhibitors among postmenopausal women [22].

A suitable strategy for women who wish to moderate toxicities from both classes of endocrine therapy is the "switch" strategy. This involves an initial 2–3 years of therapy with tamoxifen, followed by aromatase inhibitor for completion of a 5-year course of endocrine therapy. The initial decision regarding the choice of endocrine therapy for a premenopausal woman is whether to treat with gonadotropin-releasing hormone agonists, such as leuprolide acetate and goserelin for ovarian suppression, or to carry out oophorectomy to induce menopause; the subsequent decision, if inducing menopause, is whether to treat with tamoxifen or aromatase inhibitors. Patients diagnosed with HR+ breast cancer are at higher risk of recurrence even decades after the primary diagnosis. Therefore, several studies have been undertaken to examine the implications of extending therapies with tamoxifen and aromatase inhibitors beyond the recommended duration of 5 years. Two randomized controlled trials that have compared the 5- vs 10-years duration regimen of tamoxifen showed a small but significant absolute improvement (2.8%) in reduction of mortality due to breast cancer with the 10-year duration of therapy. Another randomized controlled trial examined the 5-year vs 10-year treatment regimen of aromatase inhibitor, and reported a small reduction in distant recurrences with longer duration of therapy. No significant improvement in the overall rate of survival was reported in this study with the use of extended-duration aromatase inhibitor therapy, whereas some known adverse effects of aromatase inhibitors such as new-onset osteoporosis and fracture were recorded. Therefore, it can be inferred that longer duration of endocrine therapy may provide small benefits but significantly increases toxicity, and hence warrants consideration in patients at higher risk.

For patients diagnosed with HR+/*ERBB2*− breast cancer, it is the clinicians who decide the point at which chemotherapy must be combined with endocrine therapy. The grade of the tumor and the anatomic stage are important clinicopathologic features for assessment of risk and chemosensitivity. Development of genomic risk scores, which are RNA-based, have been crucial in estimating the prognosis and benefits of chemotherapy. Based on the existing high-quality evidence, the American Society of Clinical Oncology has recommended two signature assays, the 70-gene assay and the 21-gene recurrence score, in order to aid decision-making for the administration of adjuvant chemotherapy in HR+/*ERBB2*− node-negative breast cancer patients. However, these signature assays are not considered as universally accepted indicators of benefit with chemotherapy, due to the preliminary nature of the existing data [23].

5.1.4 Chemotherapy regimen selection for ERBB2− subtypes

Chemotherapy is considered as an essential treatment to prevent the recurrence of cancer in many stage I–III breast cancer patients, despite its short- and long-term risks. It complements endocrine or *ERBB2*-directed therapies, and has proven efficacy against triple-negative breast cancers in HR+/*ERBB2*+ breast cancer patients.

A higher absolute benefit with chemotherapy is achieved in high-risk tumors, as is observed with adjuvant endocrine therapy. Several adjuvant and neoadjuvant regimens of chemotherapy are considered in patients at an early stage of breast cancer. In low-risk patients, wherein toxicity is an important factor in consideration and the overall benefits with chemotherapy are minimal, regimens including adriamycin/cyclophosphamide, docetaxel/cyclophosphamide, and cyclophosphamide/methotrexate/5-fluorouracil are considered as choice treatment regimens. On the contrary, in high-risk patients, an appropriate choice of treatment regimen is one including both taxane (e.g., adriamycin/cyclophosphamide followed by taxane) and anthracycline (e.g., adriamycin), which is found to result in the highest risk reduction. In patients diagnosed with triple-negative breast cancer having greater involvement of lymph node, the use of anthracycline is found to be most beneficial. In the case of detection of residual disease in surgery, following a complete course of neoadjuvant chemotherapy with taxane and anthracycline, the decision to start additional treatment remains uncertain. The results from a recent clinical trial shed light on the role of capecitabine in lowering the rate of recurrence and improving the rate of survival in patients detected with residual disease subsequent to neoadjuvant chemotherapy; however, further studies are recommended to provide better clarity on this approach [2].

5.1.5 Triple-negative subtype

Irrespective of the unfavorable prognosis, chemotherapy is usually administered in all cases of triple-negative breast cancer, especially when breast tumors are found to be larger than 5 mm, even when axillary nodes are negative. The only regimen approved by the US Food and Drug Administration (FDA) for triple-negative breast cancer that is nonmetastatic is with chemotherapeutic agents. Moreover, DNA-cross-linking platinum chemotherapies are being studied for triple-negative breast cancers, since it is a characteristic of some triple-negative tumors to promote damage repair for deficient DNA. Two randomized controlled trials involving triple-negative breast cancer patients receiving neoadjuvant chemotherapy with carboplatin compared to receiving neoadjuvant chemotherapy without carboplatin showed a marked improvement in the pathologic complete response (pCR) with surgery. For triple-negative tumors treated with neoadjuvant chemotherapy, pCR (all tumor gone from the breast and lymph nodes) at surgery is a highly favorable prognostic biomarker [24].

New treatment approaches for patients with triple-negative disease who do not achieve pCR are needed, and platinum chemotherapy is one strategy being evaluated in several clinical trials. To date, the only evidence-based adjuvant escalation strategy available in this setting is single-agent capecitabine, as discussed here.

5.1.6 ERBB2+ subtype

The introduction of *ERBB2*-targeted therapy for breast cancer treatment is considered as a significant development. The first clinical trials for Trastuzumab, which is a monoclonal antibody that targets the *ERBB2* extracellular domain, were conducted in the 1990s. Since then, four randomized controlled trials that enrolled patients with *ERBB2* positive breast cancer have shown that there is a significant improvement in overall survival and disease-free survival with 1 year of trastuzumab regimen taken along with a standard adjuvant chemotherapy regimen. Among the trastuzumab combination therapies that were studied in several of these trials, two neoadjuvant/adjuvant chemotherapy regimens that are recommended as per guidelines in the United States for treatment in stage II and stage III *ERBB2*+ breast cancer are docetaxel/carboplatin and adriamycin/cyclophosphamide paclitaxel. Prospective trials comparing the duration of neoadjuvant/adjuvant therapy with trastuzumab have also demonstrated that there is no difference in the efficacy when therapy is taken for 2 years compared to 1 year. However, these results highlighted the inferiority in cases where the therapy was availed for less than 1 year, compared to 1 year. Studies also demonstrated the superiority of 9 weeks of adjuvant treatment with trastuzumab when compared to no adjuvant treatment in *ERBB2*+ breast cancer patients. Subsequently, clinical trials have aimed at (1) reducing the number of adjuvant therapies along with chemotherapy in patients at low risk and (2) introduction of new adjuvant therapies in patients at high risk. Pertuzumab, which is a monoclonal antibody that targets the dimerization domain of ERBB2, and neratinib have been shown to further reduce the risk of recurrence when compared to standard regimen containing trastuzumab in patients at high risk of *ERBB2*+ breast cancer. The inclusion of pertuzumab in the treatment regimen is justifiable only in patients at a higher risk, as defined by the size of tumors and the nodal status, since there is a high risk of toxicity and additional costs are incurred for patients at a lower risk.

Another small tyrosine-kinase inhibiting molecule, neratinib, which is taken orally, is known to inhibit *ERBB2* and several other members of the HER family of genes. Presently, there is a lack of evidence to support the use of adjuvant therapy with neratinib in patients receiving pertuzumab or vice versa, since the existing trials for the two agents did not explore the use of the other agent. Hence, neither of the two agents are known to show an improved overall survival when administered as adjuvant therapies [2,25].

5.2 Systemic therapy for metastatic breast cancer

In the United States, more than 150,000 women have been diagnosed with metastatic breast cancer, with reports of approximately 41,000 deaths occurring annually. Prognostic factors at both patient level and tumor level are important with the presence of multiple metastatic sites, metastasis in brain, or visceral metastases implying a worsened prognosis. On the other hand, improved status of performance, younger age at the time of diagnosis, metastasis restricted to the bones, and longer interval with a disease-free state between the initial time of diagnosis and metastatic recurrence imply an improved prognosis.

The approach to systemic therapy for metastatic breast cancer is classified by subtypes of breast cancer, with agents used at early and later stages in the treatment regimen. Early treatment with endocrine therapy is recommended in the case of metastatic HR+/*ERBB2* – breast cancers, which are often incorporated with cyclin dependent kinase (CDK) 4/6 inhibitors (e.g., abemaciclib, palbociclib, or ribociclib) as the first-line or second-line regimen. Patients often transition to chemotherapy after developing a resistance to the initial treatment with hormones.

Several randomized controlled trials have shown that the overall rate of survival for sequential therapy with a single agent when compared to a combination chemotherapy for patients with metastatic breast cancer is associated with an improved quality of life of the patient and less adverse effects and toxicity for the single agents. Taxane with trastuzumab and pertuzumab is considered the standard first-line therapy in metastatic *ERBB2*+ breast cancer, whereas trastuzumab emtansine, which is an antibody-drug conjugate, is commonly prescribed as the second-line therapy. Subsequently, a new *ERBB2*-targeted agent for chemotherapy (or endocrine therapy-in case of HR+ patients) is known to have improved outcomes when given in continuation with an *ERBB2*-directed therapy, even in the case of progression with a history of *ERBB2*-targeted therapy [25].

Hence, for patients with metastatic triple-negative breast cancer, introduction of a new, more effective therapy is important. Data from recent studies and ongoing late-phase clinical trials that have looked at pretreated metastatic triple-negative breast cancer suggest promising effects with novel antibody-drug conjugates. Lastly, metastasis in the brain is only found in approximately 10%–20% of the patients diagnosed with metastatic breast cancer; therefore, deciding upon an appropriate treatment regimen for such patients becomes a challenge. The highest risk of developing metastasis in brain is found among patients with *ERBB2*+ and triple-negative breast cancers. Additionally, decreased permeability of several systemic therapies due to the blood–brain barrier, other unique feature of the microenvironment of the brain, and differences in brain versus nonbrain lesions, which are metastatic at a genomic level, may have led to limited progress (Tables 1 and 2).

5.3 Conventional therapies for lung cancer treatment

5.3.1 Surgery and radiotherapy

Surgical removal of a single lobe, also known as lobectomy, is a generally accepted procedure for treatment of early-stage (stage I and II) NSCLC. Imaging techniques like video-supported thoracoscopic surgery and biopsies are used to assist the surgical removal of cancer-affected areas. Retrospective clinical studies have found higher survival rates in cases undergoing lobectomy instead of limited resection of tumor [26]. Studies have found a 60%–80% survival rate among stage I NSCLC patients' postsurgical treatment and 30%–50% among stage II NSCLC patients. Surgery is also done in stage III NSCLC patients; however, it is followed by chemotherapy because of the metastatic nature of the tumor [27]. Cisplatin is mostly used as chemotherapy-based treatment postsurgery. Other than chemotherapy, radiotherapy is also widely used to elevate the efficacy of surgical treatments. Radiation therapy sessions help in the decrease of tumor size, localization of tumor cells in specific regions, and their precise resection [3].

5.3.2 Chemotherapy

Chemotherapy is effectively used at all stages of NSCLC. It is the first line of defense against progressing tumorigenesis. Chemotherapeutic drugs are classified into five broad categories: alkylating agents, antimetabolites, antibiotics, topoisomerase inhibitors, and mitotic inhibitors. Alkylating agents are used as the DNA-cross-linkers to inhibit the replication of tumor cells. Mitomycin-C, cisplatin, vindesine, ifosfamide, and vinblastine are notable examples of alkylating drugs used in the treatment of NSCLC [3].

TABLE 1 Clinical trial status of breast cancer drugs.

Drug	Status
Maytansine	Approved
Dolaflexin	Phase I
Duocarmycin	Phase III ongoing
Deruxtecan	Phase II
Auristatin	Phase II
Vedotin	Phase II
Auristatin	Phase I
Govitecan	Phase III
DM4	Phase I

Adapted from K. Barzaman, J. Karami, Z. Zarei, A. Hosseinzadeh, M.H. Kazemi, S. Moradi-Kalbolandi, E. Safari, L. Farahmand, Breast cancer: biology, biomarkers, and treatments, Int. Immunopharmacol. 84(April) (2020), https://doi.org/10.1016/j.intimp.2020.106535.

TABLE 2 FDA-approved drugs for breast cancer.

Drugs Approved to Prevent Breast Cancer

- Evista (Raloxifene Hydrochloride)
- Raloxifene Hydrochloride
- Soltamox (Tamoxifen Citrate)
- Tamoxifen Citrate

Drugs Approved to Treat Breast Cancer

- Abemaciclib
- Abraxane (Paclitaxel Albumin-stabilized Nanoparticle Formulation)
- Ado-Trastuzumab Emtansine
- Afinitor (Everolimus)
  Afinitor Disperz (Everolimus)
- Alpelisib
- Anastrozole
- Aredia (Pamidronate Disodium)
- Arimidex (Anastrozole)
- Aromasin (Exemestane)
- Atezolizumab
- Capecitabine
- Cyclophosphamide
- Docetaxel
- Doxorubicin Hydrochloride
- Ellence (Epirubicin Hydrochloride)
- Enhertu (Fam-Trastuzumab Deruxtecan-nxki)
- Epirubicin Hydrochloride
- Eribulin Mesylate
- Everolimus
- Exemestane
- 5-FU (Fluorouracil Injection)
- Fam-Trastuzumab Deruxtecan-nxki
- Fareston (Toremifene)
- Faslodex (Fulvestrant)
- Femara (Letrozole)
- Fluorouracil Injection
- Fulvestrant
- Gemcitabine Hydrochloride
- Gemzar (Gemcitabine Hydrochloride)
- Goserelin Acetate
- Halaven (Eribulin Mesylate)
- Herceptin Hylecta (Trastuzumab and Hyaluronidase-oysk)
- Herceptin (Trastuzumab)
  Ibrance (Palbociclib)

TABLE 2 FDA-approved drugs for breast cancer—cont'd

- Infugem (Gemcitabine Hydrochloride)
- Ixabepilone
- Ixempra (Ixabepilone)
- Kadcyla (Ado-Trastuzumab Emtansine)
- Keytruda (Pembrolizumab)
- Kisqali (Ribociclib)
- Lapatinib Ditosylate
- Letrozole
- Lynparza (Olaparib)
- Margenza (Margetuximab-cmkb)
- Margetuximab-cmkb
- Megestrol Acetate
- Methotrexate Sodium
- Neratinib Maleate
- Nerlynx (Neratinib Maleate)
- Olaparib
- Paclitaxel
- Paclitaxel Albumin-stabilized Nanoparticle Formulation
- Palbociclib
- Pamidronate Disodium
- Pembrolizumab
- Perjeta (Pertuzumab)
- Pertuzumab
- Pertuzumab, Trastuzumab, and Hyaluronidase-zzxf
- Phesgo (Pertuzumab, Trastuzumab, and Hyaluronidase-zzxf)
- Piqray (Alpelisib)
- Ribociclib
- Sacituzumab Govitecan-hziy
- Soltamox (Tamoxifen Citrate)
- Talazoparib Tosylate
- Talzenna (Talazoparib Tosylate)
- Tamoxifen Citrate
- Taxotere (Docetaxel)
- Tecentriq (Atezolizumab)
- Tepadina (Thiotepa)
- Thiotepa
- Toremifene
- Trastuzumab
- Trastuzumab and Hyaluronidase-oysk
- Trexall (Methotrexate Sodium)
- Trodelvy (Sacituzumab Govitecan-hziy)
- Tucatinib
- Tukysa (Tucatinib)
- Tykerb (Lapatinib Ditosylate)
- Verzenio (Abemaciclib)
- Vinblastine Sulfate
- Xeloda (Capecitabine)
- Zoladex (Goserelin Acetate)

TABLE 2 FDA-approved drugs for breast cancer—cont'd

Drug Combinations Used in Breast Cancer
• AC • AC-T • CAF • CMF • FEC • TAC

AC, doxorubicin hydrochloride (adriamycin) and cyclophosphamide; AC-T, doxorubicin hydrochloride (adriamycin), cyclophosphamide, paclitaxel (taxol); CAF, cyclophosphamide, doxorubicin hydrochloride (adriamycin), fluorouracil; CMF, cyclophosphamide, methotrexate, fluorouracil; FEC: fluorouracil, epirubicin hydrochloride, cyclophosphamide; TAC, docetaxel (taxotere), doxorubicin hydrochloride (adriamycin), cyclophosphamide.
Source: National Cancer Institute.

In addition, several drug combinations are also examined for synergistic effects in cancer treatment. Antimetabolites-based drugs act by preventing the DNA and RNA synthesis by interrupting the normal functioning of enzymes linked to the mechanism. Hence, antimetabolites work specifically during the DNA replication phase of the cell cycle. Drugs consisting of molecules like antifolates, deoxynucleotide analogs, fluoropyramidines, and thiopurines work by a similar mechanism of action. Not all drugs have proved to be effective in NSCLC treatment, except pemetrexed, a multi-targeted antifolate, which showed strong efficacy with limited side effects in randomized phase III trials [28]. Likewise, mitotic inhibitors also act by disrupting the cell proliferation process in tumor progression by stabilizing or destabilizing the microtubules during the final phase of cell mitosis. Paclitaxel is the most commonly used mitotic inhibitor. Lately, however, it has been shown to be accompanied with a severe toxicity profile. Paclitaxel is nonspecific in nature and thus disrupts the cellular mitosis in healthy cells as well as having microtubular functions in other body cells, like neurons [29]. The major drawback of chemotherapeutics is not only developing drug resistance due to the autophagy mechanism of cells but also high cytotoxic effects. Several studies are underway to overcome these drawbacks—for instance, development of nano therapies as an alternative to chemotherapy.

5.3.3 Targeted therapy

Molecular targeted therapies are based on targeting the driver mutations in oncogenes. These driver mutations are essential elements for tumor proliferation and survival. Successful approval of monoclonal antibody trastuzumab in the treatment of HER-2 positive breast cancer highlighted the potential therapeutic use of oncogenic tyrosine kinases in cancer treatment as well. Gefitinib, an EGFR inhibitor, was the first approved oncogenic tyrosine kinase inhibitor (TKI) drug in lung cancer treatment, which laid the foundation for targeted therapy in lung cancer treatment [3]. EGFR mutations are found in approximately 20% of NSCLC patients. Other than gefitinib, patients with EGFR mutations are treated with erlotinib, dacomitinib, afatinib, and osimertinib [30]. Around 5% of NSCLC patients are found to have an altered ALK gene, and 2% have ROS1 translocations, while 1%–2% of adenocarcinoma patients have BRAF mutations. Oncogenes MET and HER-2 are also found to be mutated in some NSCLC patients. These oncogenes are actively involved in the phosphorylation of tyrosine, serine, and threonine receptors, required for the process of cell proliferation pathways. Hence, these genetic targets have been an area of research for the past few years and several drugs are available to target these genetic mutations for the treatment of NSCLC [26,31] (Tables 3 and 4).

TABLE 3 Clinical trial status of drugs against oncogenic proteins in lung cancer.

Targeted oncogenic protein	Drug name	Clinical trial status
EGFR	Gefitinib	Approved
	Afatinib	
	Erlotinib	
	Osimertinib	
	Necitumumab	
ALK	Crizotinib	
	Alectinib	
	Ceritinib	
	Lorlatinib	Phase 2
	Brigatinib	
ROS1	Crizotinib	Approved
	Cabozantinib	Phase 2
	Certinib	
	Lorlatinib	
	DS-6051b	Phase 1
BRAF	Vemurafenib	Phase 2
	Dabrafenib	
MET	Crizotinib	
	Cabozantinib	
HER-2	Trastuzumab emtansine	
	Afatinib	
	Dacomitinib	

Adapted from F.R. Hirsch, G.V. Scagliotti, J.L. Mulshine, R. Kwon, W.J. Curran, Y.L. Wu, L. Paz-Ares, Lung cancer: current therapies and new targeted treatments, Lancet 389(10066) (2017) 299–311, https://doi.org/10.1016/S0140-6736(16)30958-8.

TABLE 4 FDA-approved drugs for lung cancer.

Drugs Approved for Non-Small Cell Lung Cancer

- Abraxane (Paclitaxel Albumin-stabilized Nanoparticle Formulation)
- Afatinib Dimaleate
- Afinitor (Everolimus)
- Afinitor Disperz (Everolimus)
- Alecensa (Alectinib)
- Alectinib
- Alimta (Pemetrexed Disodium)
- Alunbrig (Brigatinib)
- Atezolizumab
- Avastin (Bevacizumab)
- Bevacizumab
- Brigatinib
- Capmatinib Hydrochloride
- Carboplatin
- Cemiplimab-rwlc
- Ceritinib
- Crizotinib
- Cyramza (Ramucirumab)
- Dabrafenib Mesylate
- Dacomitinib
- Docetaxel
- Doxorubicin Hydrochloride
- Durvalumab
- Entrectinib
- Erlotinib Hydrochloride
- Everolimus
- Gavreto (Pralsetinib)
- Gefitinib
- Gilotrif (Afatinib Dimaleate)
- Gemcitabine Hydrochloride
- Gemzar (Gemcitabine Hydrochloride)
- Imfinzi (Durvalumab)
- Infugem (Gemcitabine Hydrochloride)
- Ipilimumab
- Iressa (Gefitinib)
- Keytruda (Pembrolizumab)
- Libtayo (Cemiplimab-rwlc)
- Lorbrena (Lorlatinib)
- Lorlatinib

TABLE 4 FDA-approved drugs for lung cancer—cont'd

- Lumakras (Sotorasib)
- Mekinist (Trametinib Dimethyl Sulfoxide)
- Methotrexate Sodium
- Mvasi (Bevacizumab)
- Necitumumab
- Nivolumab
- Opdivo (Nivolumab)
- Osimertinib Mesylate
- Paclitaxel
- Paclitaxel Albumin-stabilized Nanoparticle Formulation
- Paraplat (Carboplatin)
- Paraplatin (Carboplatin)
- Pembrolizumab
- Pemetrexed Disodium
- Portrazza (Necitumumab)
- Pralsetinib
- Ramucirumab
- Retevmo (Selpercatinib)
- Rozlytrek (Entrectinib)
- Selpercatinib
- Sotorasib
- Tabrecta (Capmatinib Hydrochloride)
- Tafinlar (Dabrafenib Mesylate)
- Tagrisso (Osimertinib Mesylate)
- Tarceva (Erlotinib Hydrochloride)
- Taxotere (Docetaxel)
- Tecentriq (Atezolizumab)
- Tepmetko (Tepotinib Hydrochloride)
- Tepotinib Hydrochloride
- Trametinib Dimethyl Sulfoxide
- Trexall (Methotrexate Sodium)
- Vizimpro (Dacomitinib)
- Vinorelbine Tartrate
- Xalkori (Crizotinib)
- Yervoy (Ipilimumab)
- Zirabev (Bevacizumab)
- Zykadia (Ceritinib)

Drug Combinations Used to Treat Non-Small Cell Lung Cancer

- CARBOPLATIN-TAXOL
- GEMCITABINE-CISPLATIN

Continued

TABLE 4 FDA-approved drugs for lung cancer—cont'd

Drugs Approved for Small Cell Lung Cancer
• Afinitor (Everolimus)
• Atezolizumab
• Doxorubicin Hydrochloride
• Durvalumab
• Etopophos (Etoposide Phosphate)
• Etoposide
• Etoposide Phosphate
• Everolimus
• Hycamtin (Topotecan Hydrochloride)
• Imfinzi (Durvalumab)
• Lurbinectedin
• Methotrexate Sodium
• Nivolumab
• Opdivo (Nivolumab)
• Tecentriq (Atezolizumab)
• Topotecan Hydrochloride
• Trexall (Methotrexate Sodium)
• Zepzelca (Lurbinectedin)

Source: National Cancer Institute.

6 Novel approaches for cancer therapeutics

With the advancement in cancer research, novel therapeutic approaches are emerging, involving cell cycle management, signaling pathways, monoclonal antibodies, therapies targeting molecular events, immunotherapy, and antibody-drug conjugates. Genomic makeup could change with the deregulation of signaling pathway molecules that can be linked with confrontation to conventional cancer treatment.

6.1 Repurposing of drugs

Recently, repurposing of drugs has become more popular as a novel cancer therapy with the advancement in genomics, computational biology, and proteomics. Drug repurposing offers pharmaceutical developments with safe, more effective, economic prospects, with low side effects. Generally, alkylating agents, anthracyclins, aromatase inhibitor, antimetabolites, mTOR inhibitor, CDK4/6 inhibitor, and mitotic inhibitors are considered efficacious for drug repurposing in cancer treatment. Some inhibitory drugs of CDK4 and CDK6 like palbociclib, abemaciclib, and ribociclib showed beneficial effects when given in combination with aromatase inhibitors [32]. Fulvestrant, a drug against hormone receptor (HR), increased the survival time window of BC patients when given in combination with CDK4/6 inhibitors in comparison to the treatment when given alone. The combinatorial approach of histone deacetylase (HDAC) inhibitors like entinostat and vorinostat with aromatase inhibitors such as exemestane and tamoxifen demonstrated the most significant clinical outcomes. Due to the instigation of PI3K/AkT/MTOR pathways, hormonal therapy is reported to be inefficient in 70% of cancer patients. Another FDA-approved signaling pathway inhibitor, everolimus, a mTOR inhibitor, when given in combination with exemestane or trastuzumab, reported a promising clinical outcome [1].

6.2 Immunotherapy approach

The immunotherapy-based approaches target the tumor indirectly by promoting the immune response to the cancer. In the cancer patients, the immune system gets derailed. Understanding the mechanism by which cancer cells evade and failed the patient's immune system, hence facilitate the drug development that reinstate the functioning of the immune system to the extent that it can ultimately combat cancer. One type of immune cell targeted in this therapy is regulatory T cells (Treg). As Treg is the main factor to reduce the cytotoxic immune responses and restore the tumor immunotolerance, reductions in these result in improving the natural killer (NK) and T cell functions [33].

6.2.1 Immune checkpoint inhibitors (ICIs)

ICIs are anticancer immunotherapy agents. These treatments involve the inhibition of cytotoxic T-lymphocyte-associated protein 4 (CTLA4) and programmed cell death protein 1 (PD-1) receptors, which initiates an immune response that would otherwise have been prevented by the tumor. Anti-PD1/PD-L1 directed ICIs are commonly used to cure patients with advanced NSCLC, metastases brain cancer, thymic epithelial tumors (TETs), and various other cancers. Several ICIs are already approved by the US FDA and have fruitful antitumor effects, both when given separately and in combination with local therapies. Additional studies are needed to comprehend the resistance mechanisms of tumor cells to the checkpoint inhibitors treatment to improve the therapy [34,35].

6.2.2 Immune cytokine-based strategies

On the basis that IFN-α has antitumor properties against various cancer cell lines, several clinical trials have been carried out to explore the probable antitumor properties of other cytokines, manipulating their capability to activate immune responses against cancer. As monotherapies, cytokines have a few limitations, including modest effectiveness against tumors and short half-lives. In spite of these limitations, IL2 and IFN-α are FDA-approved for the treatment of various cancer types as they have (mild) antitumor activities. It has been approved that advanced renal cell carcinoma (RCC) and metastatic melanoma are treated by IL2 treatment. Likewise, IFN-α has been approved for hairy cell leukemia, melanoma, follicular non-Hodgkin lymphoma, and AIDS-related Kaposi's sarcoma treatment [36,37].

6.2.3 Tumor-targeted superantigen/ligand-targeted superantigen

Staphylococcal enterotoxins (SEs), also known as superantigens (SAgs), are protein antigens that are highly potent and are produced by *Staphylococcus aureus*. These antigens target by nonspecifically cross-linking with the class II of the major histocompatibility complex on T cell receptors at specific V regions and antigen-presenting cells. This results in the production of huge amounts of chemokines and cytokines (e.g., tumor necrosis factor α (TNF-α), interleukin 1 (IL-1), IL-2, interferon γ (IFN- γ), and macrophage chemoattractant protein 1 (or CCL2)) due to the hyperactivation of monocytes/macrophages and T lymphocytes. The cytokines that are produced take part in the pathogenesis of various autoimmune and/or inflammatory disorders and have emerged as an attractive agent for cancer immunotherapy due to several characteristics [38]. Several studies have demonstrated the ability of immunogenic proteins to slow down the growth of cancer or kill the cancerous cells by stimulating T cells. There are two approaches by which superantigens can be delivered to the tumor. One is to link the superantigen with the cancer-specific antibody, and the second is using a ligand targeted therapeutic (also known as a tumor cell-specific peptide) in place of a full antibody to facilitate the attachment with the superantigen. The second approach is advantageous in that it provides a greater penetration for a solid tumor and reduces serious antigenicity. Additionally, the costs are also reduced in the second approach and the production becomes easier [39].

6.3 Enzyme/small molecules strategies

6.3.1 Proteolysis targeting chimera (PROTAC)

A strategy devised and developed by learning from nature, the proteolysis targeting chimera (PROTAC), draws inspiration from the mechanism of degradation of unwanted proteins in a cell. In this technique, there are two independent moieties from one molecule; one of the moieties binds to the E3 ubiquitin ligase and the other binds with the target protein. This strategy helps to bring the ubiquitinylation machinery and the target protein into proximity. After entering the cell, the molecule forms a ternary complex by binding to the ubiquitinylation machinery and the target protein [40]. The target proteins are then catalyzed to amino acids and peptides by the proteasome, and are later recycled. Then, a PROTAC molecule with two heads is formed, which consists of two-small molecule ligands. Studies have demonstrated that the two-headed PROTAC molecule provides a greater degree of degradation when compared to the one-headed molecule, and it has a better binding affinity to the concerned protein under study. The proteins containing the bromodomain and extra-terminal motif (BET), estrogen-related receptor alpha (ERRα), and RIPK2-, a serine–threonine kinase, were the first proteins to be used as targets for PROTACs. Several studies have highlighted this mechanism of targeting and degrading BET proteins, and also the MDM2 degradation by PROTAC. The proteolysis mediated by PROTAC was found to be time-dependent, resulting in a significant reduction (80%) in MDM2 after 24h. This decrease in MDM2 caused P53 and p21 to accumulate in a time-dependent manner. This strategy is currently being used to treat several diseases including cancers (in particular, the estrogen receptor-positive (ER+) breast cancer). In ER+ breast cancers and prostate cancers, it acts by carrying out ubiquitination and degradation by targeting steroid hormone receptors. The PROTAC strategy can also be applied in treating nonsmall-cell lung cancer by targeting the epidermal growth factor receptor and in degradation of anaplastic lymphoma kinase (ALK), which is associated with several human cancers [33,41].

6.3.2 Antibody-drug conjugation systems (ADCs)

Antibody-drug conjugates as targeted hybrid molecules have been identified as potential therapeutics for treating solid tumors through antibody–antigen interaction and have the ability to make a significant shift in the area of cancer therapy. These biotherapeutic systems comprise of tumor-specific mAb covalently conjugated with potent cytotoxin (payloads) through stable chemical linkers. Similarly, they conjugate covalently with the antibody through an amino acid residue side chain and alteration of surface lysine on the antibody. Furthermore, reduction of existing disulfide bonds following conjugation to the drug linker liberates the free-cysteine residues on the antibody. These are some of the strategies of chemical conjugation: the design of the molecule, type of target tumor, type of antigen (as per the differing internalization rates of antigen complex formations with ADC molecules), type of prodrug, factors related to pharmacokinetics, stability of ADC and the linker plasma at the time of circulation in bloodstream (such that the cancer cell is localized in its original formation by the ADC), immune reactivity maintenance, and efficient release of the prodrug at the favorable site [42]. Prodrugs can be converted enzymatically or chemically into active forms of drugs in order to facilitate the death of tumor cells through its ability to interfere with the cell division mechanism and/or induce irreversible damage to DNA. Furthermore, one of the important factors, linker hydrophobicity, facilitates ADC molecules aggregation, which leads to improved efficacy, reduced systemic toxicity, better coupling with hydrophobic payloads, and a significantly improved ADC development.

One of the crucial steps in designing an ADC system is to use an acceptable biological linker to connect antibodies to drugs. The mechanism of action of ADC involves three steps: (i) formation of ADC-antigen complex through the binding of mAb to target antigen; (ii) endocytosis, through which the ADC-antigen complex gets internalized into the tumor cell; and (iii) degradation of the drug and its consequent release. Antibody drug conjugation systems are innovative and important approaches in BC treatment that can provide optimal outcomes. They help to reduce toxicity due to drug and improve antitumor immunity. One of the important antibody-drug conjugates is trastuzumab emtansine (T-DM1), which has been shown to have positive effects in the case of HER2-positive BC. Moreover, it is known to have exceptional potency as a chemotherapeutic and comprises of emtansine and trastuzumab linked using a nonreducible thioester. It has also been approved in phase I, II, and III clinical trials. As per the existing evidence, the rates of survival in patients who received T-DM1 and those who received taxane with trastuzumab were similar. However, the possibilities of brain metastasis due to its ability to cross the blood–brain barrier have also been reported. Therefore, several ADC systems based on HER-2 BC have been explored in clinical trials and have been approved for treatment [1,43].

6.3.3 Antibody-directed enzyme prodrug therapy

Chemotherapy drugs are known to have a low specificity to cancer cells. On the other hand, antibodies are known to be greatly specific to their antigens and hence are preferred in diagnosis and treatment of tumors and are known to significantly reduce side effects. Another technique, which was introduced more than 30 years ago, is antibody-directed enzyme prodrug therapy (ADEPT), which helps to restrict the action around the tumor cells by giving rise to a cytotoxic drug [44]. The process takes place in two steps: firstly, an antibody, which is covalently linked to an enzyme such as the glucarpidase (or carboxypeptidase G2) delivers the enzyme to the tumor site, and then a prodrug, which is a nontoxic enzyme substrate, is injected after the antibody conjugate is cleared from the rest of the body. This enzyme is responsible for converting the prodrug into a potent cytotoxic drug in the vicinity of the tumor. An advantage of this method is that the sizes of the produced cytotoxic drugs are small, and hence they easily diffuse as compared to antibody molecules in the case of ADC. Additionally, the bystander action of this cytotoxic agent kills those cancer cells that do no express the desired antigen as well. Therefore, this approach of using an enzyme and nontoxic drugs and the prodrug protocol is an attractive option for cancer therapy. One of the first studies that used the ADEPT strategy found a significant response in advanced metastatic cancer patients who were unresponsive to all other treatments and were expected to survive for less than 8 weeks. However, the production of interstrand cross-links in the DNA of the tumor cell within 1 h of administration is a limitation of the ADEPT strategy as the cancer-cells recovered and repaired the damage within 24 h, requiring multiple cycles of ADEPT. Multiple cycles of ADEPT would only be possible if the patient had received some agent to suppress the immune responses, such as cyclosporine [45].

6.4 Nanotechnology-based therapy

Small size, high surface reactivity, and a large surface area are a few notable characteristics of nanoparticles. Moreover, they have several unique physicochemical properties, such as greater reactivity when comparted to their bulk counterparts and high surface to volume ratio. These characteristics enhance the possibility for early detection and improved treatment and diagnosis of cancer using these agents. Nanomedicines have several other advantages, including prevention of rapid clearance, enhanced bioavailability, ability to capture the drugs in nano carriers and travel through the blood stream, excellent solubility, stability, and resistance, enhanced benefits for tumor imaging, and their ability to accommodate thousands of drug molecules. Researchers have tried to overcome the challenges of existing chemotherapeutic agents with nanotechnology and establish them as tools for personalized medicine [46]. Two important approaches have been identified in developing nanoparticles: one is conjugation of active recognition moieties to the surfaces using ligands, which are biologically relevant to achieve chemotherapeutic selectivity; the other is to achieve enhanced permeability retention (EPR) of nano carriers with superior accumulation for therapeutic payloads. Some of the nanoparticles include metal-based nanoparticles, albumin nanoparticles, lipid nanoparticles, liposomes, polymeric nanoparticles, and micelles. Combining these nanoparticle therapies can help address several issues in the diagnosis and treatment of various cancers. Several important NPs are commonly used in treating and diagnosing cancer. Nanotechnology, which is new and noninvasive, is considered to be effective in preventing, detecting, and diagnosing breast cancer. Studies have failed to identify adverse complications of chemotherapy with these NPs when compared to classical treatment regimens [47].

One of the most important bio-acceptable transporter nano carriers, which has been used as a clinical excipient in several formulations, is albumin. It is biodegradable, with low antigenicity, and holds the ability to bind to several different endogenous molecules and drugs. Other important metallic nanoparticles that play an important role in cancer therapy include silver, iron oxide, and gold. Important characteristics of these NPs include exceptional thermal stability, reduced toxicity, and smaller size-to-volume ratio. These characteristics aid significantly in targeting and gene silencing diagnostic assays, as well as for drug delivery. An important strategy of these metal nanoparticles is surface functionalization with targeting ligands. These help to prolong duration of drug action and provide superior control over deposition of energy in tumors, which can significantly improve management and treatment of cancer [48,49].

Nanostructured lipid carriers (NLC) and solid lipid nanoparticles (SLN) are also lipid-based nanoparticles that have great biodegradable properties and are highly biocompatible, and hence they hold exceptional ability to treat and detect tumors. Their functions include transportation of hydrophobic and hydrophilic molecules, using a prolonged half-life to prolong the drug action time, and exceptional controlled release of the drugs incorporated. Polymeric nanoparticles, having diameters of 1–1000nm, are also important in drug targeting and diagnosis. Their several advantages include using active/passive targeting to switch the therapeutic approach target, adjusting the drug release, and minimizing the degradation in the bloodstream. Poly lactic-*co*-glycolic acid (PLGA) is an important polymeric nanoparticle, which has been approved by the FDA, with extensive biomedical applications owing to its high stability, low cytotoxicity, and favorable uptake by EPR effect [50].

Polymeric micelles are also important structures with a supramolecular core-shell. Their size ranges from 10 to 100nm, and they are formed out of amphiphilic block copolymers, which are biocompatible, in an aqueous condition. They have the ability to solubilize imaging agents as well as hydrophobic drugs and are hence considered important in the emerging techniques for imaging and diagnosing cancer. Drug stability can be achieved when the drug is encapsulated in a polymeric core, which delays inactivation and enzymatic degradation. Improvements in materials chemistry, biology, medicine, and nanoscience to accommodate the use of such nanoscale techniques have been important in changing devastating cancers to be on the verge of well-being [1].

6.5 Combinatorial cancer therapies and precision medicine

Due to the highly complex and heterogeneous nature of tumors, the treatment regimen is ideally personalized for each patient. Treatment for some cancers may require multiple strategies. Conversely, some therapies may be effective only in specific parts of the body and others may be effective in other parts of the body. Employing a combination of modern targeted therapies with traditional therapies can significantly prolong patients' lives, with the enhanced ability to overcome resistance to drugs and reduction in symptoms. Several studies have reported that improved results are observed for several FDA-approved drugs when they are used in combination (Fig. 12). In preclinical models, a combination of some drugs was observed to exhibit synergy or additivity, and showed reduction in cross-resistance, which helps to overcome patients variability. The variability between patients and the importance of precision medicine for treatment success was highlighted in this study. Higher effectiveness and high overall survival were observed for combination therapies when compared to individual treatment regimen in various cancers such as breast cancer and lung cancer. Radiotherapy, when combined with immunotherapy, is anticipated to have synergistic effects owing to the interactions between the immune system and radiation. If precision medicine is employed along with combination therapy, it might significantly improve the treatment outcomes. Combination of two or more such strategies has been explored over the years. However, the best way to use these therapies in combination is still not understood fully, and needs to be explored further in clinical trials and studies [33,51].

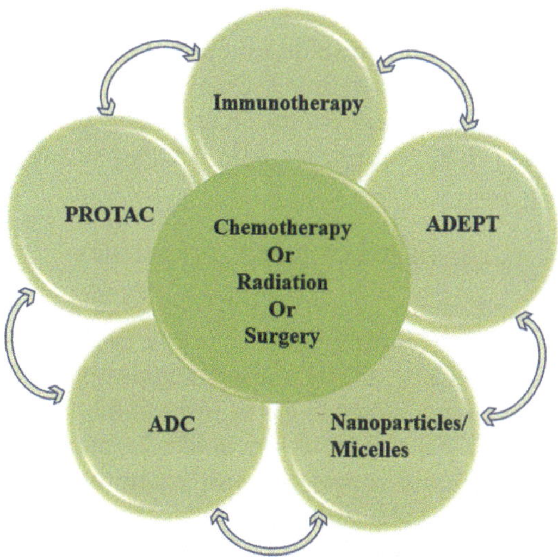

FIG. 12 Combinatorial approaches of different therapies of cancer treatment.

6.6 Targeting cancer stem cells (CSCs)

The incentive for targeting cancer stem cells came across due to its role in developing chemo-resistance and radio-resistance in tumors. Increased expression of CD44 in BCSCs implies its suitability to be used as a target. Nanomedicine is one of the important mechanisms for drug delivery in several diseases, including cancers. Drugs can be delivered to CSCs using various nanoparticles including carbon nanotubes, liposomes, PLGAs, and graphene oxide coated with specific aptamers or antibodies against markers of CSC. BCSCs were successfully targeted by nanoparticles, which were bound to anti-CD44 antibody and were loaded with chemotherapeutic agents. A new therapeutic line involving combination of chloroquine, which is an autophagy inhibiting drug, along with chemotherapeutic agents in order to target tumor CSCs has shown a significant reduction in ALDH1 of CSCs. In BC, cationic liposomes played an important role in delivering siRNA inhibitors to CSCs. However, recent evidence has indicated plasticity in CSCs and conversion of tumors of nonstem cell origin into CSCs, which poses a major challenge, suggesting the need for further research to understand these dangerous tumor leaders [52].

7 Therapeutic approaches targeting oxidative stress in cancer

Increased ROS level is responsible for sustenance of cancer progression. However, this property of carcinogenesis also provides an edge for novel therapeutic approaches in cancer treatment. Targeting ROS and its related signaling pathways and proteins has been the best approach recently in the quest for cancer therapeutics. Key novel approaches targeting oxidative stress in cancer treatment include oxidative stress-based cancer therapeutics such as radiotherapy, photodynamic therapy, antioxidant therapy, prooxidant therapy, and targeting mitochondrial cascades such as OXPHOS, electron transport chain (ETC) complexes I–IV, heme synthesis, etc. (Fig. 13).

7.1 Oxidative stress-based cancer therapeutics

Radiotherapy and photodynamic therapies were specifically designed to elevate the ROS levels in tumor cells, causing death via sudden and intense oxidative stress. Cancer cells naturally maintain a high intrinsic oxidative activity, making it easier to cause intense oxidative stress through minor elevations in ROS level. Since normal cells maintain lower levels of ROS and are endowed with powerful antioxidant mechanisms, cytotoxic levels of ROS for tumor cells do not affect the normal levels. In order to further downregulate the antioxidant response of cancer cells, inhibitors of antioxidant shielding factor Nrf2 are used that highly sensitize cancer

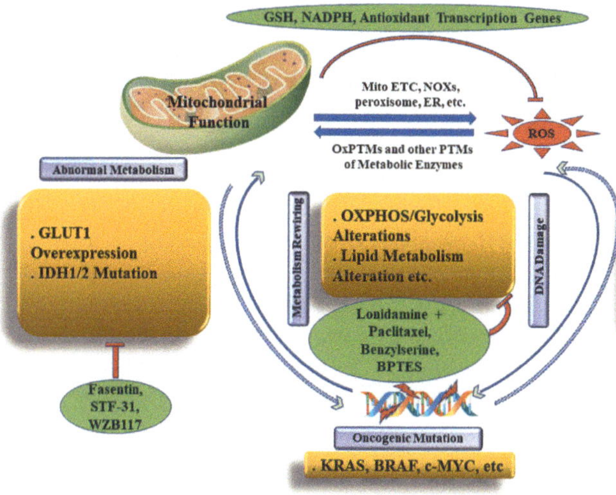

FIG. 13 Mechanistic insights on targeting oxidative stress and its related pathways in cancer therapy. *ER*, endoplasmic reticulum; *GLUT1*, glucose transporter 1; *GSH*, glutathione; *IDH1/2*, isocitrate dehydrogenase 1/2; *Mito-ETC*, mitochondrial electron transport chain; *NOX*, NADPH oxidase; *OXPHOS*, oxidative phosphorylation.

cells' response toward sudden oxidative stress. Excessive OS could produce a variety of cellular damage such as irreparable cellular lesions and commuting signaling machinery from prosurvival to apoptotic pathway and immunogenic cell death response.

Ionizing radiations (IR) can particularly disrupt the atomic structures of macro and micro molecules in cells, resulting in depleting cellular frameworks. Extensively used in radiotherapy, localized deposition of IR energy extensively generates secondary electrons and unstable species, further producing free radicals and by-products of radiolysis with heterogeneous distribution. These localized and instantaneous high ROS levels in cancer cells result in extensive clustered lesions and irreparable oxidative damage. Unlike IR, photodynamic therapy relies on the light-sensitive photosensitizers loaded in the cancer cells, which are activated by irradiation with light of a specific wavelength. Singlet oxygen burst is the major product of photodynamic irradiations, causing an intense increase in ROS levels to produce cytotoxicity in cancer cells. However, photosensitizer loading lacks specificity and intake by normal cells is unavoidable, making them vulnerable to higher side effects, hence these are designed for lower levels of cytotoxic response. In addition, the tumor-specific precision of photodynamic therapy is achieved with the help of precise light irradiation using fiber optic devices. It should be noted that the risk of genetic instability is higher in IR than in photodynamic therapy, since the latter does not target the nucleus [53].

Radiation-based therapies not only provide potential treatment for cancer but are also responsible for bystander effects. Nonirradiated cells are very much prone to exhibit similar cellular damage such as DNA strands break, point mutations, inflammation, and increased ROS levels, as those seen in tumor cells. In addition, these therapies provoke massive cell death in the biological penumbra, accounting for distant effects of OS-based therapies. Hence, there are non-radiation-based therapies that are extensively used to overcome these side effects, which include antioxidant therapy and prooxidant therapy.

Antioxidants are the first line of defense against the increased ROS levels. There are cumulative evidences suggesting the preventive effect of antioxidants against tumorigenesis. As a result, they are often used as supplements in combination with chemotherapy for potential reduction in dose-limiting toxicities among cancer patients [7]. Treatment of cancer cells, particularly with polyphenol-based antioxidants, is highly effective in combating the disease. Synthetic and metal-based antioxidants with biologically active ligands are also of great help in scavenging ROS molecules and curbing tumor growth. Furthermore, overexpressing antioxidant codifying genes also has promising effects in cancer therapy. Additionally, Nrf2 activators are also potential approaches for increasing the antioxidative mechanism of cells in cancer state.

Opposed to the antioxidative approach to treating cancer, prooxidant therapy works on a similar conceptual framework to radiation therapy, i.e., instantly elevating the levels of ROS in cancer cells, sufficient to cause cancer cell annihilation. However, instead of radiation, prooxidant therapy relies on prooxidant agents, which help in generating free radicals and unstable molecules in cancer cells, thus helping to decrease tumor growth. Nathan and Cohn used glucose oxidase for the first time in 1981 to report the effectiveness of prooxidant therapy. Glucose oxidase is a precursor of hydrogen peroxide, thus generating peroxide radicals in cancer cells and downregulating tumor growth [54]. Additionally, certain chemotherapeutic agents are also observed to induce ROS, leading to oxidative stress in the cancer microenvironment. Some of the chemotherapeutic drugs exhibiting such ROS-inducing effects include 5-fluorouracil, oxaliplatin, and paclitaxel. In the case of breast cancer, treatment with paclitaxel translocate Rac1 (positive regulatory protein of Nox) and enhanced ROS levels are found to be accumulated in the intercellular matrix instead of the intracellular matrix, thus causing lethal damage to bystander cancer cells that are not exposed to the drug treatment. This effect is also observed for some other anticancer agents, such as vincristine and taxotere. In most cases, a combinatorial approach provides the highest efficacy against tumor cells and cancer treatment [55].

7.2 Mitochondria as a target for cancer therapeutics

Mitochondria plays a crucial role in cancer pathogenesis. It is actively involved in the oxidative metabolism and generation of ROS. Hence, mitochondria is a novel therapeutic approach targeting oxidative stress in cancer treatment. In recent years, several anticancer mitochondrial agents have been classified on the basis of their mechanism of action, where they target either mitochondrial genetic alterations or mitochondrial signaling cascades [56]. Recent evidence suggests that mitochondrial enzymes involved in aerobic glycolysis are potential targets for anticancer therapy. For instance, glucose transporter GLUT1 is the potential target of anticancer drugs such as fasentin, STF-31, and WZB117. These drugs inhibit the glucose uptake by cancer cells, leading to cytotoxic effects [57]. Similarly, lonidamine and 2-deoxy-D-glucose (2-DG) inhibit another key enzyme involved in aerobic glycolysis, hexokinase II. In addition, lonidamine has also been used in combination with paclitaxel and epidermal growth factor receptor targeted nano carrier system to treat multidrug-resistant breast cancer [58]. Furthermore, a combinatorial approach has also exhibited synergistic therapeutic effects of 2-DG when used with PI3K or mTOR inhibitors, which shifted the aerobic glycolysis toward OXPHOS, causing strong cytotoxicity in primary effusion lymphoma cells [59].

In order to sustain the Krebs cycle, high demand of biosynthetic precursors, and NADPH, tumor cells rely on glutaminolysis. Glutamine uptake is regulated by plasma membrane transporters such as SLC1A5 (ASCT2) and SLC7A5. After uptake, it is converted to glutamate with the help of glutaminase (GLS) in the mitochondria. Glutamate is further broken down into α-ketoglutarate by glutamate dehydrogenase (GDH) or α-keto-acids. Hence, upregulated levels of membrane transporters, GLS, and GDH are seen in tumor cells, making them potential targets for cancer treatment. Lung cancer cells exhibit higher levels of SLC1A5 as a primary glutamine membrane transporter. Hence, inhibition of SLC1A5 with benzylserine or glutamyl-p-nitroanilide (GPNA) has occluded tumor growth [60]. Benzophenanthridinone 968 and BPTES are the most suitable GLS inhibitors available, and are known to suppress cancer progression in several cancer types. A BPTES derivative, CB-839, is undergoing phase I clinical trials for hematological and solid malignancies. Specific GDH inhibitors are also extensively researched, like purpurin and its derivative, R162. These inhibitors are known to increase the ROS levels in cancer cells and suppress tumor growth [61]. Mitochondrial retrograde signaling is activated in the case of mitochondrial dysfunction. This signaling activates the transcription factors, creating resistance to apoptosis and cancer progression. The retrograde signaling involves several proteins and pathways that are targetable for cancer treatment. These include upregulation of antiapoptotic Bcl-2 family proteins, PI3K/Akt, ERK, and Ras pathways.

Several anticancer drugs are used to inhibit the Bcl-2 family proteins by acting on their BH3 domains. Navitoclax, gossypol, and ABT-737 are extensively used BH3 inhibitors, promoting membrane channel formation by Bax and Bak, and initiating apoptosis [56].

In order to fulfill their bioenergetic needs, cancer cells exhibit metabolic plasticity, making it easy to shift from aerobic glycolysis to OXPHOS. This characteristic feature of cancer cells has been extensively explored in recent years to target the mitochondrial respiratory complexes along with mitochondrial dysfunction to produce lethal cytotoxic effects on cancer cells. OXPHOS is found to be upregulated in several cancer types including breast and lung cancer. Such upregulation is known to elevate the ROS levels, tumorigenesis, and resistance to chemotherapy [62]. Breast cancer and lung cancer types have recently been found to exhibit metabolic heterogeneity, i.e., elevated rates of glycolysis and oxidative metabolism can be observed in these cancer types. Along with OXPHOS, lung cancer cells are also found to exhibit higher levels of heme, a key molecule in mitochondrial OXPHOS [63]. Thus, in light of such evidence, several drugs are being screened for their impairing effect on mitochondrial respiratory complexes I–IV, which are the backbone of the ETC, causing reduction in flow of electrons, increased ROS production, and activation of proapoptotic process in cancer cells (Table 5). MitoTam is a novel mitochondrial therapeutic agent and complex I inhibitor that increases ROS production and promotes cell death in breast cancer. Similarly, MitoVes is a complex II inhibitor and apoptotic promoter in breast and lung cancer cells [64].

Iron protoporphyrin IX, heme, is an iron-containing key molecule in mitochondrial OXPHOS. Heme can be synthesized by human cells and can even be consumed in dietary intake. Epidemiological studies indicate an association of higher amount of heme intake with increased risk of several types of cancer including lung, breast, pancreatic, and colorectal. Heme serves as a cofactor for enzymes involved in oxygen transportation and is crucial for complexes II, III, and IV to function [64]. Elevated levels of heme flux are observed in NSCLC cell lines when compared to nontumorigenic cells. These elevations are further corroborated with overexpression of heme biosynthesis enzyme ALAS1 and heme uptake protein

TABLE 5 Approved mitochondrial complexes inhibitor drugs in cancer treatment.

Mitochondrial Complex	*Inhibitor Anticancer Drug*
Complex I	Rotenone, Piericidine A, Capsaicin, MitoTam, Metformin, Tocotrienol
Complex II	α-tocopheryl succinate, Gracillin, Atpenins, Lonidamine, MitoVes
Complex III	Atovaquone
Complex IV	Mitotane, Arsenic Trioxide,

SLC48A1, thus establishing its crucial role in lung cancer progression [65]. Therefore, limiting the heme availability or targeting its biosynthesis enzyme is an effective strategy against lung tumor growth. Succinyl acetone and cyclopamine tartrate (CycT) have been tested for their anticancer efficacy by inhibiting heme biosynthesis and suppressing proliferation, colony formation, and evasion in NSCLC cells. Apart from synthetic drugs, bacterial secretions like hemophores are also evaluated in some studies for their inhibitory effects on uptake of heme in NSCLC cells [66].

8 Conclusion

Cancer development and sustenance is promoted by a vast array of external and internal factors, majorly by persisting oxidative stress. Remarkably, cancer cells thrive on higher levels of ROS for their survival and affect the cellular microenvironment accordingly. Invading apoptotic pathways, autocrine signaling mechanisms, and mitochondrial disruption are some of the ways by which tumorigenesis persist in the body. Several conventional therapies such as surgery, chemotherapy, radiotherapy, and targeted molecular therapy are used for the treatment of several types of cancer, including lung and breast cancer. However, higher cytotoxicity and lower specificity of conventional therapies have led to the development of novel therapeutic approaches for cancer treatment, such as nanotherapy, combinational therapy, precision medicine, immunotherapy, etc. Apart from this, oxidative stress-based therapies have now emerged as novel and potential therapeutic approaches for cancer treatment. In order to achieve this, several oxidative stress-related biomarkers, enzymes, and mitochondrial molecular markers have been exploited to find the best target to inhibit cancer growth. Novel therapeutic approaches not only have been shown to be highly effective against cancer development, but also have limiting side effects, making further studies and drug development essential.

9 Future perspective

Years of cancer research have provided answers to several questions regarding cancer mechanistic and biology. However, we still lack a great deal of knowledge related to functionalities of cancer cells. Development of therapies for cancer treatment will only yield effective results when we evaluate cancer biology with precision. Hence, application of genomics and proteomics holds great significance in future cancer studies. Additionally, future studies are required to validate specific oxidative stress-based biomarkers for targeted therapy, as well as predicting the response of anti- and pro-oxidants in cancer treatment.

Acknowledgments

The authors would like to acknowledge the support of the Department of Science and Technology FIST (No. SR/FST/LS-I/2017/05[C]) grant to Department of Toxicology, Jamia Hamdard. Prof. S.P. is thankful for the DST PURSE grant (No. SR/PURSE Phase 2/39[C]). The Research Associate Fellowship granted from the Indian Council of Medical Research (ICMR, Govt. of India) (No. 45/41/2019-NAN-BMS) to Dr. P.K. is acknowledged. Ms. M.K. gratefully acknowledges the support of Senior Research Fellowship from ICMR, Govt. of India (No. 3/1/2/Neuro/136/2020-NCD-I).

References

[1] K. Barzaman, J. Karami, Z. Zarei, A. Hosseinzadeh, M.H. Kazemi, S. Moradi-Kalbolandi, E. Safari, L. Farahmand, Breast cancer: biology, biomarkers, and treatments, Int. Immunopharmacol. 84 (April) (2020), https://doi.org/10.1016/j.intimp.2020.106535.

[2] A.G. Waks, E.P. Winer, Breast cancer treatment: a review, JAMA 321 (3) (2019) 288–300, https://doi.org/10.1001/jama.2018.19323.

[3] C. García-Fernández, C. Fornaguera, S. Borrós, Nanomedicine in non-small cell lung cancer: from conventional treatments to immunotherapy, Cancers 12 (6) (2020) 1–26, https://doi.org/10.3390/cancers12061609.

[4] T. Ismail, Y. Kim, H. Lee, D.S. Lee, H.S. Lee, Interplay between mitochondrial peroxiredoxins and ROS in cancer development and progression, Int. J. Mol. Sci. 20 (18) (2019), https://doi.org/10.3390/ijms20184407.

[5] F. Hecht, C.F. Pessoa, L.B. Gentile, D. Rosenthal, D.P. Carvalho, R.S. Fortunato, The role of oxidative stress on breast cancer development and therapy, Tumor Biol. 37 (4) (2016) 4281–4291, https://doi.org/10.1007/s13277-016-4873-9.

[6] X. Liu, Z. Chen, The pathophysiological role of mitochondrial oxidative stress in lung diseases, J. Transl. Med. 15 (1) (2017) 1–13, https://doi.org/10.1186/s12967-017-1306-5.

[7] V. Sosa, T. Moliné, R. Somoza, R. Paciucci, H. Kondoh, LLeonart, M. E., Oxidative stress and cancer: An overview, Ageing Res. Rev. 12 (1) (2013) 376–390, https://doi.org/10.1016/j.arr.2012.10.004.

[8] E.A.C. Wiemer, Stressed tumor cell, chemosensitized cancer, Nat. Med. 17 (12) (2011) 1552–1554, https://doi.org/10.1038/nm.2593.

[9] E. Omori, M. Inagaki, Y. Mishina, K. Matsumoto, J. Ninomiya-Tsuji, Epithelial transforming growth factor β-activated kinase 1 (TAK1) is activated through two independent mechanisms and regulates reactive oxygen species, Proc. Natl. Acad. Sci. U. S. A. 109 (9) (2012) 3365–3370, https://doi.org/10.1073/pnas.1116188109.

[10] T.O. Khor, M.T. Huang, K.H. Kwon, J.Y. Chan, B.S. Reddy, A.N. Kong, Nrf2-deficient mice have an increased susceptibility to dextran sulfate sodium-induced colitis, Cancer Res. 66 (24) (2006) 11580–11584, https://doi.org/10.1158/0008-5472.CAN-06-3562.

[11] M. Benezra, N. Chevallier, D.J. Morrison, T.K. MacLachlan, W.S. El-Deiry, J.D. Licht, BRCA1 augments transcription by the NF-κB transcription factor by binding to the rel domain of the p65/RelA subunit, J. Biol. Chem. 278 (29) (2003) 26333–26341, https://doi.org/10.1074/jbc.M303076200.

[12] P. Rai, J.J. Young, D.G.A. Burton, M.G. Giribaldi, T.T. Onder, R.A. Weinberg, Enhanced elimination of oxidized guanine nucleotides inhibits oncogenic RAS-induced DNA damage and premature senescence, Oncogene 30 (12) (2011) 1489–1496, https://doi.org/10.1038/onc.2010.520.

[13] F. Rothé, M. Ignatiadis, C. Chaboteaux, B. Haibe-Kains, N. Kheddoumi, S. Majjaj, B. Badran, H. Fayyad-Kazan, C. Desmedt, A.L. Harris, M. Piccart, C. Sotiriou, Global microRNA expression profiling identifies MiR-210 associated with tumor proliferation,

invasion and poor clinical outcome in breast cancer, PLoS One 6 (6) (2011), https://doi.org/10.1371/journal.pone.0020980.

[14] E. Provenzano, G.A. Ulaner, S.F. Chin, Molecular classification of breast cancer, PET Clinics 13 (3) (2018) 325–338, https://doi.org/10.1016/j.cpet.2018.02.004.

[15] M. Thomas, E.D. Kelly, J. Abraham, M. Kruse, Invasive lobular breast cancer: a review of pathogenesis, diagnosis, management, and future directions of early stage disease, Semin. Oncol. 46 (2) (2019) 121–132, https://doi.org/10.1053/j.seminoncol.2019.03.002.

[16] S. Gilad, G. Lithwick-Yanai, I. Barshack, S. Benjamin, I. Krivitsky, T.B. Edmonston, M. Bibbo, C. Thurm, L. Horowitz, Y. Huang, M. Feinmesser, J. Steve Hou, St. Cyr, B., Burnstein, I., Gibori, H., Dromi, N., Sanden, M., Kushnir, M., & Aharonov, R., Classification of the four main types of lung cancer using a microRNA-based diagnostic assay, J. Mol. Diagn. 14 (5) (2012) 510–517, https://doi.org/10.1016/j.jmoldx.2012.03.004.

[17] M. Zheng, Classification and pathology of lung Cancer, Surg. Oncol. Clin. N. Am. 25 (3) (2016) 447–468, https://doi.org/10.1016/j.soc.2016.02.003.

[18] T.A. King, M. Morrow, Surgical issues in patients with breast cancer receiving neoadjuvant chemotherapy, Nat. Rev. Clin. Oncol. 12 (6) (2015) 335–343, https://doi.org/10.1038/nrclinonc.2015.63.

[19] S.A. McLaughlin, Surgical Management of the Breast. Breast conservation therapy and mastectomy, Surg. Clin. N. Am. 93 (2) (2013) 411–428, https://doi.org/10.1016/j.suc.2012.12.006.

[20] M. Donker, G. Tlenhoven, van, M. Straver, P. Meljnen, C.J.H. van de Velde, R.E. Mansel, L. Cataliotti, A.H. Westenberg, J.H.G. Klinkenbijl, S.C. Veltkamp, L. Slaets, N.J. Duez, P.W. De Graaf, T. Van Dalen, H. Rijna, M. Snoj, N.J. Bundred, J.W.S. Merkus, Y. Belkacemi, J. Bogaerts, AMAROS trial, Lancet Oncol. 15 (12) (2015) 1303–1310, https://doi.org/10.1016/S1470-2045(14)70460-7.Radiotherapy.

[21] A.S. Coates, E.P. Winer, A. Goldhirsch, R.D. Gelber, M. Gnant, M.J. Piccart-Gebhart, B. Thürlimann, H.J. Senn, F. André, J. Baselga, J. Bergh, H. Bonnefoi, H. Burstein, F. Cardoso, M. Castiglione-Gertsch, M. Colleoni, G. Curigliano, N.E. Davidson, A. Di Leo, B. Xu, Tailoring therapies-improving the management of early breast cancer: St Gallen international expert consensus on the primary therapy of early breast cancer 2015, Ann. Oncol. 26 (8) (2015) 1533–1546, https://doi.org/10.1093/annonc/mdv221.

[22] E.J. Blok, M.G.M. Derks, J.J.M. van der Hoeven, C.J.H. van de Velde, J.R. Kroep, Extended adjuvant endocrine therapy in hormone-receptor positive early breast cancer: current and future evidence, Cancer Treat. Rev. 41 (3) (2015) 271–276, https://doi.org/10.1016/j.ctrv.2015.02.004.

[23] M. Dowsett, J. Cuzick, J. Ingle, A. Coates, J. Forbes, J. Bliss, M. Buyse, M. Baum, A. Buzdar, M. Colleoni, C. Coombes, C. Snowdon, M. Gnant, R. Jakesz, M. Kaufmann, F. Boccardo, J. Godwin, C. Davies, R. Peto, Meta-analysis of breast cancer outcomes in adjuvant trials of aromatase inhibitors versus tamoxifen, J. Clin. Oncol. 28 (3) (2010) 509–518, https://doi.org/10.1200/JCO.2009.23.1274.

[24] G. Von Minckwitz, A. Schneeweiss, S. Loibl, C. Salat, C. Denkert, M. Rezai, J.U. Blohmer, C. Jackisch, S. Paepke, B. Gerber, D.M. Zahm, S. Kümmel, H. Eidtmann, P. Klare, J. Huober, S. Costa, H. Tesch, C. Hanusch, J. Hilfrich, M. Untch, Neoadjuvant carboplatin in patients with triple-negative and HER2-positive early breast cancer (GeparSixto; GBG 66): a randomised phase 2 trial, Lancet Oncol. 15 (7) (2014) 747–756, https://doi.org/10.1016/S1470-2045(14)70160-3.

[25] N. Harbeck, M. Gnant, Breast cancer, Lancet 389 (10074) (2017) 1134–1150, https://doi.org/10.1016/S0140-6736(16)31891-8.

[26] N. Duma, Non-small cell lung cancer: epidemiology, Mayo Clin. Proc. 94 (August) (2019) 1623–1640.

[27] O.H. Hassan Lemjabbar-Alaouia, Y.-W. Yanga, P. Buchanana, Lung cancer: biology and treatment options, Physiol. Behav. 176 (5) (2017) 139–148, https://doi.org/10.1016/j.bbcan.2015.08.002.Lung.

[28] J.H. Schiller, Okcolo g. i s t Three Emerging New Drugs for NSCLC, 2005, pp. 282–291.
[29] M.R. Green, G.M. Manikhas, S. Orlov, B. Afanasyev, A.M. Makhson, P. Bhar, M.J. Hawkins, Abraxane®, a novel Cremophor®-free, albumin-bound particle form of paclitaxel for the treatment of advanced non-small-cell lung cancer, Ann. Oncol. 17 (8) (2006) 1263–1268, https://doi.org/10.1093/annonc/mdl104.
[30] K.C. Arbour, G.J. Riely, Systemic therapy for locally advanced and metastatic non-small cell lung cancer: a review, JAMA 322 (8) (2019) 764–774, https://doi.org/10.1001/jama.2019.11058.
[31] F.R. Hirsch, G.V. Scagliotti, J.L. Mulshine, R. Kwon, W.J. Curran, Y.L. Wu, L. Paz-Ares, Lung cancer: current therapies and new targeted treatments, Lancet 389 (10066) (2017) 299–311, https://doi.org/10.1016/S0140-6736(16)30958-8.
[32] G.N. Hortobagyi, S.M. Stemmer, H.A. Burris, Y.-S. Yap, G.S. Sonke, S. Paluch-Shimon, M. Campone, K.L. Blackwell, F. André, E.P. Winer, W. Janni, S. Verma, P. Conte, C.L. Arteaga, D.A. Cameron, K. Petrakova, L.L. Hart, C. Villanueva, A. Chan, J. O'Shaughnessy, Ribociclib as first-line therapy for HR-positive, advanced breast Cancer, N. Engl. J. Med. 375 (18) (2016) 1738–1748, https://doi.org/10.1056/nejmoa1609709.
[33] S.S. Bashraheel, A. Domling, S.K. Goda, Update on targeted cancer therapies, single or in combination, and their fine tuning for precision medicine, Biomed. Pharmacother. 125 (February) (2020), https://doi.org/10.1016/j.biopha.2020.110009, 110009.
[34] A. Jeanson, P. Tomasini, M. Souquet-Bressand, N. Brandone, M. Boucekine, M. Grangeon, S. Chaleat, N. Khobta, J. Milia, L. Mhanna, L. Greillier, J. Biemar, I. Nanni, L. Ouafik, S. Garcia, J. Mazières, F. Barlesi, C. Mascaux, Efficacy of immune checkpoint inhibitors in KRAS-mutant non-small cell lung Cancer (NSCLC), J. Thorac. Oncol. 14 (6) (2019) 1095–1101, https://doi.org/10.1016/j.jtho.2019.01.011.
[35] J. Mazieres, A. Drilon, A. Lusque, L. Mhanna, A.B. Cortot, L. Mezquita, A.A. Thai, C. Mascaux, S. Couraud, R. Veillon, M. Van Den Heuvel, J. Neal, N. Peled, M. Früh, T.L. Ng, V. Gounant, S. Popat, J. Diebold, J. Sabari, O. Gautschi, Immune checkpoint inhibitors for patients with advanced lung cancer and oncogenic driver alterations: results from the IMMUNOTARGET registry, Ann. Oncol. 30 (8) (2019) 1321–1328, https://doi.org/10.1093/annonc/mdz167.
[36] J.E. Groopman, D.T. Scadden, Interferon therapy for Kaposi sarcoma associated with the acquired immunodeficiency syndrome (AIDS), Ann. Intern. Med. 110 (5) (1989) 335–337, https://doi.org/10.7326/0003-4819-110-5-335.
[37] P. Solal-Celigny, E. Lepage, N. Brousse, F. Reyes, C. Haioun, M. Leporrier, M. Peuchmaur, A. Bosly, Y. Parlier, P. Brice, B. Coiffier, C. Gisselbrecht, Recombinant interferon alfa-2b combined with a regimen containing doxorubicin in patients with advanced follicular lymphoma, N. Engl. J. Med. 329 (22) (1993) 1608–1614, https://doi.org/10.1056/nejm199311253292203.
[38] I.V. Pinchuk, E.J. Beswick, V.E. Reyes, Staphylococcal enterotoxins, Toxins 2 (8) (2010) 2177–2197, https://doi.org/10.3390/toxins2082177.
[39] X. Liu, M. Ling, C. Chen, F. Luo, P. Yang, D. Wang, X. Chen, H. Xu, J. Xue, Q. Yang, L. Lu, J. Lu, Q. Bian, A. Zhang, Q. Liu, Impaired autophagic flux and p62-mediated EMT are involved in arsenite-induced transformation of L-02 cells, Toxicol. Appl. Pharmacol. 334 (2017) 75–87, https://doi.org/10.1016/j.taap.2017.09.004.
[40] Y. Zou, D. Ma, Y. Wang, The PROTAC technology in drug development, Cell Biochem. Funct. 37 (1) (2019) 21–30, https://doi.org/10.1002/cbf.3369.
[41] C. Zhang, X.R. Han, X. Yang, B. Jiang, J. Liu, Y. Xiong, J. Jin, Proteolysis targeting chimeras (PROTACs) of anaplastic lymphoma kinase (ALK), Eur. J. Med. Chem. 151 (2018) 304–314, https://doi.org/10.1016/j.ejmech.2018.03.071.
[42] S.L. Kitson, D.J. Quinn, T.S. Moody, D. Speed, W. Watters, D. Rozzell, Antibody-drug conjugates (ADCs) – biotherapeutic bullets, Chemistry Today 31 (August) (2013) 30–36.
[43] N. Pondé, P. Aftimos, M. Piccart, Antibody-drug conjugates in breast Cancer: a comprehensive review, Curr. Treat. Options in Oncol. 20 (5) (2019), https://doi.org/10.1007/s11864-019-0633-6.

[44] R.F. Sherwood, A cytotoxic agent sites, Cancer (1988) 700–703.

[45] R.F. Sherwood, R.G. Melton, S.M. Alwan, P. Hughes, Purification and properties of carboxypeptidase G2 from Pseudomonas sp. strain RS-16: use of a novel triazine dye affinity method, Eur. J. Biochem. 148 (3) (1985) 447–453, https://doi.org/10.1111/j.1432-1033.1985.tb08860.x.

[46] F. Din, ud, W, Aman, A., Ullah, I., Qureshi, O. S., Mustapha, O., Shafique, S., & Zeb, A., Effective use of nanocarriers as drug delivery systems for the treatment of selected tumors, Int. J. Nanomedicine 12 (2017) 7291–7309.

[47] H. Kobayashi, R. Watanabe, P.L. Choyke, Improving conventional enhanced permeability and retention (EPR) effects; what is the appropriate target? Theranostics 4 (1) (2014) 81–89, https://doi.org/10.7150/thno.7193.

[48] F.F. An, X.H. Zhang, Strategies for preparing albumin-based nanoparticles for multifunctional bioimaging and drug delivery, Theranostics 7 (15) (2017) 3667–3689, https://doi.org/10.7150/thno.19365.

[49] A. Sharma, A.K. Goyal, G. Rath, Recent advances in metal nanoparticles in cancer therapy, J. Drug Target. 26 (8) (2018) 617–632, https://doi.org/10.1080/1061186X.2017.1400553.

[50] B. García-Pinel, C. Porras-Alcalá, A. Ortega-Rodríguez, F. Sarabia, J. Prados, C. Melguizo, J.M. López-Romero, Lipid-based nanoparticles: application and recent advances in cancer treatment, Nano 9 (4) (2019) 1–23, https://doi.org/10.3390/nano9040638.

[51] A.C. Palmer, P.K. Sorger, Combination Cancer therapy can confer benefit via patient-to-patient variability without drug additivity or synergy, Cell 171 (7) (2017) 1678–1691.e13, https://doi.org/10.1016/j.cell.2017.11.009.

[52] P. Dey, M. Rathod, A. De, Targeting stem cells in the realm of drug-resistant breast cancer, Breast Cancer 11 (2019) 115–135, https://doi.org/10.2147/BCTT.S189224.

[53] G. Manda, G. Isvoranu, M.V. Comanescu, A. Manea, B. Debelec Butuner, K.S. Korkmaz, The redox biology network in cancer pathophysiology and therapeutics, Redox Biol. 5 (2015) 347–357, https://doi.org/10.1016/j.redox.2015.06.014.

[54] B.Y.C.F. Nathan, Z.A. Cohn, Effects of Hydrogen Peroxide In Vivo, vol. 10021. 154(November), From The Rockefeller University, New York, 1981, pp. 1539–1553.

[55] S. Afzal, S.A. Jensen, J.B. Sørensen, T. Henriksen, A. Weimann, H.E. Poulsen, Oxidative damage to guanine nucleosides following combination chemotherapy with 5-fluorouracil and oxaliplatin, Cancer Chemother. Pharmacol. 69 (2) (2012) 301–307, https://doi.org/10.1007/s00280-011-1700-2.

[56] J. Neuzil, L.F. Dong, J. Rohlena, J. Truksa, S.J. Ralph, Classification of mitocans, anticancer drugs acting on mitochondria, Mitochondrion 13 (3) (2013) 199–208, https://doi.org/10.1016/j.mito.2012.07.112.

[57] A. Nomura, P. Dauer, V. Gupta, O. McGinn, N. Arora, K. Majumdar, C. Uhlrich, J. Dalluge, V. Dudeja, A. Saluja, S. Banerjee, Microenvironment mediated alterations to metabolic pathways confer increased chemo-resistance in CD133+ tumor initiating cells, Oncotarget 7 (35) (2016) 56324–56337, https://doi.org/10.18632/oncotarget.10838.

[58] K. Nath, D.S. Nelson, D.F. Heitjan, D.B. Leeper, R. Zhou, J.D. Glickson, Lonidamine induces intracellular tumor acidification and ATP depletion in breast, prostate and ovarian cancer xenografts and potentiates response to doxorubicin, NMR Biomed. 28 (3) (2015) 281–290, https://doi.org/10.1002/nbm.3240.

[59] D. Zhang, J. Li, F. Wang, J. Hu, S. Wang, Y. Sun, 2-deoxy-D-glucose targeting of glucose metabolism in cancer cells as a potential therapy, Cancer Lett. 355 (2) (2014) 176–183, https://doi.org/10.1016/j.canlet.2014.09.003.

[60] M. Hassanein, J. Qian, M.D. Hoeksema, J. Wang, M. Jacobovitz, X. Ji, F.T. Harris, B.K. Harris, K.L. Boyd, H. Chen, R. Eisenberg, P.P. Massion, Targeting SLC1a5-mediated glutamine dependence in non-small cell lung cancer, Int. J. Cancer 137 (7) (2015) 1587–1597, https://doi.org/10.1002/ijc.29535.

[61] L. Jin, G.N. Alesi, S. Kang, Glutaminolysis as a target for cancer therapy, Oncogene 35 (28) (2016) 3619–3625, https://doi.org/10.1038/onc.2015.447.
[62] L. Frattaruolo, M. Brindisi, R. Curcio, F. Marra, V. Dolce, A.R. Cappello, Targeting the mitochondrial metabolic network: a promising strategy in cancer treatment, Int. J. Mol. Sci. 21 (17) (2020) 1–21, https://doi.org/10.3390/ijms21176014.
[63] S.A. Swenson, C.M. Moore, J.R. Marcero, A.E. Medlock, A.R. Reddi, O. Khalimonchuk, From synthesis to utilization: the ins and outs of mitochondrial heme, Cell 9 (3) (2020) 10–19, https://doi.org/10.3390/cells9030579.
[64] P. Ghosh, C. Vidal, S. Dey, L. Zhang, Mitochondria targeting as an effective strategy for cancer therapy, Int. J. Mol. Sci. 21 (9) (2020) 1–19, https://doi.org/10.3390/ijms21093363.
[65] S. Sohoni, P. Ghosh, T. Wang, S.P. Kalainayakan, C. Vidal, S. Dey, P.C. Konduri, L. Zhang, Elevated heme synthesis and uptake underpin intensified oxidative metabolism and tumorigenic functions in non-small cell lung cancer cells, Cancer Res. 79 (10) (2019) 2511–2525, https://doi.org/10.1158/0008-5472.CAN-18-2156.
[66] M.M. Alam, S. Sohoni, S.P. Kalainayakan, M. Garrossian, L. Zhang, Cyclopamine tartrate, an inhibitor of hedgehog signaling, strongly interferes with mitochondrial function and suppresses aerobic respiration in lung cancer cells, BMC Cancer 16 (1) (2016) 1–10, https://doi.org/10.1186/s12885-016-2200-x.

CHAPTER

11

Advancements in novel drug delivery systems: Providing a challenge and threat to oxidative stress in various diseases

Sumira Malik[a], *Shilpa Prasad*[a], *and Archna Dhasmana*[b]

[a]Amity Institute of Biotechnology, Amity University Jharkhand, Ranchi, Jharkhand, India, [b]Department of Biotechnology, School of Applied and Life Sciences, Uttaranchal University Dehradun, Uttarakhand, India

1 Introduction

Drug delivery systems are comprised of a pathway that can deliver a particular drug to its target efficiently. Each drug is delivered via a specific delivery system and the effect of that drug in treatment of disease specifically depends on the efficacy of that delivery system. To date, various advancements have been made in the field of drug research as well as delivery, but progress is still required to improve novel drug delivery systems. Novel drug delivery has proved beneficial in dealing with oxidative stress-related issues. Oxidative stress is a condition led by overproduction of reactive oxygen species (ROS) and shifting of the balance between oxidant and anti-oxidant statuses [1]. ROS are produced by various metabolic activities that happen at cellular level, as well as due to several environmental factors, e.g., air pollutants, cigarette smoke. As the name suggests, ROS are highly reactive molecules as they are composed of unpaired electrons that enable them to react with several biological macromolecules present in cells, such as carbohydrates, proteins, lipids, and nucleic acids,

https://doi.org/10.1016/B978-0-323-90905-1.00011-0

thereby altering their functions. ROS can also further affect the expression of genes as they can upregulate redox-sensitive transcription factors and chromatin remodeling.

An appropriate level of ROS is indispensable for cell viability, activation, proliferation, and function [2]. ROS works as a double-edged sword as it can be beneficial in modulating various physiological functions that are important for an organism, but overproduction of ROS is associated with various diseases such as atherosclerosis, neurodegenerative diseases like Alzheimer's and Parkinson's, cancer, diabetes mellitus, and inflammatory diseases; it is also associated with accelerated aging processes [1,3]. The deleterious effects of ROS can be controlled through accurate administration of anti-oxidants [1]. Conventional anti-oxidative agents have shown positive effects in the treatment of oxidative stress, but in recent decades, significant achievements have been made in the development of antioxidant nano therapies to enhance the outcome of this novel approach, and these have shown multiple advantages in this field, such as such as pharmacokinetics, stable anti-oxidative activity, and intrinsic ROS-scavenging properties [3].

Anti-oxidant supplementation is one of the novel strategies to treat oxidative stress and other pathological conditions associated with it, as it can detoxify the reactive intermediates and can also repair the damage caused by toxic effects of peroxides and free radicals [1,3]. Anti-oxidants can be natural or synthetic depending on their occurrence, as shown in Fig. 1 [4,5]. Natural anti-oxidants are present in various dietary sources such as different fruits and vegetables and possess the ability to prevent and/or detain specific types of cell damage [6]. Certain commonly used anti-oxidants include beta-vitamin C, vitamin A, vitamin E, carotene, lycopene, lutein, selenium, alpha-tocopherol, ascorbate, anthocyanin, cysteine, carotenoid, curcumin, flavonoid, flavonol, hydrogen peroxide, kaempferol, uric acid, *N*-acetylcysteine, phenolic acid, nitric oxide, and

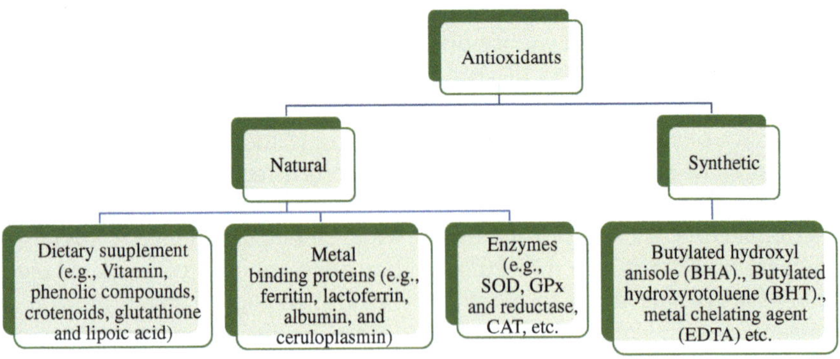

FIG. 1 Anti-oxidants classifications.

resveratrol [6]. Polyphenols constitute a heterogeneous sub-group of phenolic compounds, including flavonoids, which serve as a group of natural antioxidants [7,8].

Antioxidants that are derived from plant extracts are mostly used in supplements, and phyto-bioactive compounds such as curcumin, resveratrol, catechins, and quercetin are natural compounds used as potential modulators to improve cellular damages caused by oxidative stress [9]. The low bioavailability of natural anti-oxidants limits their successful therapeutic activity at the targeted site and causes poor solubility, stability, and absorbance of anti-oxidants in the gastrointestinal tract [9,10]. Delivery of drugs to targeted cells in oxidative microenvironments effectively, together with proper anti-oxidant activity, can be obtained with advancements in drug delivery processes. The incorporation of anti-oxidant compounds within implanted biomaterials should make it possible to retain the anti-oxidant activity [10]. The developments in nanotechnology-based systems to develop bioengineered anti-oxidants have been found beneficial for development of novel therapeutic approaches over the conventional preparations in different parameters such as superior solubility and stability, extended half-life, improved epithelium permeability and bioavailability, enhanced tissue targeting, and minimized side effects [9]. The recently developed bioengineered anti-oxidant poly (ethylene glycol)-functionalized hydrophilic carbon clusters (PEG-HCCs) have been reported to neutralize ROS superoxide anions and hydroxyl radicals. A perfectly designed novel drug delivery system can be an effective approach to solve the problems related to the release of the drug at a specific site at a specific rate. The present review summarizes recent advances in drug delivery systems with anti-oxidants to combat oxidative stress issues in various diseases.

2 Anti-oxidants working principle and pathways

Biological systems always experience oxidative stress, and have developed defense mechanisms, both physiological and biochemical, to manage it. Anti-oxidants work as the first line of defense through different pathways against the ill effects of oxidative stress. The physiological level maintains the levels of O_2 in the tissues, and the biochemical level comprises the anti-oxidant enzymatic or non-enzymatic defense mechanism [11]. Physiologically, ROS generated by different metabolic processes works as a signaling messenger, but it also shows certain deleterious effect depending on its level in an individual. To maintain a proper level of ROS and prevent oxidative stress, removal of ROS is required by using supplementation of anti-oxidants [12,13]. There are different types of ROS free radicals that can increase oxidative stress in biological systems, but O_2^- is the most active oxidizing agent [11].

Anti-oxidants can be classified as direct or indirect, depending on pathways. Direct anti-oxidants are those that are able to neutralize ROS through simple and direct reactions, and indirect anti-oxidants work as activators of certain cell components, i.e., the Kelch-like ECH-associated protein 1 (Keap1)/nuclear factor erythroid 2-related factor 2 (Nrf2)/anti-oxidant response element (ARE) pathway, which is responsible for inducing phase 2 enzymes and elevating the cytoprotective defenses of the cell [14]. Direct anti-oxidants can regenerate with respective activity but, on the other hand, indirect anti-oxidants can or cannot be redox active, enabling a pro-oxidant effect [15]. Anti-oxidant insufficiency can be developed because of decreased anti-oxidant intake, synthesis of endogenous enzymes, or increased anti-oxidant utilization. Different types of anti-oxidants are available that may be generated through endogenous body cells or exogenous agents such as dietary supplements.

There are various dietary anti-oxidants that can be obtained from citrus fruits, vegetables, etc., but they have two advantages, i.e., bioavailability and stability. During in vivo delivery, reactivity of anti-oxidants is an important issue as they can get attracted toward their opposite ROS and control oxidative stress. This can lead to low bioavailability and stability of anti-oxidants, making it difficult to store them without instant oxidation [16]. Natural anti-oxidants that are orally administered can get activated only after enzymatic cleavage by human microbiota and their bioavailability primarily depends on dietary intake, digestion, gut absorption, metabolism, and clearance [10]. It is not possible for the active form of these anti-oxidants to be delivered followed by implantation to the specific tissue target due to its intrinsic physicochemical properties [10]. To overcome the instability of a single anti-oxidant, multiple anti-oxidants can be developed using novel approaches to form a complex network as this synergizes their effect in terms of cytoprotection [10], regeneration [17], and healing [18]. The limitations for single anti-oxidant treatment pave the way for a new strategy of bioengineering to be accompanied with natural as well as synthetic anti-oxidants, thus enhancing stability and bioavailability.

The cellular level of free radicals may be decreased by anti-oxidants, either by inhibiting the activities or expression of free radical generating enzymes such as NAD(*P*)H oxidase and xanthine oxidase (XO), and by promoting the activities and expression of anti-oxidant enzymes such as superoxide dismutase (SOD), catalase (CAT), and glutathione peroxidase (GPx) [19]. Anti-oxidant classification can be done in various ways based on different criteria [20,21]. Based on activity, anti-oxidants can be classified as enzymatic and non-enzymatic anti-oxidants and, depending on their solubility, they can be classified as water-soluble or lipid-soluble anti-oxidants. Vitamin C is a type of water-soluble vitamin found in cellular fluids such as cytosol or cytoplasmic matrix. Small- or large-molecule anti-oxidants are classified based on their respective size. The small-molecule anti-oxidants neutralize the ROS in a process named radicals scavenging and carry them away. Vitamin C, vitamin E, carotenoids, and glutathione (GSH) are the main anti-oxidants in this category. Large-molecule anti-oxidants include enzymes (SOD, CAT, and GPx) and proteins such as albumin that can absorb ROS and prevent initiation of oxidative stress.

Kinetically active anti-oxidants can be categorized based on the different pathways that they follow. Anti-oxidants like phenol, naphthol, hydroquinone, aromatic amines, and aminophenols that can break chains through reacting with peroxyl radicals containing weak O–H or N–H bonds, anti-oxidants with a capability to break chains by reacting with alkyl radicals such as quinines, nitrones, and iminoquinones. Another group are anti-oxidants that terminate cyclic chains such as aromatic amines, nitroxyl radicals, and variable valence metal compounds, as well as those anti-oxidants that are able to decompose hydroperoxide. Anti-oxidants such as sulfide, phosphide, and thiophosphate, and metal-deactivating anti-oxidants, include diamines, hydroxyl acids, and bifunctional compounds.

Chain-breaking natural anti-oxidants can react with radicals and convert them into more stable products. They are phenolic in structure and include anti-oxidant minerals, i.e., cofactors (selenium, copper, iron, zinc, and manganese), vitamins (C, E, and B), phytochemicals (phenolic compounds derivatives such as flavonoids, catechins, carotenoids, carotene, and lycopene), as well as herbs and spices (diterpene, rosmariquinone, thyme, nutmeg, clove, black pepper, ginger, garlic, and curcumin). Synthetic anti-oxidants are compounds that include mainly butylated hydroxyl anisole (BHA), butylated hydroxyltoluene (BHT), propyl gallate (PG), metal chelating agent (EDTA), tertiary butyl hydroquinone (TBHQ), and nordihydroguaiaretic acid (NDGA). These compounds carry out the role of scavenging free radicals and stopping the chain reaction shown in Fig. 1 [19].

In biological systems, if a cell gets damaged by oxidative stress and ROS then there are also certain enzymes constituting the repair systems that are able to repair or remove the biomolecules, such as lipids, proteins, and DNA, that have been damaged due to oxidative stress. These enzymes mainly include DNA repair enzymes such as polymerases, glycosylases, and nucleases, and proteolytic enzymes such as proteinases, proteases, and peptidases. They are found in both the cytosol and the mitochondria of mammalian cells and act as intermediaries in the repair process against oxidative stress [11].

3 Types of anti-oxidants and their mode of action

In our body, an excessive amount of free radical or ROS generation in different metabolic pathways such as aerobic oxidation, pathogenic degradation, cytochrome P450 enzymes, and peroxisomes activity produces an oxidative burst and releases oxide (NO), O_2^-, H_2O_2, and OCl. Thus, the oxidative stress induced in the cells can be prevented by the biological anti-oxidant defense system by neutralizing the radicals or oxidant activity including enzymes, and metal chelating and free radical scavenging molecules. The dietary anti-oxidants have a synergistic effect along with the cells' own anti-oxidants and thereby maintain the anti-oxidant level in the body.

Free radicals contribute to many different diseases. In a biological system, production of energy molecule ATP (adenosine triphosphate) by electron transport chain (ETC) in the mitochondria results in generation of free radical formation. This will result in the formation of ROS (reactive oxygen species) such as superoxide anion (O_2^-) and hydroxyl radicals (HO.) and singlet oxygen (O_{21}), which results in degeneration of macromolecules and cell aging. In addition, harmful radiation, such as UV (ultraviolet) light, and inhaled air pollutants also generate free radicals.

In our dietary sources, a wide variety of anti-oxidants are present, such as phytonutrients, vitamins A, E, and C, quercetin, flavonoids, etc. Besides these, different enzymes present in a biological system have anti-oxidant activity. They all either prevent the oxidant of macromolecules or block the chain-breaking pathway from occurring due to the generation of free radicals in the system [6]. Natural components such as vitamins, flavonoids, and β-carotene scavenge the free radicals to block the oxidation chain reaction as well as stabilize the transition of metal radicals.

Broadly, the mode of action of anti-oxidant involves (i) the inhibition of free alkyl radicals generation, (ii) intruding on the free radical chain generating molecules, (iii) slowing the degeneration or oxidation of lipids, and (iv) donation of free elections and hydrogen and formation of metal chelates [22].

Anti-oxidants are classified as any bio-molecules, metal ions, or peptides that at very low concentration and suppress the oxidants activity and act as oxidation pathway chain-inhibitors. Therefore, anti-oxidants can be classified (i) based on origin, either natural (primary) or synthetic (secondary), and (ii) by chemical nature, i.e., enzymatic or non-enzymatic [23].

The natural anti-oxidants are broadly classified into two categories based on their activity: enzymatic and non-enzymatic. In the case of enzymatic pathways, different enzymes or co-factors metabolically scavenge the free radicals or degenerate oxidants [24]. However, the non-enzymatic anti-oxidants involve vitamins, phytochemicals such as lipoic acid, polyphenols, and carotenoids that are derived from dietary sources and produced in minute amounts by the cells. They may act directly or indirectly as chelating agents by binding with the redox metals to stop free radical generation.

There are several types of chemically synthesized components, e.g., butylated hydroxytoluene (BHT) and butylated hydroxy anisole (BHA), called synthetic anti-oxidants, which are traditionally used as food preservatives. It was reported that in small amounts they have anti-oxidative properties, but high consumption results in chronic toxicity and tumor formation (Table 1) [25].

TABLE 1 Types of anti-oxidants and their modes of action.

Anti-oxidant type	Sources	Mode of action
Natural		
Vitamin A (*retinoids*)	Liver, sweet potatoes, carrots, milk, egg yolks, and mozzarella cheese	Interact with free radicals and prohibit cell lipid peroxidation
Vitamin C (*ascorbic acid*)	Citrus fruits and vegetables such as cabbage, broccoli, peas, tomatoes, potatoes and sprouts	Acts as electron donor and chelating agent, stabilizes and excretes folate, enhances non-haem iron absorption, prevents the formation of mutagenic *N*-nitroso compounds, and protects low-density lipoproteins oxidation
Vitamin E (*tocopherol (Toc) and tocotrienol (Toc-3) derivatives*)	Tocopherols: polyunsaturated vegetable oils, the cereal seeds germ Tocotrienols: aleurone and subaleurone layers of cereal seeds and in palm oils	α-Toc induced lipid peroxidation in microsomes, α-Toc-3 peroxyl radical scavenging potency, reacts with fatty acid peroxyl radicals in liposomal membranes to form stable radicals
β-Carotene	Fruits and vegetables	Reacts with oxygen in the interception of the autoxidation radical chain process, which is not perpetuated by oxygen but by the fatty acid that quenches singlet oxygen double bonds to donate electron density
Phenolic compounds (*volatile oils, phenolic acids, diterpenes flavonoids*)	Vegetables, fruits	Traps free radicals, metal chelates, and scavenge ROS due to the presence of side chain OH, R groups
Selenium (*selenoproteins or selenoenzymes*)	Seafood, meat, eggs, grains	Acts as co-factor with antioxidant enzymes, such as glutathione peroxidase (GPx), thioredoxinreductase (TrxR), and iodothyroninedeiodinases (IDD) to inhibit the free radicals and ROS generation
Superoxide dismutase (SOD)	Cabbage, Brussels sprouts, wheat grass, barley grass, and broccoli	Donates electrons, inhibits free radicals (N_3^-, CN^-, superoxide (O_2^-), produces the stable and less toxic components hydrogen peroxide (H_2O_2) and oxygen (O_2)
Catalase (CAT)	Animal tissue	Degrades the H_2O_2 to water and oxygen, protects from oxidative stress.

Continued

TABLE 1 Types of anti-oxidants and their modes of action—cont'd

Anti-oxidant type	Sources	Mode of action
Glutathione peroxidase (GPx)	Isoenzymes found in mammals	Prevents free radicals synthesis of hydrogen and lipid peroxides, detoxification of endogenous compounds
Glutathione reductase (GR)		Maintains intracellular glutathione (GSH) level and prevents oxidative damage
Synthetic		
Butylated hydroxytoluene (BHT)		Used as primary food antioxidant preservative, inhibits the free radical formation, suppresses autoxidation and autocatalytic reaction by converting peroxy radicals to hydroperoxides
Butylated hydroxyanisole (BHA)		Scavenges and stabilizes the free radicals and prevents rancidity of lipids

Natural antioxidants release free electrons either from hydroxyl groups or by double bonds by chemical reaction and prevent oxidative damage. Thus, the consumption of these natural anti-oxidants suppresses free radicals, cell-aging, and cancer formation, and prevents cardiovascular, neurodegenerative, and several other metabolic diseases.

4 Traditional approaches

Traditionally, Ayurveda herbal plant extract is used as medicine to cure different diseases. Different types of fruits, vegetables, and medicinal herbs enriched with different types of compounds have anti-oxidant properties, such as endogenous metabolites, vitamins, phenolic acids, flavonoids, quinones, coumarins, lignans, stilbenes, tannins alkaloids, amines, betalains, terpenoids, etc. [26]. All these components, along with anti-oxidative properties, have anti-inflammatory, anti-atherosclerotic, anti-tumor, anti-mutagenic, anti-carcinogenic, anti-bacterial, and anti-viral activities [27].

Therefore, these sources are used both as dietary supplements and food preservatives to prevent oxidative damage to cells and food items. They may be taken directly in food items or a bio-chemical extract collected by different chemical reactions for enhanced activity. Many studies reported that the anti-oxidants modulate the cellular pathways and regulate gene expression, which may result in protective shielding of the immune system.

The natural sources having anti-oxidative properties can potentially be used as dietary, nutritious, and healing agents to suppress the effects of free radical-generated cellular deformities. A wide-range of plant extracts are either consumed directly or purified to obtain phenolic therapeutic compounds. In the ancient literature, different herbal formulations, such as Ayurveda and charaka samhita, are used for the treatment of chronic diseases caused by free radical formation [28].

In food and beverages industries, different types of natural (ascorbic acid and tocopherols) and synthetic anti-oxidants (BHA and BHT) are used for the long-term preservation and storage of food items. There are different cosmetic products containing fat compounds that have anti-oxidant preservatives to prevent rancidity.

Different types of phytochemicals are used as therapeutic or chemoprotective agents to protect the cell from oxidative stress. Vitamin E blocks the apoptosis by controlling the cancer gene expression and delays the cell aging process by inducing faster metabolic lipid degradation [29]. Another dietary supplement, lycopene, found in red vegetables and fruits such as tomatoes, acts as a potential anti-cancer agent, having chemo-protective properties against the attack of (ROS)-mediated chronic diseases such as cardiovascular diseases [30]. Lipoic acid in combination with other anti-oxidants such as glutathione, coenzyme Q10, vitamin C, and vitamin E provides a protective shield to the cell against free radicals and is used for the treatment of aging stroke, heart attack, cataracts, and dementia, such as Alzheimer's disease [31].

Recently, it has been found that certain microbial species such as actinomycetes, bacteria algae, fungi, and lichens, release secondary metabolites having potential anti-oxidative properties [32]. They are also rich in micro-bioactive agents potentially applicable for industrial purposes.

The health benefits and lower toxicity of natural anti-oxidants play a significant role. However, the synthetic anti-oxidants' effectiveness in small amounts makes them applicable, but the negative impacts of high doses results in major safety issues. Although the natural anti-oxidants are less toxic and bio-compatible, their direct intake through traditional methods results in low efficiency, degradation, loss of activity in different biological systems, and limited shelf life, which limits their applicability. Therefore, there is a need for better drug delivery agents for better efficiency and effect.

5 Innovative approaches in nano delivery of natural and synthetic anti-oxidants

The dietary anti-oxidants are not in favor due to their low solubility, bioavailability, and stability in the fluids of the human body [9]. These problems are addressed through the development of new technology-based innovative approaches. The nanotechnological innovative approach helps in enhancing bioavailability and therapeutic effectiveness of natural and synthetic anti-oxidants. The recent advancements in the application of natural and synthetic anti-oxidant-based nano delivery systems may aid in regulation of several diseases [33]. Many diseases that are generated through degenerated ROS system or atherosclerosis are treated by anti-oxidants; therefore, dietary supplements are surface substituents of synthetic oxidants [34,35].

5.1 Natural antioxidants

The phytochemicals like curcumins, resveratrol, catechines, and genistein regenerate inflammation and stress related to oxidative situations [36,37]. The phytotherapeutic compounds are discussed in Table 2.

5.2 Synthetic antioxidants

The vitamins such as vitamin A, E, and beta carotene synthetic anti-oxidants are potentially lethal and may aid in progression of death rate in a healthy population [39]. Furthermore, many clinical trials have also reported failure in treatment of cardiovascular disease and inflammation [40,41]. The nanostructures can be highly active or passive for the targeted therapeutic delivery [42]. The different systems involved in therapeutic nano delivery systems are extended half-life, increased bioavailability and solubility, stability, and permeability, with enhanced specific tissue targeting and low side effects [43]. Various nano drug delivery systems provide health benefits, which depend on desired therapeutic application, agents to be loaded, and different chemical properties [44].

The details of different nano drug delivery systems involved in therapeutic applications are explained in Table 2.

6 Limitations and future perspectives of novel drug delivery systems at clinical translation level

There are several challenges in the emerging field of therapeutics-based nanotechnology that are found to be associated with uncertain possibilities. They have been observed without threats but does not provide promising results. This has been shown in Fig. 2. Even though these nanoparticles in delivery system have specific functions with specific ligands for successful drug targeting, still the challenges of limited drug delivery and loading, aggregation of particles, and high clearance rate in diagnosis exist. Furthermore, factors such as biodegradability, biocompatibility, safety, and delivery of nano-drugs are also critical issues to be addressed, which

TABLE 2 Different nano drug delivery systems involved in therapeutic applications.

S. No.	Nano drug delivery system	Size	State	Disease treatment	Advantage	Reference
1	Polymeric NPs	10–1000nm	Solid colloidal	Osteoarthritis, SCID, multiple sclerosis, prostate cancer, hepatitis B and C, acromegaly, rheumatoid arthritis, Crohn's disease	Increased immunogenicity, increased stability in vivo, high protein stability, extended release, extended circulation time, longer half-life, controlled clearance	[9,38]
2	Nanocrystal NPs	10–800nm	Solid	Hyper lipedema, bone substitute, muscle relaxant, psychostimulant, schizophrenia, immunosuppressant	Mimic bone structure, slow release, high drug loading, simple administration, increased bioavailability	
3	Solid liquid NPs	50–1000nm	Solid	Brain tumor, HIV encephalitis	Increased permeability, increased surface area, increased loading efficiency, low toxicity, large-scale production	
4	Nano liposomes	50–450nm	Liquid	Pancreatic cancer, respiratory syndrome, respiratory distress, ovarian cancer, fungal and protozoan syndrome	Increased systemic toxicity, increased delivery to diseased site, decreased toxicity	
5	Nano emulsions	20–200nm	Liquid	CNS disorder, neurological disorder	Improved oral bioavailability	
6	Nano polymerosome	1–100nm	Liquid	Tissue cancer	Increased oral bioavailability, increased anti-tumorigenic tendency, longer half-life	
7	Protein NPs	NA	Solid	Cutaneous lymphoma, breast and pancreatic cancer	Increased delivery to tumor site, increased solubility, targeted cell therapy	
8	Inorganic NPs	NA	Solid	Iron deficiency CKD	Prolonged release	
9	Dendrimers	100nm	Solid	Multiple drug molecules, diagnostic agents, and/or targeting ligands	Increased distribution, increased retention at site of inflammation, long-lasting systemic exposure	

Continued

TABLE 2 Different nano drug delivery systems involved in therapeutic applications—cont'd

S. No.	Nano drug delivery system	Size	State	Disease treatment	Advantage	Reference
10	Block copolymer micelles	NA	Liquid	Vasomotor disorder, neurological disorder	Increased bioavailability, increased systemic exposure	
11	Metallic NPs	1–100 nm	Liquid	Biomedical application	Biocompatibility, easy chemical modification, tunable biophysical character	
12	Nano resveratrol	103–304 nm, 206–217 nm	Liquid	Type 2 disease, obesity, chronic kidney, dementia, colorectal cancer	Increased bioavailability and anti-oxidative potential, decreased toxicity	
	Nano curcumin	200 nm	Liquid	Anti-cancer, anti-inflammatory, anti-neurodegenerative	Increased bioavailability and water solubility	
	Nano quercetin	180 nm	Liquid	Anti-diabetic	Increased bioavailability	
	Nano genistein	33 nm	Liquid	Anti-atherogenic, type 2 diabetes, anti-cancer	Increased bioavailability	
	Nano epigallocatechin 3 galate	100 nm	Liquid	Anti-tumor, CNS treatment	Increased bioavailability and stability	

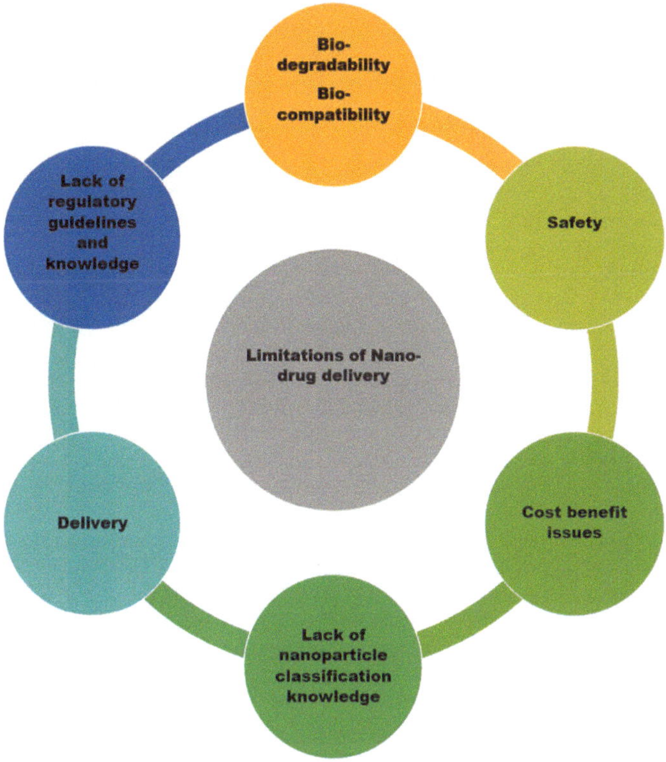

FIG. 2 Limitations of nano drug delivery systems.

invite future research in successful clinical translation of novel drug delivery systems from bench to bedside. For the future perspective, factors such as biodegradability, non-toxicity, newer non-invasive drug-administration routes, and an eco-friendly approach, and factors like charge, moiety, size, and shape attached to nanomaterials should be well explicated and evaluated. This is essential for the development of novel targeted drug delivery systems in the future.

7 Summary

The ongoing development of the nano-based drug delivery approaches is mostly focused on the formation of new approaches for drug delivery systems. These nanoparticles carriers-based nano delivery systems have enabled targeted drug delivery in in vivo systems and in vitro systems, and could offer promising successful treatment of different diseases such as cancer, neurological disorders, and aging disorders.

References

[1] S.M. Al-Dalaen, A.I. Al-Qtaitat, Review article: oxidative stress versus antioxidants, Am. J. Biosci. Bioeng. 2 (5) (2014) 60–71, https://doi.org/10.11648/j.bio.20140205.11.

[2] E. Birben, U.M. Sahiner, C. Sackesen, S. Erzurum, O. Kalayci, Oxidative stress and antioxidant defense, World Allergy Organ. J. 5 (1) (2012) 9–19, https://doi.org/10.1097/WOX.0b013e3182439613.

[3] C.W. Li, L.L. Li, S. Chen, J.X. Zhang, W.L. Lu, Antioxidant nanotherapies for the treatment of inflammatory diseases, Front. Bioeng. Biotechnol. 8 (2020) 200, https://doi.org/10.3389/fbioe.2020.00200.

[4] R.F. Hurrell, Influences of vegetable protein sources on trace element and mineral bioavailability, J. Nutr. 133 (9) (2003) 2973S–2977S, https://doi.org/10.1093/jn/133.9.2973S.

[5] B.B. Mathew, A. Tiwari, S.K. Jatawa, Free radicals and antioxidants: a review, J. Pharm. Res. 4 (12) (2011) 4340–4343.

[6] A.W.K. Yeung, N.T. Tzvetkov, O.S.E. Tawil, S.G. Bungău, M.M. Abdel-Daim, A.G. Atanasov, Antioxidants: scientific literature landscape analysis, in: Oxidative Medicine and Cellular Longevity, 2019, https://doi.org/10.1155/2019/8278454. 11 p.

[7] C.G. Fraga, K.D. Croft, D.O. Kennedy, F.A. Tomás-Barberán, The effects of polyphenols and other bioactives on human health, Food Funct. 10 (2) (2019) 514–528, https://doi.org/10.1039/c8fo01997e.

[8] M. Malaguti, C. Angeloni, S. Hrelia, Polyphenols in exercise performance and prevention of exercise-induced muscle damage, Oxidative Med. Cell. Longev. 2013 (2013) 825928, https://doi.org/10.1155/2013/825928.

[9] A. Vaiserman, A. Koliada, A. Zayachkivska, O. Lushchak, Nanodelivery of natural antioxidants: an anti-aging perspective, Front. Bioeng. Biotechnol. 7 (2020) 447, https://doi.org/10.3389/fbioe.2019.00447.

[10] P. Marrazzo, C. O'Leary, Repositioning natural antioxidants for therapeutic applications in tissue engineering, Bioengineering 7 (3) (2020) 104, https://doi.org/10.3390/bioengineering7030104.

[11] N.F. Santos-Sanchez, R. Salas-Coronado, C. Villanueva-Canongo, B. Hernandez-Carlos, Antioxiant compounds and their antioxidant mechanism, in: E. Shalaby (Ed.), Antioxidants, IntechOpen, 2019, https://doi.org/10.5772/intechopen.85270. Available from: https://www.intechopen.com/chapters/66259.

[12] E. Niki, Antioxidants: basic principles, emerging concepts, and problems, Biom. J. 37 (3) (2014) 106–111, https://doi.org/10.4103/2319-4170.128727.

[13] R. Sarangarajan, S. Meera, R. Rukkumani, P. Sankar, G. Anuradha, Antioxidants: friend or foe? Asian Pac J Trop Med 10 (12) (2017) 1111–1116, https://doi.org/10.1016/j.apjtm.2017.10.017.

[14] L. Jin, K. Ren, Q. Xu, T. Hong, S. Wu, Y. Zhang, Z. Wang, Multifunctional carbon dots for live cell staining and tissue engineering applications, Polym. Compos. 39 (1) (2018) 73–80, https://doi.org/10.1002/pc.23903.

[15] A.T. Dinkova-Kostova, P. Talalay, Direct and indirect antioxidant properties of inducers of cytoprotective proteins, Mol. Nutr. Food Res. 52 (2008) S128–S138, https://doi.org/10.1002/mnfr.200700195.

[16] K. Goszcz, S.J. Deakin, G.G. Duthie, D. Stewart, S.J. Leslie, I.L. Megson, Antioxidants in cardiovascular therapy: panacea or false hope? Front. Cardiovasc. Med. 2 (2015) 29, https://doi.org/10.3389/fcvm.2015.00029. eCollection 2015.

[17] S. Shafi, H.R. Ansari, W. Bahitham, S. Aouabdi, The impact of natural antioxidants on the regenerative potential of vascular cells, Front. Cardiovasc. Med. 6 (2019) 28, https://doi.org/10.3389/fcvm.2019.00028.

[18] C.K. Sen, S. Khanna, G. Gordillo, D. Bagchi, M. Bagchi, S. Roy, Oxygen, oxidants, and antioxidants in wound healing, Ann. N. Y. Acad. Sci. 957 (2002) 239–249, https://doi.org/10.1111/j.1749-6632. 2002.tb02920. x.

[19] M.A. Aziz, A.S. Diab, A.A. Mohammed, Antioxidant categories and mode of action, in: E. Shalaby (Ed.), Antioxidants, IntechOpen, 2019, https://doi.org/10.5772/intechopen.83544. Available from: https://www.intechopen.com/chapters/65225.
[20] S.B. Nimse, D. Pal, Free radicals, natural antioxidants, and their reaction mechanisms, Roy. Soc. Chem. Adv. 5 (35) (2015) 27986–28006, https://doi.org/10.1039/C4RA13315C.
[21] F. Shahidi, Y. Zhong, Novel antioxidants in food quality preservation and health promotion, Eur. J. Lipid Sci. Technol. 112 (9) (2010) 930–940, https://doi.org/10.1002/ejlt.201000044.
[22] M. Antolovich, P.D. Prenzler, E. Patsalides, S. McDonald, K. Robards, Methods for testing antioxidant activity, Analyst 127 (1) (2002) 183–198.
[23] A. Augustyniak, G. Bartosz, A. Čipak, G. Duburs, L.U. Horáková, W. Łuczaj, M. Majekova, A.D. Odysseos, L. Rackova, E. Skrzydlewska, M. Stefek, Natural and synthetic antioxidants: an updated overview, Free Radic. Res. 44 (10) (2010) 1216–1262, https://doi.org/10.3109/10715762.2010.508495.
[24] Y. Gilgun-Sherki, E. Melamed, D. Offen, Oxidative stress induced-neurodegenerative diseases: the need for antioxidants that penetrate the blood brain barrier, Neuropharmacology 40 (8) (2001) 959–975, https://doi.org/10.1016/s0028-3908(01)00019-3.
[25] B. Poljsak, D. Šuput, I. Milisav, Achieving the balance between ROS and antioxidants: when to use the synthetic antioxidants, Oxidative Med. Cell. Longev. 2013 (2013), https://doi.org/10.1155/2013/956792.
[26] B. Dimitrios, Sources of natural phenolic antioxidants, Trends Food Sci. Technol. 17 (9) (2006) 505–512, https://doi.org/10.1016/j.tifs.2006.04.004.
[27] Z.T. Schafer, A.R. Grassian, L. Song, Z. Jiang, Z. Gerhart-Hines, H.Y. Irie, S. Gao, P. Puigserver, J.S. Brugge, Antioxidant and oncogene rescue of metabolic defects caused by loss of matrix attachment, Nature 461 (7260) (2009) 109–113, https://doi.org/10.1038/nature08268.
[28] A. Das, D. Chaudhuri, R. Sarkar, N.B. Ghate, S. Panja, N. Mandal, Plants of Indian traditional medicine with antioxidant activity, in: Nutritional Antioxidant Therapies: Treatments and Perspectives, Springer, Cham, 2017, pp. 27–64, https://doi.org/10.1007/978-3-319-67625-8_2.
[29] A. Rietveld, S. Wiseman, Antioxidant effects of tea: evidence from human clinical trials, J. Nutr. 133 (10) (2003) 3285S–3292S, https://doi.org/10.1093/jn/133.10.3285S.
[30] M. Kelkel, M. Schumacher, M. Dicato, M. Diederich, Antioxidant and anti-proliferative properties of lycopene, Free Radic. Res. 45 (8) (2011) 925–940, https://doi.org/10.3109/10715762.2011.564168.
[31] A.A. Hamid, O.O. Aiyelaagbe, L.A. Usman, O.M. Ameen, A. Lawal, Antioxidants: its medicinal and pharmacological applications, Afr. J. Pure Appl. Chem. 4 (8) (2010) 142–151. http://www.academicjournals.org/AJPAC.
[32] P. Chandra, R.K. Sharma, D.S. Arora, Antioxidant compounds from microbial sources: a review, Food Res. Int. 129 (2020) 108849, https://doi.org/10.1016/j.foodres.2019.108849.
[33] P. Toledo-Ibelles, J. Mas-Oliva, Antioxidants in the fight against atherosclerosis: is this a dead end? Curr. Atheroscler. Rep. 20 (7) (2018) 36, https://doi.org/10.1007/s11883-018-0737-7.
[34] K. Neha, M.R. Haider, A. Pathak, M.S. Yar, Medicinal prospects of antioxidants: a review, Eur. J. Med. Chem. 178 (2019) 687–704, https://doi.org/10.1016/j.ejmech.2019.06.010.
[35] A. Serino, G. Salazar, Protective role of polyphenols against vascular inflammation, aging and cardiovascular disease, Nutrients 11 (1) (2018) 53, https://doi.org/10.3390/nu11010053.
[36] R. Corrêa, R.M. Peralta, C. Haminiuk, G.M. Maciel, A. Bracht, I. Ferreira, New phytochemicals as potential human anti-aging compounds: reality, promise, and challenges, Crit. Rev. Food Sci. Nutr. 58 (6) (2018) 942–957, https://doi.org/10.1080/10408398.2016.1233860.

[37] J. Martel, D.M. Ojcius, Y.F. Ko, C.J. Chang, J.D. Young, Antiaging effects of bioactive molecules isolated from plants and fungi, Med. Res. Rev. 39 (5) (2019) 1515–1552, https://doi.org/10.1002/med.21559.

[38] S. Naqvi, A. Panghal, S. Flora, Nanotechnology: a promising approach for delivery of neuroprotective drugs, Front. Neurosci. 14 (2020) 494, https://doi.org/10.3389/fnins.2020.00494.

[39] G. Bjelakovic, D. Nikolova, C. Gluud, Antioxidant supplements and mortality, Curr. Opin. Clin. Nutr. Metabol. Care 17 (1) (2014) 40–44, https://doi.org/10.1097/MCO.0000000000000009.

[40] J. Gruber, B. Halliwell, Approaches for extending human healthspan: from antioxidants to healthspan pharmacology, Essays Biochem. 61 (3) (2017) 389–399, https://doi.org/10.1042/EBC20160091.

[41] S.K. Myung, W. Ju, B. Cho, S.W. Oh, S.M. Park, B.K. Koo, B.J. Park, Korean Meta-Analysis Study Group, Efficacy of vitamin and antioxidant supplements in prevention of cardiovascular disease: systematic review and meta-analysis of randomised controlled trials, BMJ (Clin. Res. ed.) 346 (2013) f10, https://doi.org/10.1136/bmj.f10.

[42] J. Kydd, R. Jadia, P. Velpurisiva, A. Gad, S. Paliwal, P. Rai, Targeting strategies for the combination treatment of cancer using drug delivery systems, Pharmaceutics 9 (4) (2017) 46, https://doi.org/10.3390/pharmaceutics9040046.

[43] A. Kermanizadeh, L.G. Powell, V. Stone, P. Møller, Nanodelivery systems and stabilized solid-drug nanoparticles for orally administered medicine: current landscape, Int. J. Nanomedicine 13 (2018) 7575–7605, https://doi.org/10.2147/IJN.S177418.

[44] P. Ganesan, P. Ramalingam, G. Karthivashan, Y.T. Ko, D.K. Choi, Recent developments in solid lipid nanoparticle and surface-modified solid lipid nanoparticle delivery systems for oral delivery of phyto-bioactive compounds in various chronic diseases, Int. J. Nanomedicine 13 (2018) 1569–1583, https://doi.org/10.2147/IJN.S155593.

Further reading

A.A. Date, J. Hanes, L.M. Ensign, Nanoparticles for oral delivery: design, evaluation and state-of-the-art, J. Control. Release 240 (2016) 504–526, https://doi.org/10.1016/j.jconrel.2016.06.016.

L. Goth, P. Rass, A. Páy, Catalase enzyme mutations and their association with diseases, Mol. Diagn. 8 (3) (2004) 141–149, https://doi.org/10.2165/00066982-200408030-00001.

P. Marrazzo, C. Angeloni, S. Hrelia, Combined treatment with three natural antioxidants enhances neuroprotection in a SH-SY5Y 3D culture model, Antioxidants 8 (10) (2019) 420, https://doi.org/10.3390/antiox8100420.

A. Yadav, R. Kumari, A. Yadav, J.P. Mishra, S. Srivatva, S. Prabha, Antioxidants and its functions in human body-a review, Res. Environ. Life Sci. 9 (11) (2016) 1328–1331.

Index

Note: Page numbers followed by *f* indicate figures and *t* indicate tables.

Kolibri 360 GmbH
Ferdinand-Jühlke-Straße 7
99095 Erfurt, Deutschland
produktsicherheit@kolibri360.de